Mechatronics and Intelligent Systems for Off-road Vehicles

Francisco Rovira Más · Qin Zhang · Alan C. Hansen

Mechatronics and Intelligent Systems for Off-road Vehicles

Springer

Francisco Rovira Más, PhD
Polytechnic University of Valencia
Departamento de Ingeniería Rural
46022 Valencia
Spain
frovira@dmta.upv.es

Alan C. Hansen, PhD
University of Illinois at Urbana-Champaign
Agricultural Engineering Sciences Building
360P AESB, MC-644
1304 W. Pennsylvania Avenue
Urbana, IL 61801
USA
achansen@illinois.edu

Qin Zhang, PhD
Washington State University
Center for Automated Agriculture
Department of Biological Systems Engineering
Prosser Campus
Prosser, WA 99350-9370
USA
qinzhang@wsu.edu

ISBN 978-1-4471-5709-0 ISBN 978-1-84996-468-5 (eBook)
DOI 10.1007/978-1-84996-468-5
Springer London Dordrecht Heidelberg New York

British Library Cataloguing in Publication Data
A catalogue record for this book is available from the British Library

Cover design: eStudioCalmar, Girona/Berlin

Printed on acid-free paper

Springer is part of Springer Science+Business Media (www.springer.com)

Contents

Chapter 1
Introduction

1.1 Evolution of Off-road Vehicles Towards Automation: the Advent of Field Robotics and Intelligent Vehicles

Following their invention, engine-powered machines were not immediately embraced by the agricultural community; some time was required for further technical developments to be made and for users to accept this new technology. One hundred years on from that breakthrough, field robotics and vehicle automation represent a second leap in agricultural technology. However, despite the fact that this technology is still in its infancy, it has already borne significant fruit, such as substantial applications relating to the novel concept of precision agriculture. Several developments have contributed to the birth and subsequent growth over time of the field of intelligent vehicles: the rapid increase in computing power (in terms of speed and storage capacity) in recent years; the availability of a rich assortment of sensors and electronic devices, most of which are relatively inexpensive; and the popularization of global localization systems such as GPS. A close look at the cabin of a modern tractor or harvester will reveal a large number of electronic controls, signaling lights, and even flat touch screens. Intelligent vehicles can already be seen as agricultural and forestry robots, and they constitute the new generation of off-road equipment aimed at *delivering power with intelligence*.

The birth and development of agricultural robotics was long preceded by the nascency of general robotics, and the principles of agricultural robotics obviously need to be considered along with the development of the broader discipline, and particularly mobile robots. Robotics and automation are intimately related to artificial intelligence. The foundations for artificial intelligence, usually referred as "AI," were laid in the 1950s, and this field has been expanding ever since then. In those early days, the hardware available was no match for the level of performance already shown by the first programs written in Lisp. In fact, the bulkiness, small memory capacities, and slow processing speeds of hardware prototypes often discouraged researchers in their quest to create mobile robots. This early software–hardware developmental disparity certainly delayed the completion of robots with the degree of

F. Rovira Más, Q. Zhang, A.C. Hansen, *Mechatronics and Intelligent Systems for Off-road Vehicles*. © Springer 2010

autonomy predicted by the science fiction literature of that era. Nevertheless, computers and sensors have since reached the degree of maturity necessary to provide mobile platforms with a certain degree of autonomy, and a vehicle's ability to carry out computer reasoning efficiently, that is, its *artificial intelligence*, defines its value as an intelligent off-road vehicle.

In general terms, AI has divided roboticists into those who believe that a robot should *behave like humans*; and those who affirm that a robot should *be rational* (that is to say, it should do the right things) [1]. The first approach, historically tied to the Turing test (1950), requires the study of and (to some extent at least) an understanding of the human mind: the enunciation of a model explaining how we think. Cognitive sciences such as psychology and neuroscience develop the tools to address these questions systematically. The alternative tactic is to base reasoning algorithms on *logic rules* that are independent of emotions and human behavior. The latter approach, rather than implying that humans may behave irrationally, tries to eliminate systematic errors in human reasoning. In addition to this philosophical distinction between the two ways of approaching AI, intelligence can be directed towards *acting* or *thinking*; the former belongs to the *behavior domain*, and the latter falls into the *reasoning domain*. These two classifications are not mutually exclusive; as a matter of fact, they tend to intersect such that there are four potential areas of *intelligent behavior design: thinking like humans*, *acting like humans*, *thinking rationally*, and *acting rationally*. At present, design based on rational agents seems to be more successful and widespread [1].

Defining intelligence is a hard endeavor by nature, and so there is no unique answer that ensures universal acceptance. However, the community of researchers and practitioners in the field of robotics all agree that *autonomy* requires some degree of intelligent behavior or ability to handle knowledge. Generally speaking, the grade of autonomy is determined by the *intelligence* of the device, machine, or living creature in question [2]. In more specific terms, three fundamental areas need to be adequately covered: *intelligence*, *cognition*, and *perception*. Humans use these three processes to navigate safely and efficiently. Similarly, an autonomous vehicle would execute reasoning *algorithms* that are programmed into its intelligence unit, would make use of knowledge stored in *databases* and lookup tables, and would constantly perceive its surroundings with *sensors*. If we compare artificial intelligence with human intelligence, we can establish parallels between them by considering their principal systems: the *nervous system* would be represented by architectures, processors and sensors; *experience and learning* would be related to algorithms, functions, and modes of operation. Interestingly enough, it is possible to find a reasonable connection between the nervous system and the artificial system's *hardware*, in the same way that experience and learning is naturally similar to the system's *software*. This dichotomy between software and hardware is actually an extremely important factor in the constitution and behavior of intelligent vehicles, whose reasoning capacities are essential for dealing with the unpredictability usually encountered in open fields.

Even though an approach based upon rational agents does not necessarily require a deep understanding of intelligence, it is always helpful to get a sense of its inner workings. In this context, we may wonder how we can estimate the *capacity* of an

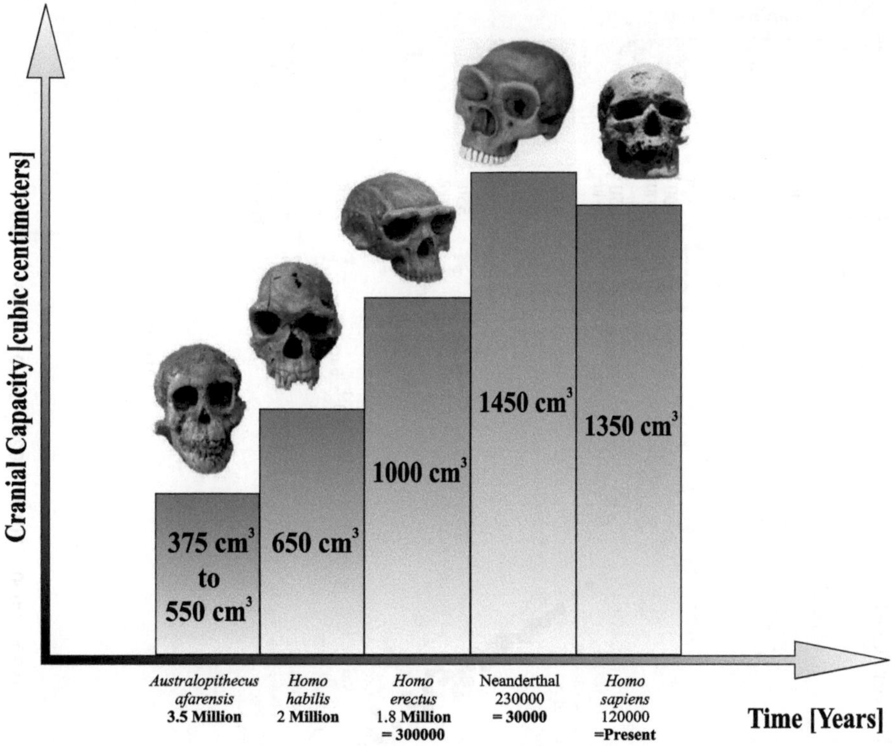

Figure 1.1 Brain capacity and degree of sophistication over the course of evolution

intelligent system, as many things seem easier to understand when we can measure and classify them. A curious fact, however, is shown in Figure 1.1, which depicts how the degree of sophistication of humans over the course of evolution has been directly related to their brain size. According to this "evolutionary stairway," we generally accept that a bigger brain will lead to a higher level of society. However, some mysteries remain unsolved; for example, Neanderthals had a larger cranial capacity than we do, but they became extinct despite their high potential for natural intelligence.

It is thus appealing to attempt to quantify intelligence and the workings of the human mind; however, the purpose of learning from natural intelligence is to extract knowledge and experience that we can then use to furnish computer algorithms, and eventually off-road vehicles, with reliable and robust artificial thinking. Figure 1.1 provides a means to estimate brain capacity, but is it feasible to compare brain power and computing power? Hans Moravec has compared the evolution of computers with the evolution of life [3]. His conclusions, graphically represented in Figure 1.2, indicate that contemporary computers are reaching the level of intelligence of small mammals. According to his speculations, by 2030 computing power could be comparable to that of humans, and so robots will compete with humans;

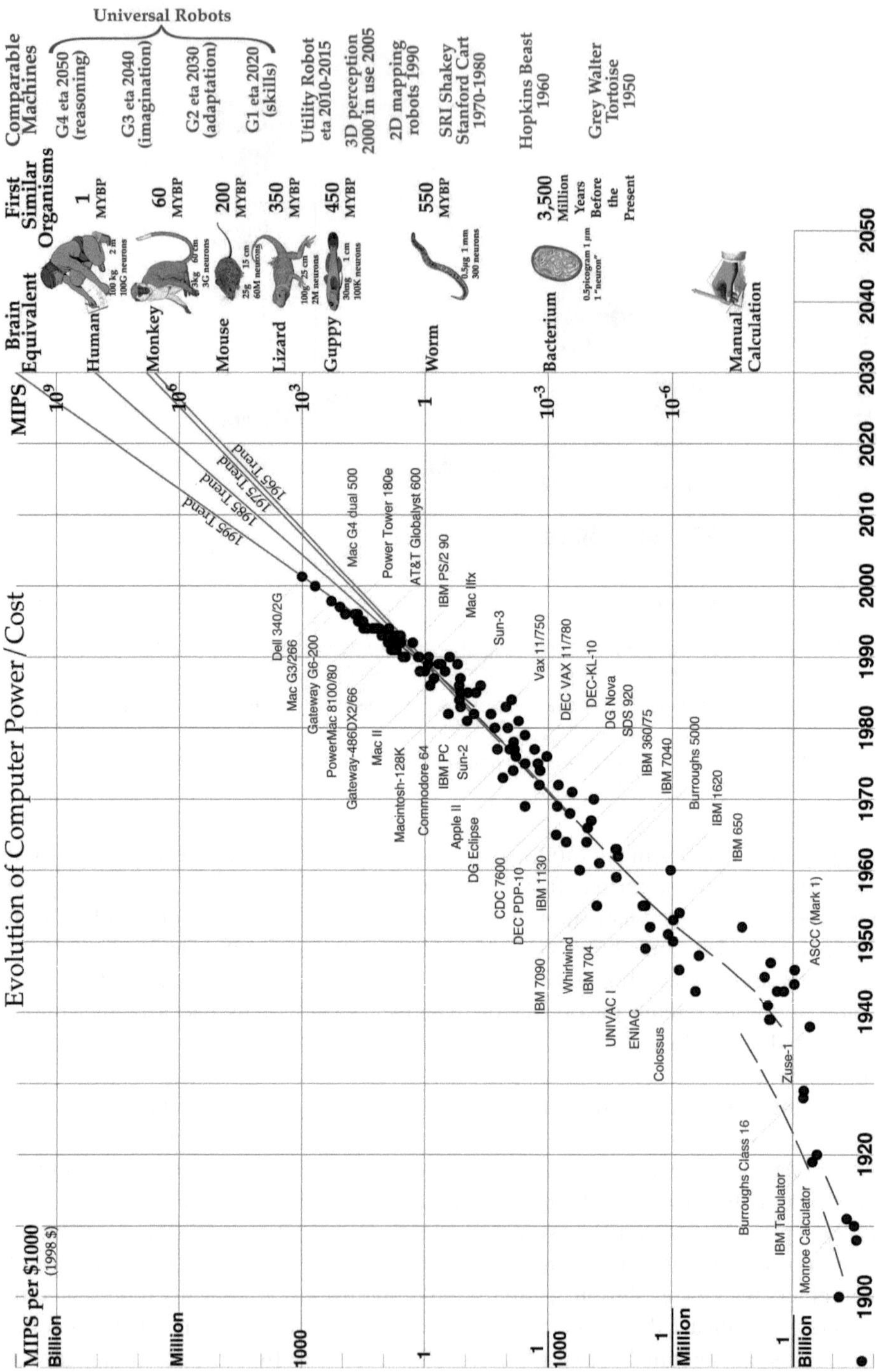

Figure 1.2 Hans Moravec's comparison of the evolution of computers with the evolution of life [3]

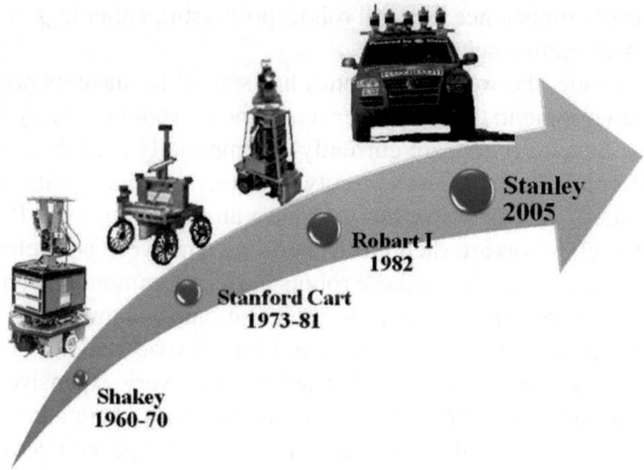

Figure 1.3 Pioneering intelligent vehicles: from laboratory robots to off-road vehicles

in other words, a fourth generation of universal robots may abstract and reason in a humanlike fashion.

Many research teams and visionaries have contributed to the field of mobile robotics in the last five decades, and so it would be impractical to cite all of them in this introductory chapter. Nevertheless, it is interesting to mention some of the breakthroughs that trace the trajectory followed by field robotics from its origins. *Shakey* was a groundbreaking robot, developed at the Stanford Research Institute (1960–1970), which solved simple problems of perception and motion, and demonstrated the benefits of artificial intelligence and machine vision. This pioneering work was continued with the *Stanford Cart* (1973–1981), a four-wheeled robot that proved the feasibility of stereoscopic vision for perception and navigation. In 1982, *ROBART I* was endowed with total autonomy for random patrolling, and two decades later, in 2005, *Stanley* drove for 7 h autonomously across the desert to complete and win Darpa's Grand Challenge. Taking an evolutionary view of the autonomous robots referred to above and depicted in Figure 1.3, successful twenty-first century robots might not be very different from off-road vehicles such as *Stanley*, and so agricultural and forestry machines possess a typology that makes them suited to robotization and automation.

In order to move autonomously, vehicles need to follow a *navigation model.* In general, there are two different architectures for such a model. The *traditional model* requires a cognition unit that receives perceptual information on the surrounding environment from the sensors, processes the acquired information according to its intelligent algorithms, and executes the appropriate actions. This model was implemented, for instance, in the robot *Shakey* shown in Figure 1.3. The alternative model, termed *behavior-based robotics* and developed by Rodney Brooks [4], eliminates the cognition box by merging perception and action. The technique used to apply this approach in practice is to implement sequential layers of control that have different levels of competence. Several robots possessing either legs or wheels have followed this architecture successfully.

In the last decade, the world of robotics has started to make its presence felt in the domestic environment: there has been a real move from laboratory prototypes to retail products. Several robots are currently commercially available, although they look quite differently from research prototypes. Overall, commercial solutions tend to be well finished, very task-specific, and have an appealing look. Popular examples of off-the-shelf robots are vacuum cleaners, lawn mowers, pool-cleaning robots, and entertainment mascots. What these robots have in common are a small size, low power demands, no potential risks from their use, and a competitive price. These properties are just the opposite of those found for off-road vehicles, which are typically enormous, actuated by powerful diesel engines, very expensive and – above all – accident-prone. For these reasons, even though they share a common ground with general field robotics, off-road equipment has very special needs, and so it is reasonable to claim a distinct technological niche for it within robotics: agricultural robotics.

1.2 Applications and Benefits of Automated Machinery

Unlike planetary rovers (the other large group of vehicles that perform autonomous navigation), which wander around unstructured terrain, agricultural vehicles are typically driven in fields arranged into crop rows, orchard lanes or greenhouse corridors; see for example the regular arrangement of the vineyard and the ordered rows of orange trees in Figure 1.4 (a and b, respectively). These man-made structures provide features that can assist in the navigation of autonomous vehicles, thus facilitating the task of auto-steering. However, as well as the layout of the field, the nature of agricultural tasks makes them amenable to automation too. Farm duties such as planting, tilling, cultivating, spraying, and harvesting involve the execution of repetitive patterns where operators need to spend many hours driving along farming rows. These long periods of time repeating the same task often result in tiredness and fatigue that can lead to physical injuries in the long run. In addition, a sudden lapse in driver concentration could result in fatalities.

Figure 1.4 Vineyard in Northern California (**a**) and an orange grove in Valencia, Spain (**b**)

One direct benefit of automating farming tasks is a gain in ergonomics: when the farmer does not need to hold the steering wheel for 8 h per day, but can instead check the vehicle's controls, consult a computer, and even answer the phone, individual workloads clearly diminish. The vehicle's cabin can then be considered a working office where several tasks can be monitored and carried out simultaneously. The machine may be driven in an autopilot mode — similar to that used in commercial aircraft – where the driver has to perform some turns at the ends of the rows, engage some implements, and execute some maneuvers, but the autopilot would be in charge of steering inside the field (corresponding to more than 80% of the time).

Vehicle automation complements the concept of *precision agriculture* (PA). The availability of large amounts of data and multiple sensors increases the accuracy and efficiency of traditional farming tasks. Automated guidance often reaches sub-inch accuracies that only farmers with many years of experience and high skill levels can match, and not even expert operators can reach such a degree of precision when handling oversized equipment. Knowledge of the exact position of the vehicle in real time reduces the amount of overlapping between passes, which not only reduces the working time required but decreases the amount of chemicals sprayed, with obvious economic and environmental benefits. Operating with information obtained from updated maps of the field also contributes to a more rational use of resources and agricultural inputs. For instance, an autonomous sprayer will shut off the nozzles when traversing an irrigation ditch since the contamination of the ditch could have devastating effects on cattle or even people. A scouting camera may stop fertilization if barren patches are detected within the field.

As demonstrated in the previous paragraphs, the benefits and advantages of off-road vehicle automation for agriculture and forestry are numerous. However, safety, reliability and robustness are always concerns that need to be properly addressed before releasing a new system or feature. Automatic vehicles have to outperform humans because mistakes that people would be willing to accept from humans will never be accepted from robotic vehicles. Safety is probably the key factor that has delayed the desired move from research prototypes to commercial vehicles in the field of agricultural intelligent vehicles.

1.3 Automated Modes: Teleoperation, Semiautonomy, and Full Autonomy

So far, we have been discussing vehicle automation without specifying what that term actually means. There are many tasks susceptible to automation, and multiple ways of automating functions in a vehicle, and each one demands a different level of intelligence. As technology evolves and novel applications are devised, new functions will be added to the complex design of an intelligent vehicle, but some of the functions that are (or could be) incorporated into new-generation vehicles include:

- automated navigation, comprising guidance visual assistance, autosteering, and/or obstacle avoidance;
- automatic implement control, including implement alignment with crops, smart spraying, precise planting/fertilizing, raising/lowering the three-point hitch without human intervention, *etc.*;
- mapping and monitoring, gathering valuable data in a real-time fashion and properly storing it for further use by other intelligent functions or just as a historical data recording;
- automatic safety alerts, such as detecting when the operator is not properly seated, has fallen asleep, or is driving too fast in the vicinity of other vehicles or buildings;
- routine messaging to send updated information to the farm station, dealership, loading truck, or selling agent about crop yields and quality, harvesting conditions, picking rates, vehicle maintenance status, *etc.*

Among these automated functions, navigation is the task that relieves drivers the most, allowing them to concentrate on other managerial activities while the vehicle is accurately guided without driver effort. There are different levels of navigation, ranging from providing warnings to full vehicle control, which evidently require different complexity levels. The most basic navigation kit appeared right after the popularization of the global positioning system (GPS), and is probably the most extended system at present. It is known as a *lightbar* guidance assistance device, and consists of an array of red and green LEDs that indicate the magnitude of the offset and the orientation of the correction, but the steering is entirely executed by the driver who follows the lightbar indications. This basic system, regardless of its utility and its importance as the precursor for other guidance systems, cannot be considered an automated mode *per se* because the driver possesses full control over the vehicle and only receives advice from the navigator. The next grade up in complexity is represented by *teleoperated* or *remote-controlled* vehicles. Here, the vehicle is still controlled by the operator, but in this case from outside the cabin, and sometimes from a remote position. This is a hybrid situation because the machine is moving driverless even though all of its guidance is performed by a human operator, and so little or no intelligence is required. This approach, while utilized for planetary rovers (despite frustrating signal delays), is not attractive for off-road equipment since farm and forestry machines are heavy and powerful and so the presence of an operator is normally required to ensure safety. Wireless communications for the remote control of large machines have still not yet reached the desired level of reliability. The next step is, at present, the most interesting for intelligent off-road vehicles, and can be termed *semiautonomy*. It constitutes the main focus of current research into autonomous navigation and corresponds to the autopilots employed in airplanes: the operator is in place and in control, but the majority of time – along the rows within the field – steering is performed automatically. Manual driving is typically performed from the machinery storage building to the field, to engage implements, and in the headlands to shift to the next row. The majority of the material presented in this book and devoted to autonomous driving and autoguidance will refer to semiautonomous applications. The final step in the evolutionary path for autonomous navigation is represented by *full autonomy*. This is the stage that has long been dreamed of by visionaries. In full autonomy, a herd of completely autonomous machines farm the field by themselves and return to the farm after the task is done without human intervention. The current state of technology and even human mentality are not ready for such an idyllic view, and it will certainly take some years, probably decades, to fulfill that dream. System reliability and safety is surely the greatest obstacle to achieving full autonomy (although accomplishing semiautonomy is also a remarkable advance that is well worth pursuing). The future – probably the next two decades – will reveal when this move should be made, if it ever happens.

1.4 Typology of Field Vehicles Considered for Automation

When confronted with the word *robot*, our minds typically drift to the robots familiar to us, often from films or television watched during childhood. Hence, well-known robots like *R2-D2*, *HAL-9000*, or *Mazinger Z* can bias our opinions of what a robot actually is. As a matter of fact, a robotic platform can adopt any configuration that serves a given purpose, and agricultural and forestry production can benefit for many types of vehicles, from tiny scouting robots to colossal harvesters. The rapid development of computers and electronics and the subsequent birth of agricultural robotics have led to the emergence of new vehicles that will coexist with conventional equipment. In general, we can group off-road field vehicles into two categories: conventional vehicles and innovative platforms.

Conventional vehicles are those traditionally involved in farming tasks, such as all types of tractors, grain harvesters, cotton and fruit pickers, sprayers, self-propelled forage harvesters, *etc.* Robotized machines differ from conventional vehicles in that they incorporate a raft of sensors, screens, and processors, but the actual chassis of the vehicle is the same, and so they are also massive, powerful and usually expensive. These vehicles, which we will term robots from now on, are radically different from the small rovers and humanoids that take part in planetary explorations or dwell in research laboratories. Farm equipment moving in (semi)autonomous mode around fields typically frequented by laborers, machines, people or livestock poses acute challenges in terms of liability; mortal accidents are unlikely to occur in extraterrestrial environments, research workshops, or amusement parks, but they do happen in rural areas where off-road equipment is extensively used. Since the drivers of these vehicles need special training and to conduct themselves responsibly, automated versions of these vehicles will have to excel in their precaution and safeguarding protocols. A great advantage of robotized conventional off-road vehicles over typical small mobile robots is the durability of the energy source. One of the known problems with domestic and small-scale robots is their autonomy, due to the limited number of operating hours afforded by their power sources. Most of them are powered by solar cells (planetary rovers) or lithium batteries (humanoids, vacuum cleaners, entertainment toys, *etc.*). This serious inconvenience is nonexistent in farming vehicles, since they are usually powered by potent diesel engines, meaning that the energy requirements of onboard computers, flat screens, and sensors are insignificant.

The quest for updated field data, precision in the application of farming inputs, and the rational adoption of information technology methods has led to a number of novel and unusual vehicles that can be grouped under the common term of *innovative vehicles*. These platforms follow an unconventional design which is especially tailored to the specific task that it is assigned to carry out. Most of them are still under development, or only exist as research prototypes, but the numbers and varieties of innovative vehicles will probably increase in the future as more robotic solutions are incorporated into the traditional farm equipment market. Among these innovative vehicles, it is worth mentioning legged robots capable of climbing steep mountains for forestry exploitation, midsized robotic utility vehicles (Figure 1.5a), localized remote-controlled spraying helicopters (Figure 1.5b), and small scouting robots (Figure 1.5c) that can operate individually or implement swarm intelligence strategies.

Figure 1.5 Innovative field vehicles: (**a**) utility platform; (**b**) spraying helicopter; (**c**) scouting robot (courtesy of Yoshisada Nagasaka)

1.5 Components and Systems in Intelligent Vehicles

Despite of the lure of innovative unconventional vehicles, most of today's intelligent off-road vehicles are conventional agricultural vehicles, and probably most of tomorrow's will be too. These machines possess special characteristics that place them among the largest and most powerful mobile robots. For instance, a common tractor for farming corn and soybeans in the American Midwest can weigh 8400 kg, incorporates an engine of 200 HP, and has an approximate price of $100,000. A wheat harvester that is frequently used in Northern Europe might weigh 16,000 kg, be powered by a 500 HP engine, and have a retail value of $300,000. A self-propelled sprayer for extensive crops can feature a 290 HP engine, weigh 11,000 kg, and cost $280,000. All of these figures indicate that the off-road vehicles that will be robotized for deployment in agricultural fields will not have any trouble powering their sensors, the cost of the sensors and ancillary electronics will represent a modest percentage of the machine's value, and the weight of the "brain" (the hardware and architecture that supports the intelligent systems onboard) will be insignificant compared to the mass of the vehicle. On the other hand, reliability and robustness will be major concerns when automating these giants, so the software used in them will need to be as heavyweight as the vehicle itself – meaning that such machines can be thought of as "smart dinosaurs."

1.5.1 Overview of the Systems that Comprise Automated Vehicles

Given the morphology of the vehicles under consideration, the design of the system architecture must take the following aspects into account (in order of priority): robustness and performance; cost; size; power requirements; weight. An individual description of each sensing system is provided subsequently, but regardless of the specific properties of each system, it is important to consider the intelligent vehicle as a whole rather than as an amalgamation of sensors (typical of laboratory prototyping). In this regard, a great deal of thought must be devoted early in the design process to how all of the sensors and actuators form a unique body, just like the human body. Although field vehicles can be very large, cabins tend to be full of devices, levers and controls without much room to spare, so it is essential to plan efficiently and, for example, merge the information from several sources into a single screen

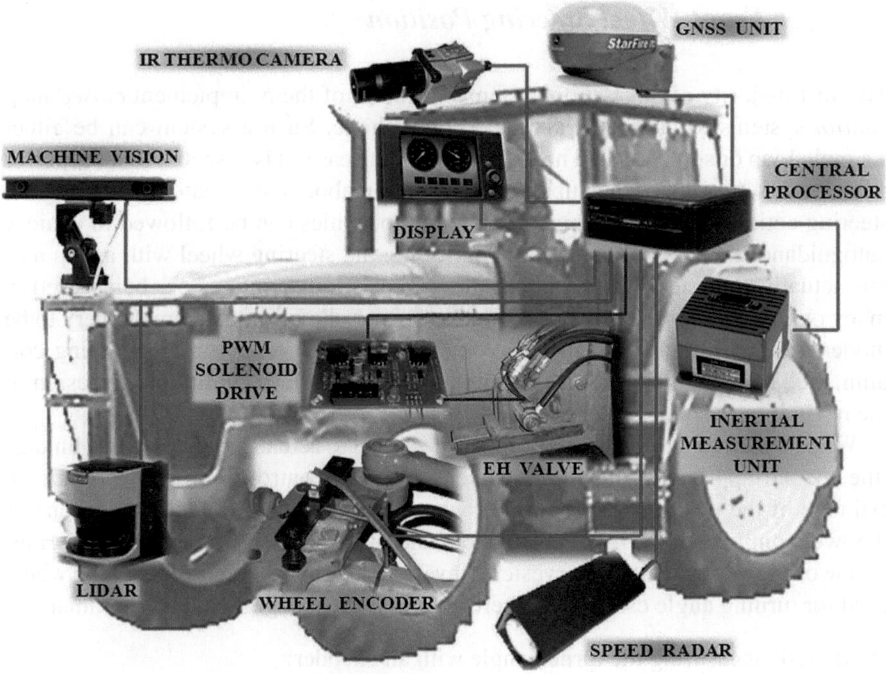

Figure 1.6 General architecture for an intelligent vehicle

with a clear display and friendly interfaces. The complexity of the intelligent system does not necessarily have to be translated into the cabin controls, and it should never be forgotten that the final user of the vehicle is going to be a professional farmer, not an airliner pilot. The physical positions of the sensors and actuators are also critical to ensuring an efficient design. One of the main mistakes made when configuring a robotized vehicle that is intended to roam in the open field is a lack of consideration of the harsh environment to which the vehicle can be exposed: freezing temperatures in the winter, engine and road vibrations, abrasive radiation and temperatures in the summer, strong winds, high humidity, dew and unpredicted rains, dust, friction from branches, exposure to sprayed chemicals, *etc.* These conditions make off-road vehicle design special, as it diverges from classic robotic applications where mobile robots are designed to work indoors (either in offices or in manufacturing buildings). If reliability is the main concern, as previously discussed, hardware endurance is then a crucial issue. Not only must the devices used be of high quality, but they must also have the right protection and be positioned optimally. In many cases, placing a delicate piece in an appropriate position can protect it from rough weather and therefore extend its working life. Figure 1.6 shows a robotized tractor with some of the usual systems that comprise intelligent vehicles.

1.5.2 Flow Meters, Encoders, and Potentiometers for Front Wheel Steering Position

The vast majority of navigation systems, if not all of them, implement *closed loop control* systems to automatically guide the vehicle. Such a system can be either a simple loop or sophisticated nested loops. In any case, it is essential to incorporate a feedback sensor that sends updated information about the actuator generating the steering actions. Generally speaking, two philosophies can be followed to achieve autoguidance in terms of actuation: controlling the steering wheel with a step motor; actuating the steering linkage of the vehicle. Both solutions are being used in many ongoing research projects. While the former allows outdated machinery to be modernized by mounting a compact autosteering kit directly on the steering column, the latter keeps the cabin clearer and permits more flexibility in the design of the navigation system.

When the automatic steering system is designed to actuate on the steering linkage (the second approach), the feedback sensor of the control loop must provide an estimate of the position of the turning wheel. This wheel will generally be one of the two front wheels on tractors, sprayers and utility vehicles (Ackerman steering) or one of the rear wheels on harvesters (inverse Ackerman). Regardless of the wheel used for turning angle estimation, there are three ways to get feedback commands:

1. directly measuring the turned angle with an encoder;
2. indirectly measuring the angle by estimating the displacement of the hydraulic cylinder actuating the steering linkage;

3. indirectly measuring the wheel angle by monitoring the flow traversing the steering cylinder.

Estimating the turning angle through the linear displacement of the cylinder rod of the steering linkage requires sensor calibration to relate linear displacements to angles. As described in Chapter 2, when steering is achieved by turning the front or rear wheels (that is, for non-articulated geometries), the left and right wheels of the same axle do not turn the same amount for a given extension of the cylinder rod. Thus, the nonlinear relationship between both wheels must be established, as the sensor will usually estimate the angle turned by one of them. The sensor typically employed to measure rod displacements is a *linear potentiometer*, where changes in electrical resistivity are converted into displacements. This sort of sensor yields a linear response inside the operating range, and has been successfully used with off-road vehicles, although the potentiometer assemblage is sometimes difficult to mount on the steering mechanism. The position of the rod can also be calculated from the flow rate actuating the cylinder. In this case, the accuracy of the flow meter is vital for accomplishing precise guidance.

An alternative to a linear potentiometer is to use *optical encoders* to estimate the angle turned by one or both of the turning wheels. These electromechanical devices usually consist of a disc with transparent and opaque areas that allow a light beam to track the angular position at any time. Such rotary encoders are preferably mounted on the king pin of the wheel whose angle is being recorded. Assembly is difficult in this case, since it is necessary to fix either the encoder's body or the encoder's shaft to the vehicle's chassis so that relative movements can be tracked and wheel angles measured. King pins are not easy to access, and encoders require a customized housing to keep them or their shafts affixed to the vehicle while protecting them from the harsh surroundings of the tire. The calibration of optical encoders is straightforward (see Section 7.1), and establishes a relationship between output voltage and angle turned. Encoders, as well as potentiometers, require an input voltage, which has to be conducted to the wheels through the appropriate wires. Figure 1.7 shows the assembly of encoders for tractors with two different wheel-types.

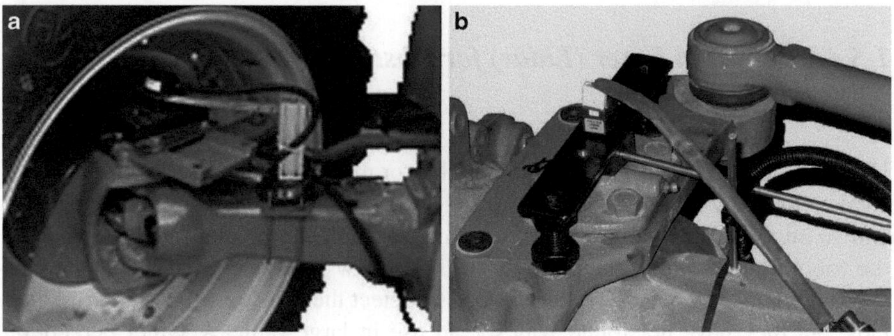

Figure 1.7 Assembly of optical encoders on two robotized tractors with different wheel-types

1.5.3 Magnetic Pulse Counters and Radars for Theoretical and Ground Speed

Automatic navigation can be achieved by implementing a great variety of algorithms, from simplistic reactive feelers to sophisticated trajectory planners. Most of the strategies that have actually been used require the estimation of the vehicle forward velocity, as it is incorporated into models that predict and trace trajectories. Knowledge of the speed is indispensable for adjusting the steering angle appropriately as the vehicle increases speed or, for instance, for calculating states in the Kalman filter-based sensor fusion employed by a navigation planner. Dead reckoning is a navigation technique that is used to estimate the current position of a vehicle based on its speed of travel and the time elapsed from a previous position. While it is used in many robotic applications, it is never recommended for off-road vehicles because *wheel slip* is a common phenomenon when traversing off-road terrains, and when such slippage occurs, errors in the positions estimated through dead reckoning grow considerably. The slippage can however be calculated when the theoretical speed of the vehicle and the actual speed can be measured.

The *theoretical forward speed* of a vehicle can be calculated if the number of revolutions made by the wheel in a certain time and the diameter of the wheel are known. The angular speed of the wheel can easily be measured by a magnetic pulse counter installed directly in the wheel or axle shaft. The counter needs a set of stripes or some other means of marking angular positions and a timer.

Although the theoretical speed is necessary to estimate wheel slip, the *real speed* is more important for navigational purposes, as it is the parameter used in most models. Other automated functions aside from navigation also make use of it; for instance, it is used to estimate the changes in nozzle actuation required during intelligent spraying according to the speed. The forward speed can be measured with devices based on the principle of time-of-flight calculations, such as radar. Vehicles equipped with global navigation satellite systems such as GPS can also estimate the forward speed from messages sent to the receiver from satellites, since position and time are acquired in real time.

1.5.4 Sonar and Laser (Lidar) for Obstacle Detection and Navigation

Ultrasonic distance sensing became popular for mobile robotics due to a sonar sensor developed by Polaroid for camera range-finding. These sensors were inexpensive and so an affordable solution was to arrange a matrix of them around the body of the robot, thus avoiding the problem of the narrow field of each sensor. This idea worked well for small robots that needed to detect the walls of offices and research labs, but they have not found widespread use in large vehicles. Other perception sensors, such as vision and laser devices, have been found to be more efficient for outdoor applications.

Lidar (light detection and ranging) is an optical device that is used to find ranges or distances to objects and surfaces. Different light sources can be used to find ranges, but the prevalent trend is to use laser pulses, and therefore a lidar and a laser rangefinder will be assumed to be equivalent devices hereafter, unless otherwise specified. Lasers are ideal for vehicle navigation because the beam density and coherency are excellent. However, lasers possess a very narrow beam, which forces the emitter to rotate to cover the field of view in front of the vehicle. The high resolutions of lidars have made them popular for obstacle detection and avoidance in field robots, such as the participants in the Grand Challenge competition for unmanned vehicles, where most of the off-road vehicles featured one – and often several – lidar heads [5]. Figure 1.3 shows *Stanley the Robot*, an intelligent vehicle off-road with five lidars on its roof.

1.5.5 GNSS for Global Localization

The tremendous boost given by GPS to the automation of agricultural vehicles, especially with regards to automatic guidance, has had a very positive effect on the development of agricultural robotics. The cancellation of selective availability by the United States Department of Defense in 2000 marked the beginning of an wave of commercial products and research projects that took advantage of the availability of real-time vehicle localization. While farm equipment firms have directed significant effort toward global navigation systems, other electronics and communications manufacturers have also expanded their market share to include agricultural applications. At the present time, most of the leading manufacturers of agricultural machinery include navigation assistance systems among their advanced products.

Even though GPS triggered the growth of satellite-based navigation, it is more appropriate to consider a general term under which other similar systems can be grouped: *global navigation satellite systems*, often referred to as *GNSS*. Under the umbrella of GNSS, we will consider GPS (USA), Galileo (Europe), GLONASS (Russia), Beidou (China), and other satellite localization systems that may appear in the future. Currently only GPS is fully operational, and so all commercial navigation assistance applications currently rely on it.

In spite of the extensive use and clear benefits of global positioning, it has some important drawbacks that need to be addressed by autonomous navigation applications. The main disadvantage of global sensing is a lack of local awareness. Any unpredicted event that occurs in the vicinity of the vehicle will always remain unnoticed in a global frame. Such events include small trajectory corrections and real-time changes that affect the robot's predetermined course. Another difficulty that is of great importance for orchards and greenhouses is related to double-path errors and signal drops. Tall trees create tunnel-like inter-row lanes where GNSS signals from satellites tend to be inconsistent. The hazards caused by unreliable navigation commands directing a massive off-road vehicle demand sensor redundancy, including local perception, which is usually achieved with lidars or imaging sensors. The tractor depicted in Figure 1.6 features a GPS receiver that is customized for agricultural production needs. The different GNSS solutions that are available for agriculture are discussed in Chapter 3.

It is important to establish a distinction between a GNSS receiver and a complete GNSS-based navigation system. The receiver provides real-time geodesic coordinates and the velocity of the vehicle, and it is the navigation algorithm's task to process these data, often by fusing them with data from other redundant sensors to generate guidance commands. When we refer to a GNSS-based navigation system, we mean a complete system that utilizes global positioning data to feed a controller whose output instructions steer the vehicle. This approach is followed by some manufacturers, and in this case the whole system must be considered a black box with limited or no access to the internals of the controller.

1.5.6 Machine Vision for Local Awareness

It has been noticed by some advanced farmers (early adopters of GNSS and related technologies) that, unless high-accuracy systems such as RTK-GPS are used in the field, the coordinates of crop rows recorded during planting are not the same as the positions of the same rows detected during harvesting. Unless proper corrections are made before harvesting, automated machines could cause irreversible damage to the valuable crops if only global localization is used. A solution to this serious problem can be found in local perception sensors; among them, machine vision probably has the greatest potential due to its ability to "see" ahead of the vehicle. The slight correction that needs to be done to adjust the harvester head to the crop rows can be performed with an onboard camera. These corrections often change over time, and so a fixed offset is not a robust solution to the problem. A camera with a fast frame rate of up to 30 images/s and an adjustable field of view and resolution can calculate the small tolerances that a robotic vehicle needs to navigate without damaging the crops.

Instantaneous rectifications are not the only benefit of image sensors. Moreover, they do not represent their most important advantage over global sensing. The main reason for incorporating video cameras into the perception systems of autonomous robots is usually the advantages of computer vision for safeguarding and obstacle detection. Agricultural vehicles operate in fields where other workers, vehicles, and even livestock move around without following predetermined paths. An obstacle can interfere with the vehicle's trajectory at any time, and there is a need to detect it in real time so that the vehicle can be halted or detoured to avoid collision.

The rich visual information made available by this technique can also be employed for other uses besides vehicle navigation. As features from the field environment are grabbed in a continuous sequence of images, mapping and monitoring algorithms can recreate the field scene and estimate growth status or maturity.

The range of imaging sensors available is diverse, and each concrete application demands a different solution. The detection of plant health for automated fertilizing has been successfully realized with *hyperspectral* and *multispectral cameras*. Smart spraying has been achieved with *monocular cameras*. Autonomous guidance can make use of both monocular cameras and binocular stereovision rigs. Three-dimensional maps can be assembled from images obtained with a *stereoscopic camera*. All of the sensors require appropriate calibration. The vehicle shown in Figure 1.6 features a stereo camera located on the front of the tractor for 3D imaging. Chapters 4 and 5 provide a more detailed description of vision sensors.

There is no perfect sensor that can fulfill all of the requirements of a robotic vehicle, and from that point of view sensor fusion and redundancy is more than necessary. Just like other sensors, vision systems also have weaknesses, such as the computational load associated with many vision algorithms, the amount of data that some processes need to handle (especially for stereo images), and the dependency of such systems on lighting conditions, with notable consequences for reliability and robustness.

1.5.7 Thermocameras and Infrared for Detecting Living Beings

Vision sensors, lidars (lasers), and ultrasonic devices cannot penetrate through a thick layer of densely planted crops at the time of harvesting. Corn, for example, can easily reach over six feet at the end of its vegetative cycle. In this situation, the safeguarding engine of an automated machine cannot detect the presence of living beings standing within the crop if it is exclusively based on local perception sensors such as cameras and optical rangefinders. There have been reports of cattle being run over, and even negligent laborers being injured by (manually operated) farm machines. Accidents caused by agricultural vehicles are not uncommon, and when they do happen they are shocking and are quickly reported by the media. In order to avoid this situation, some sophisticated vehicles incorporate *infrared thermocameras* that are capable of generating a thermographic map of a given scene.

Thermocameras, also called FLIR (forward-looking infrared), are cameras that form images based on the infrared radiation emitted by objects. During low-intensity illumination at, say, dawn or dusk, or even for nighttime tasks, when an operator would found it more difficult to distinguish a person or animal concealed by plants, the temperatures of living beings are significantly superior to those of the surrounding plants and soil. Such temperature differences can be identified on the thermographic profile of the scene, and the safeguarding algorithm can, after conducting a *thermographic analysis* of the infrared image, output warning messages that the vehicle should be detoured or stopped. These sensors have not been widely exploited so far for civil applications, although they have been used for defense purposes for a long time. As the cost of FLIR sensors decreases, more intelligent vehicles will incorporate them into their perception units. Figure 1.8 shows a thermographic map (a) of an agricultural scene (b) that can be used to analyze the water content of the soil.

Figure 1.8 Thermographic map (**b**) of a Japanese rural area (**a**) (courtesy of Noboru Noguchi)

1.5.8 Inertial and Magnetic Sensors for Vehicle Dynamics: Accelerometers, Gyroscopes, and Compasses

Feedback control systems are a set of techniques that are universally utilized to achieve automation in field robotics. The basic idea is to estimate the difference (*i.e.*, the error) between the desired state and the actual state and use it to control the vehicle in subsequent motion orders. The specific way to do this is defined by the design of the controller algorithm and control loop. This procedure requires the instantaneous estimation of the states of the vehicle, in other words its position, linear velocity, angular rate (angular velocity), acceleration, pitch, roll, and heading angle. These states are measured by inertial sensors and are essential for assessing the vehicle's dynamic behavior. The dynamics of the motion are related to the response of the vehicle to the navigational commands sent by the intelligence unit, and are usually included in the motion equations of the dynamic model, such as state space control models and Kalman filters.

Inertial measurement units (IMU) are motion sensors created from a combination of accelerometers and gyroscopes. The accelerometers of the IMU detect the acceleration (the change in velocity over time) of the vehicle. Once the acceleration is known, integrating it gives an estimate of the velocity, and integrating it again allows the position to be evaluated. Similarly, the gyroscopes can detect the angular rates turned by the vehicle; integrating these leads to roll, pitch and yaw values. Typical inertial measurement units comprise three accelerometers and three gyroscopes assembled along three perpendicular axes that reproduce a Cartesian coordinate system. With this configuration, it is possible to calculate the three components of acceleration and speed in Cartesian coordinates as well as Euler angles. New IMU designs are smaller and less expensive, which is favorable for multiple and more accurate estimates of vehicle states. The rate of reduction is such that the sizes of some IMUs are similar to those of small devices such as microelectromechanical systems (MEMS).

The principal disadvantage of inertial measurement units is drift – the accumulation of error with time. This problem is caused by the way that measurements are carried out, with previous values being used to calculate current ones, following the same philosophy applied in dead reckoning. The effects of drift on navigation are significant, and other sensing devices need to be taken into account to complement IMU readings. Traditionally, position or motion data from different sources are combined through sensor fusion techniques such as Kalman filtering or fuzzy logic. A common approach to making IMU navigation (also known as inertial navigation systems, or INS) more robust is integrate it with GNSS, which can provide regular corrections of position and speed. GPS, in particular, provides the speed and heading when the roving vehicle is in motion.

Most of the navigation models that are currently implemented require knowledge of the *vehicle's heading*, which gives the orientation of the vehicle with respect to the direction reference, usually north in a local tangent plane system of coordinates. Correction errors for automatic guidance are based on two key parameters: offset and heading, meaning that knowledge of the heading tends to be indispensable. The problem of drift in IMU-based navigation systems practically rules out gyroscopes for heading estimations. Two alternatives are eligible: GPS to determine the vehicle's heading course, although several points are needed for reliability, and the vehicle needs to be in motion; and a magnetic compass, to find the orientation with respect to the magnetic North Pole. Intelligent vehicles utilize a modern version of the conventional magnetic compass: the electronic or fluxgate compass. This device can output electronic measurements that are easier for the vehicle's circuitry to handle.

1.5.9 Other Sensors for Monitoring Engine Functions

Complete automation of an agricultural vehicle may involve many different functions, such as raising the implement in the headlands, slowing down at the ends of rows, accelerating and up-shifting at the beginning of a row, engaging the power take-off, locking the differential or one of the traction wheels for turning, and so on. The implementation of these functions requires the proper actuators and a number of feedback sensors, either to directly assist in the operation of the vehicle or to send the information to a data-acquisition station through a wireless network. Typical sensors may include engine tachometers, speedometers, gear lever position, three-point-hitch position, fuel consumption and tank usage, brake position, differential lock–unlock position, *etc.* A successful way to coordinate these sensors is via an onboard computer area network (CAN), an information bus that links all of the sensors in the vehicle and facilitates its overall control.

References

1. Russell S, Norvig P (2002) Artificial intelligence: a modern approach. Prentice Hall, Upper Saddle River, NJ

2. Meystel A (1993) Autonomous mobile robots: vehicles with cognitive control. World Scientific Publishing Co., Singapore
3. Moravec H (2000) Robot: mere machine to transcendent mind. Oxford University Press, New York
4. Brooks R (1999) Cambrian intelligence: the early history of the new AI. MIT Press, Cambridge, MA
5. Gibbs WW (2008) Innovations from a robot rally. Sci Am Rep 18:80–88

Chapter 2
Off-road Vehicle Dynamics

2.1 Off-road Vehicle Dynamics

A thorough understanding of vehicle dynamics is essential when designing high performance navigation systems for off-road vehicles. This section intends to provide readers with a comprehensive framework of the dynamics involved with wheel-type off-road vehicles. For a theoretical analysis of vehicle dynamics, it is a common practice to define the motion equations in reference to the body of the vehicle, and so vehicle-fixed coordinate systems are often used to describe the fundamental dynamics of vehicles [1]. As depicted in Figure 2.1, a conventional vehicle coordinate system consists of *body-fixed coordinates* (hereafter the *body* coordinates) and *steering wheel-fixed coordinates* (hereafter the *wheel* coordinates). The origin of *body coordinates* is often defined as the center of gravity (CG) of the vehicle, with

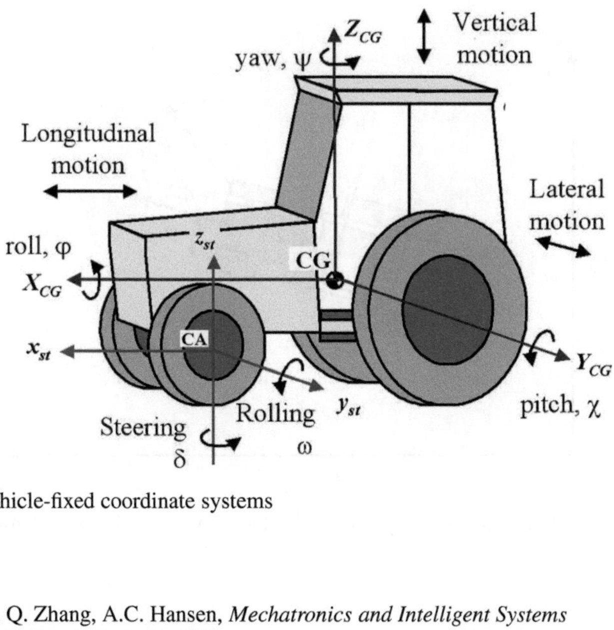

Figure 2.1 Vehicle-fixed coordinate systems

its X_{CG} direction pointing in the direction of travel (or longitudinal motion), its Y_{CG} coordinate following the left-side direction (also denoted the lateral motion), and its Z_{CG} direction indicating the vertical direction. The *wheel coordinates*, also represented in Figure 2.1, are defined with their origin at the center of the axle connecting both front wheels (sometimes referred to as the center of the axle, CA), its x_{st} direction pointing in the forward longitudinal direction, and the y_{st} and z_{st} coordinates defined similarly to those of the body coordinates. The origin of the wheel coordinates is offset from that of the vehicle coordinates by a fixed distance determined by the physical separation of the vehicle's center of mass (CG) from the axle center of the front (steered) wheels (CA).

An essential task of automated off-road vehicles that move around sites of operation is to guide the vehicle in a path that allows it to perform its assigned functions. The path information may be presented as a series of location data points P_i in a site-specific coordinate system, the *site coordinates*, whereas the parameters defining the motion of the vehicle are often expressed in body coordinates. Therefore, it is necessary to update the vehicle positioning information, usually given in site coordinates, in terms of the vehicle motion information expressed in vehicle-fixed coordinates. In most cases, even when off-road vehicles traverse uneven and unpredictable terrains, the amplitude of vertical motion is normally negligible compared to the magnitude of horizontal displacement. Consequently, it is reasonable to assume that, in general, off-road vehicles move on 2D surfaces. As illustrated in Figure 2.2, the expected trajectory of a vehicle can be estimated via its motion status given in body coordinates (X_V, Y_V). When the current position and orientation of a vehicle is known in site coordinates, the expected trajectory along the working site can be traced to provide the information needed for navigation.

If the forward velocity of the vehicle in body coordinates at time instant t is given by the derivative of x_b with respect to time, and its lateral velocity is given by the derivative of y_b with respect to time, the expected current position of the vehicle

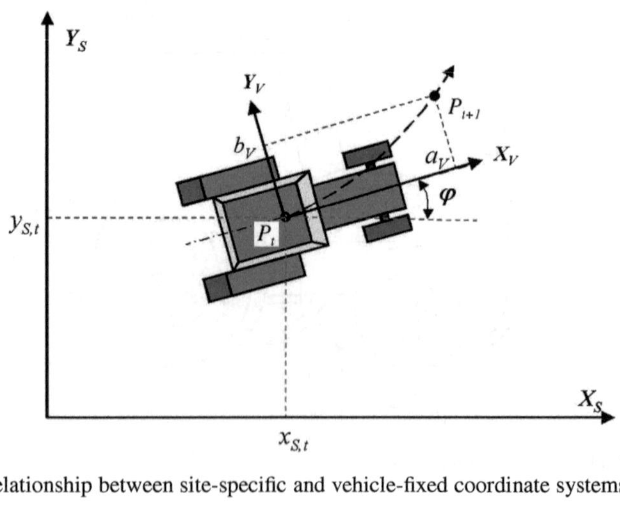

Figure 2.2 Relationship between site-specific and vehicle-fixed coordinate systems

$(t+1)$ in relation to its previous position (t) after a finite time interval Δt has elapsed can be calculated with Equation 2.1.

$$\left.\begin{aligned} x_{b,t+1} &= \dot{x}_b \cdot \Delta t \\ y_{b,t+1} &= \dot{y}_b \cdot \Delta t \end{aligned}\right\} \tag{2.1}$$

When the current position $(x_{s,t}, y_{s,t})$ and the heading angle φ of the vehicle, both expressed in site coordinates, are known, the expected position of the vehicle given in the body coordinates of Equation 2.1 can be transformed to site coordinates through Equation 2.2, where $[x_{b,t+1}, y_{b,t+1}]^{\mathrm{T}}$ is the expected vehicle position in body coordinates, $[x_{s,t}, y_{s,t}]^{\mathrm{T}}$ is the current position in site coordinates, $[x_{s,t+1}, y_{s,t+1}]^{\mathrm{T}}$ is the expected position of the vehicle expressed in site coordinates, and φ is the heading angle of the vehicle, also represented in site coordinates. Site coordinates are usually globally referenced, given the popularity of satellite positioning systems. The sequence of expected positions for the vehicle in site coordinates defines the resulting trajectory of the vehicle for a given dynamic status, and usually provides the basis for making navigational decisions in the field.

$$\begin{bmatrix} x_{s,t+1} \\ y_{s,t+1} \end{bmatrix} = \begin{bmatrix} \cos\varphi & -\sin\varphi \\ \sin\varphi & \cos\varphi \end{bmatrix} \cdot \begin{bmatrix} x_{b,t+1} \\ y_{b,t+1} \end{bmatrix} + \begin{bmatrix} x_{s,t} \\ y_{s,t} \end{bmatrix} \tag{2.2}$$

The *wheel tractive force* is a function of the axle load and the wheel slip of the vehicle, and is considered one of the main parameters affecting motion dynamics of off-road vehicles, in addition to accelerating, braking, steering and hauling dynamics. The analysis of axle loads can provide a basic, but effective, means of quantifying the tractive efforts of a vehicle under a multitude of working conditions. Figure 2.3 depicts the force balance of a conventional off-road vehicle in a generic situation. Based upon the assumption that the vehicle is climbing a flat (2D) slope, and considering that all of the mass of the vehicle is acting on its center of mass (CG), the force balance for this vehicle in the longitudinal (x) direction can be expressed as Equation 2.3, where all of the parameters involved in the equation are listed in Table 2.1. Notice that the lateral dimension (y) and the turning moments should also be taken into account to show the effects of the drawbar load and the slope angle θ on the tractive effort. Equation 2.3 can be further simplified to Equation 2.4 if F_{t} is the total tractive force and R_{r} the total wheel–ground friction.

$$m \cdot \ddot{x} = F_{\mathrm{f}} + F_{\mathrm{r}} - R_{\mathrm{a}} - R_{\mathrm{rf}} - R_{\mathrm{rr}} - R_{\mathrm{d}} - R_{\mathrm{g}} \tag{2.3}$$

$$F_{\mathrm{t}} = m \cdot \ddot{x} + R_{\mathrm{a}} + R_{\mathrm{r}} + R_{\mathrm{d}} + R_{\mathrm{g}} \tag{2.4}$$

The tractive effort is defined as the ability of a powered wheel to grip the ground underneath, generating a torque on the wheel and thus moving the vehicle. The maneuverability of an off-road vehicle is limited by the maximum tractive effort on the wheels. Thus, for instance, the steering performance of a vehicle is greatly affected by the tractive effort on the front wheels (for front-wheel-steered vehicles), and the acceleration–braking performance is determined by the tractive effort on the

Table 2.1 Fundamental forces acting on off-road vehicles, as represented in Figure 2.3

Parameter	Definition
$\ddot{x}$	Linear acceleration of the vehicle in the longitudinal direction
m	Mass of the vehicle
F_f	Tractive effort acting on the front wheel for front wheel drive vehicles
F_r	Tractive effort acting on the rear wheel
R_a	Aerodynamic resistance acting on the vehicle
R_{rf}	Rolling resistance between the front wheel and the ground surface
R_{rr}	Rolling resistance between the rear wheel and the ground surface
R_d	Drawbar load acting on the vehicle
R_g	Grade resistance acting on the vehicle

driven wheels. The normal loads acting on the axles can be determined from a force balance analysis conducted on the vehicle at any arbitrary position. For a vehicle climbing up a slope, as illustrated in Figure 2.3, the normal loads on the front and rear wheels can be evaluated by summing up the moments at the wheel–ground contact point of the rear wheels (point A in Figure 2.3) or the wheel–ground contact point of the front wheels (point B). Because off-road vehicles usually travel slowly while they are operating, it is reasonable to ignore the aerodynamic resistance in the dynamic equations presented above. This simplification results in the expressions for the normal axle loads shown in Equations 2.5 (front) and 2.6 (rear), where W is the total weight of the vehicle, W_f and W_r are the normal loads on the front and rear wheel axles, h and h_d represent the elevation of the center of mass and the drawbar load acting on the vehicle, L_f and L_r are the respective distances of the front and rear axles from the vehicle's center of mass, L is the distance between the front and

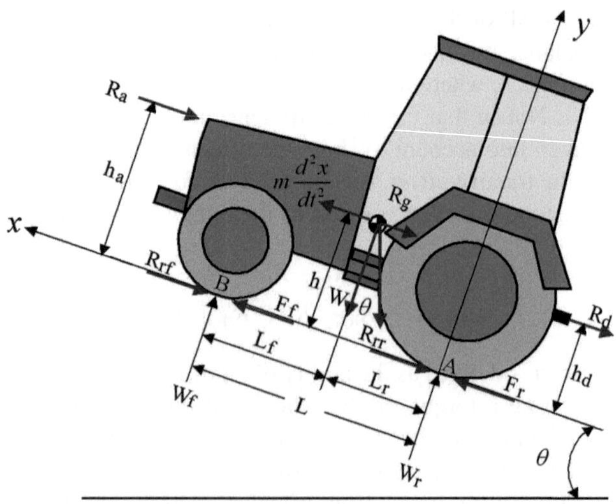

Figure 2.3 Dynamic analysis of a generic off-road vehicle

rear axles (wheel base), and θ is the slope of the ground.

$$W_{\mathrm{f}} = \frac{W \cdot L_{\mathrm{r}} \cdot \cos\theta - W \cdot h \cdot \sin\theta - m \cdot \ddot{x} \cdot h - R_{\mathrm{d}} \cdot h_{\mathrm{d}}}{L} \tag{2.5}$$

$$W_{\mathrm{r}} = \frac{W \cdot L_{\mathrm{f}} \cdot \cos\theta + W \cdot h \cdot \sin\theta + m \cdot \ddot{x} \cdot h + R_{\mathrm{d}} \cdot h_{\mathrm{d}}}{L} \tag{2.6}$$

When off-road vehicles operate on flat terrains where slopes are very small, it is reasonable to assume that $\cos\theta \approx 1$ and $\sin\theta \approx \theta$. Applying this simplification to Equations 2.5 and 2.6, and introducing F_{t} from Equation 2.4 to replace the inertial force term when $h \approx h_{\mathrm{d}}$, which is often true for many off-road vehicles, the normal loads on the front and rear axles can be rewritten as Equations 2.7 and 2.8. Note that the first terms on the right side of the equality represent the static load on the axle when the vehicle is at rest on horizontal ground, while the second terms provide the dynamic components of the normal loads on the wheel axles.

$$W_{\mathrm{f}} = \frac{L_{\mathrm{r}}}{L} \cdot W - \frac{h}{L} \cdot (F_{\mathrm{t}} - R_{\mathrm{r}}) \tag{2.7}$$

$$W_{\mathrm{r}} = \frac{L_{\mathrm{f}}}{L} \cdot W + \frac{h}{L} \cdot (F_{\mathrm{t}} - R_{\mathrm{r}}) \tag{2.8}$$

The friction between the wheels and the ground is determined by the friction coefficient and the normal static load acting on the wheels. The *friction coefficient* f is defined as the ratio of the frictional force acting on the wheel and the wheel–ground contact force [2]. Introducing the friction coefficient f into Equations 2.7 and 2.8 leads to Equations 2.9 and 2.10.

$$W_{\mathrm{f}} = \left(\frac{L_{\mathrm{r}}}{L} + \frac{h}{L} \cdot f\right) \cdot W - \frac{h}{L} \cdot F_{\mathrm{t}} \tag{2.9}$$

$$W_{\mathrm{r}} = \left(\frac{L_{\mathrm{f}}}{L} - \frac{h}{L} \cdot f\right) \cdot W + \frac{h}{L} \cdot F_{\mathrm{t}} \tag{2.10}$$

Once the normal loads acting on the wheel axles have been calculated, the tractive effort can be obtained by multiplying the coefficient of traction by the normal load acting on the wheel axle [3]. Equation 2.11 defines the maximum tractive effort F_{max} of a rear-wheel-driven vehicle, where μ is the wheel's *tractive coefficient*, which varies with wheel type, traveling speed, and the nature of the soil. The tractive coefficient μ is determined experimentally.

$$F_{\mathrm{max}} = \mu \cdot W_{\mathrm{r}} \tag{2.11}$$

The maximum tractive effort of a rear-wheel-driven vehicle can be estimated by combining Equations 2.10 and 2.11, leading to Equation 2.12.

$$F_{\mathrm{max}} = \frac{\mu \cdot W \cdot (L_{\mathrm{f}} - f \cdot h)}{L - \mu \cdot h} \tag{2.12}$$

In the case of a four-wheel-drive (4WD) vehicle, the maximum tractive effort is limited by the normal load on the front axle, and can be calculated via Equation 2.13. This mathematical expression assumes that the right and left wheels give identical performance. The tractive forces play a critical role in the steering performance of an off-road vehicle, and so it is essential to estimate them when modeling the steering dynamics. Equations 2.11 and 2.12 provide approximations for vehicles traveling slowly on flat terrains; however, for higher operational speeds or significant slopes, the loads caused by aerodynamics and terrain inclination will also have to be taken into account.

$$F_{\max -4WD} = \frac{\mu \cdot W \cdot (L_r + f \cdot h)}{L + \mu \cdot h} \tag{2.13}$$

2.2 Basic Geometry for Ackerman Steering: the Bicycle Model

The design of high-performance control systems that assist in the navigation of agricultural vehicles requires a comprehensive understanding of steering dynamics, given that they reveal the vehicle's response to a steering action. Given that this response depends on the method of steering and the chassis of the vehicle (that is, whether a rigid-body vehicle or an articulated vehicle is considered), this section will focus on the steering dynamics of rigid-body vehicles, also known as *Ackerman steering*, as they constitute the most generalized configuration. An important issue when studying the steering dynamics of rigid-body off-road vehicles is the *cornering behavior* of the vehicle when traveling on level terrain at low speeds, thus neglecting the effects of slopes and centrifugal forces on the vehicle's responses to steering actions. The schematic of Figure 2.4 shows the typical steering geometry

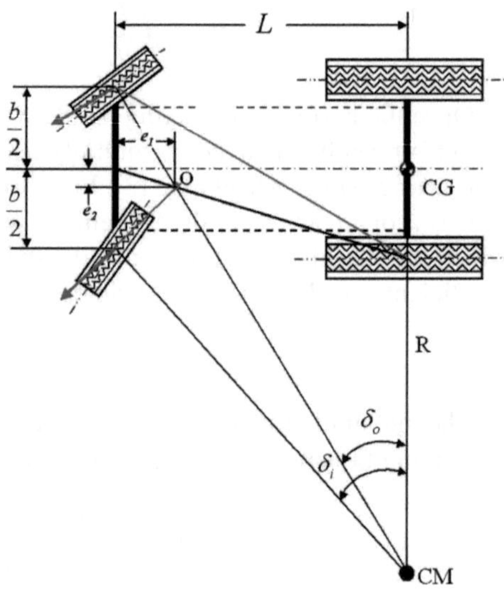

Figure 2.4 Steering geometry for rigid-body off-road vehicles

of rigid-body off-road vehicles, where turning is achieved by changing the heading of the front wheels after actuating the steering mechanism. In order to simplify this analysis without losing generality, the center of mass (CG) of the vehicle is placed at the center of the rear axle, as shown in the diagram.

The first step in the analysis of the steering performance of rigid-body vehicles is to determine the nonlinear relationship between the turning angles of the inside front wheel and the outside front wheel of the vehicle for any demanding steering action. Assuming slow movements, wheel slippages can be ignored for the calculation of this angular relation. According to Figure 2.4, the geometrical relationship between the inside steering angle δ_i and the outside steering angle δ_o can be expressed via Equations 2.14, 2.15, and 2.16, where L is the wheelbase (front–rear axle distance), b is the separation of the front wheels, and R is the radius of the turn made by the vehicle.

$$\cot(\delta_o) = \frac{R + \frac{b}{2}}{L} \tag{2.14}$$

$$\cot(\delta_i) = \frac{R - \frac{b}{2}}{L} \tag{2.15}$$

$$\cot(\delta_o) - \cot(\delta_i) = \frac{b}{L} \tag{2.16}$$

The relationship established in Equation 2.16 implies that the turning radius of the outside front wheel intersects at O with the straight line connecting the midpoint of the front axle and the center of the inside rear wheel, as drawn in Figure 2.4. Point O is denoted the *vehicle turning center*, and this steering geometry is usually referred to as the *Ackerman steering geometry*. Notice that the vehicle turning center O generates the two additional parameters e_1 and e_2, which provide a means to accurately determine its location in the vehicle. Furthermore, the general relationships described by Equations 2.14 to 2.16 can be expressed in terms of e_1 and e_2, as given in Equations 2.17–2.20. The Ackerman steering geometry provides a theoretical framework to establish a mathematical relationship between the steering angles of the inside and outside front wheels. In practice, however, it is customary to use the average value of the two turning angles δ_i and δ_o as the commanded steering angle; this is generally known as the *Ackerman steering angle*.

$$\cot(\delta_o) = \frac{e_2 + \frac{b}{2}}{e_1} \tag{2.17}$$

$$\cot(\delta_i) = \frac{\frac{b}{2} - e_2}{e_1} \tag{2.18}$$

$$\cot(\delta_o) - \cot(\delta_i) = 2 \cdot \frac{e_2}{e_1} \tag{2.19}$$

$$\frac{e_2}{e_1} = \frac{b}{2 \cdot L} \tag{2.20}$$

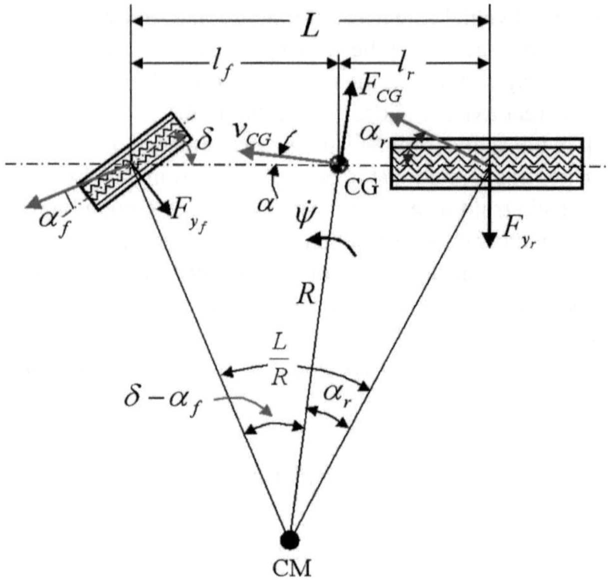

Figure 2.5 Bicycle model for analyzing the steering systems of rigid-body vehicles

The final objective of a navigation control system is to deduce the Ackerman angle from the relative position of the vehicle in relation to its desired path, so that proper commanded angles can guide the vehicle accurately. In an attempt to simplify the analysis of the steering system, the four-wheel geometrical model of Figure 2.4 is normally condensed into the *bicycle model* of Figure 2.5, which facilitates the calculation of the Ackerman angle when the vehicle is subjected to centrifugal and cornering forces caused by turning maneuvers. Following the geometry proposed in Figure 2.5, the steering angle of the front wheel can be determined with Equation2.21, where δ is the Ackerman angle, R is the turning radius, L is the wheel base, α_{f} is the front wheel slip angle, and α_{r} is the rear wheel slip angle.

$$\delta = \frac{L}{R} + \alpha_{\mathrm{f}} - \alpha_{\mathrm{r}} \tag{2.21}$$

Equation 2.21 indicates that wheel slippage plays an important role in determining the commanded steering angle δ. The slip angles α are the result of the cornering forces that act on the wheels to balance the centrifugal force of the vehicle provoked by the turning maneuver. The cornering forces induced in the vehicle can be calculated by establishing a dynamic equilibrium in the lateral direction. For small steering angles, the cornering forces acting on the front and rear wheels can be estimated with Equations 2.22 and 2.23 [3], where $F_{y\mathrm{f}}$ is the cornering force acting on the front wheel, $F_{y\mathrm{r}}$ is the cornering force acting on the rear wheel, W_{f} is the part of the total weight resting on the front axle, W_{r} is the weight on the rear axle, and v is the traveling velocity of the vehicle.

$$F_{y\mathrm{f}} = 2 \cdot \frac{W_\mathrm{f} \cdot v^2}{g \cdot R} \tag{2.22}$$

$$F_{y\mathrm{r}} = 2 \cdot \frac{W_\mathrm{r} \cdot v^2}{g \cdot R} \tag{2.23}$$

The cornering forces $F_{y\mathrm{f}}$ and $F_{y\mathrm{r}}$ are acting as lateral tractive forces that prevent the wheels from slipping under centrifugal forces. The cornering behavior of the tires is usually determined by the *cornering stiffness*, an experimental coefficient defined as the derivative of the cornering force with respect to the slip angle before slippage occurs. In essence, the cornering stiffness is a linear approximation of the lateral force versus slip angle curve at low slip angles. The cornering stiffness for a given type of tire must be determined empirically, and it is necessary to estimate the front and rear slip angles with Equations 2.24 and 2.25, where C_f is the cornering stiffness of the front wheel and C_r is the cornering stiffness of the rear wheel [3]. The substitution of Equations 2.24 and 2.25 into Equation 2.21 gives the maximum steering angle that can be implemented to perform a turn free of slippage at a traveling speed v, as shown in Equation 2.26. It is important to keep in mind that the cornering stiffness can vary with changes in velocity, axle load, or the ground surface. For this reason, the maximum slipless turning angle is always a dynamic magnitude that depends on shifting traveling conditions.

$$\alpha_\mathrm{f} = \frac{F_{y\mathrm{f}}}{C_\mathrm{f}} = \frac{2 \cdot W_\mathrm{f} \cdot v^2}{C_\mathrm{f} \cdot g \cdot R} \tag{2.24}$$

$$\alpha_\mathrm{r} = \frac{F_{y\mathrm{r}}}{C_\mathrm{r}} = \frac{2 \cdot W_\mathrm{r} \cdot v^2}{C_\mathrm{r} \cdot g \cdot R} \tag{2.25}$$

$$\delta = \frac{L}{R} + \left(\frac{W_\mathrm{f}}{C_\mathrm{f}} - \frac{W_\mathrm{r}}{C_\mathrm{r}}\right) \cdot \frac{2 \cdot v^2}{g \cdot R} \tag{2.26}$$

Since off-road vehicles often operate within specific sites, their navigational parameters, such as target path waypoints, are often expressed in site coordinates. Consequently, before calculating the commanded steering angles, it is convenient to define the bicycle model in site coordinates, as shown in Figure 2.6 [4]. According to this figure, when the vehicle turns at slow speeds, its velocity presents an angle α with the heading direction of the vehicle, pointing in the turning direction. This angle α is caused by the slippage of the wheel during the turn, and is often regarded as the vehicle *sideslip angle*. In actual situations, off-road vehicles may start at any location in the site and travel in any possible direction. To simplify the computation of Ackerman angles, a starting point for the vehicle can be set in order to construct the bicycle model. A conventional approach is based on the assumption that the vehicle always starts at the origin of the site coordinates, which in practice leads us to define the origin of the coordinates as the center of gravity of the vehicle (CG), and the vehicle heading direction as the X direction of the site coordinates.

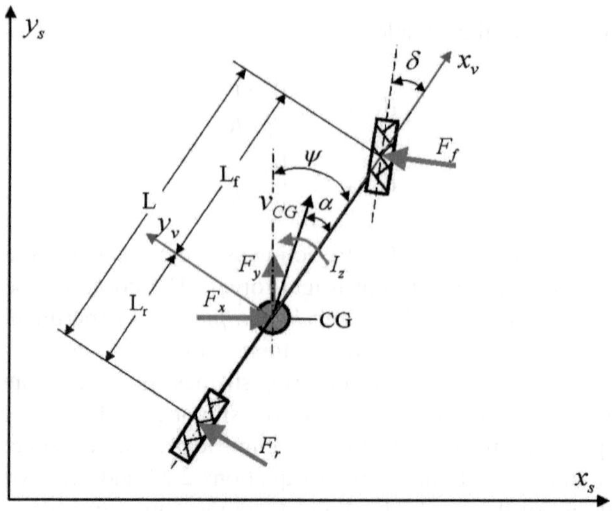

Figure 2.6 Definition of the bicycle model in site coordinates

The motion dynamics of a vehicle can be defined mathematically according to Newton's second law, which takes the form of Equations 2.27 and 2.28 when applied to the bicycle model shown in Figure 2.6, where I_z is the moment of inertia with respect to the center of mass CG, ψ is the yaw (or heading) angle of the vehicle, F_y is the centrifugal force in the Y direction expressed in site coordinates, M is the mass of the vehicle, y_s is its longitudinal displacement represented in site coordinates, F_f is the cornering force on the front wheel, and F_r is the cornering force on the rear wheel.

$$I_z \cdot \ddot{\psi} = -F_r \cdot L_r + F_f \cdot L_f \cdot \cos\delta \tag{2.27}$$

$$M \cdot \ddot{y}_s = F_y - F_r \cdot \cos\psi - F_f \cdot \cos\psi \cdot \cos\delta \tag{2.28}$$

Both the lateral (X direction) and the longitudinal (Y direction) velocities of the vehicle play important roles in the turning dynamics. Equations 2.29 and 2.30 provide an expression to estimate both velocities from the velocity of the center of mass of the vehicle v_{CG}, where α is the slip angle.

$$\dot{x}_s = v_{CG} \cdot \cos(\psi - \alpha) \tag{2.29}$$

$$\dot{y}_s = v_{CG} \cdot \sin(\psi - \alpha) \tag{2.30}$$

The application of the desired steering angle δ requires the actuation of a hydraulic steering cylinder in coordination with the articulation of a mechanical linkage. The sketch shown in Figure 2.7 models a generic steering mechanism for an off-road vehicle, following the traditional four-bar assemblage of tractors. The hydraulic cylinder extends its rod to pull or push bar "*d*," whose movement is immediately transmitted to bar "*b*" by the connecting bar "*c*." The main geometrical

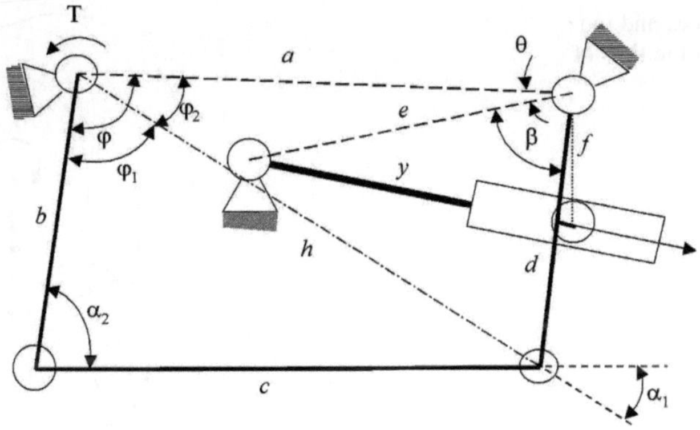

Figure 2.7 Four-bar steering mechanism used in agricultural tractors

relationship for the particular case of Figure 2.7 provides the angle of kingpin arm φ in Equation 2.31, where the intermediate distance $h(\beta)$ can be calculated with Equation 2.32 and the angle $\beta(y)$ can be estimated through Equation 2.33. The rest of the geometrical parameters "e," "f," and "a," and the extension of the rod "y" are included in Figure 2.7. Although the control gain corresponding to the steering linkage is nonlinear, it can be considered linear for part of the steering angle range, and fortunately this part includes most of the angles turned while navigating along straight or slightly curved rows. However, the steering torque applied to front wheel kingpins is highly nonlinear.

$$\varphi = \varphi_1 + \varphi_2 = \arccos\left(\frac{b^2 + (h(\beta))^2 - c^2}{2 \cdot b \cdot h(\beta)}\right) + \arccos\left(\frac{a^2 + (h(\beta))^2 - d^2}{2 \cdot a \cdot h(\beta)}\right) \tag{2.31}$$

$$h(\beta) = \sqrt{a^2 + d^2 - 2 \cdot a \cdot d \cdot \cos(\beta(y) - \theta)} \tag{2.32}$$

$$\beta(y) = \arccos\left(\frac{e^2 + f^2 - y^2}{2 \cdot e \cdot f}\right) \tag{2.33}$$

2.3 Forces and Moments on Steering Systems

The forces and moments exerted on turning wheels originate from the ground's reaction to the steering action generated at the *tire–ground interface* [5]. The main forces and moments on the steering tires of two-wheel-drive (2WD) vehicles are illustrated in Figure 2.8, and comprise three reaction forces (known as the normal force F_z, the tractive force F_x, and the lateral force F_y), and three reaction moments (called the aligning torque M_z, the rolling resistance moment M_y, and the overturn-

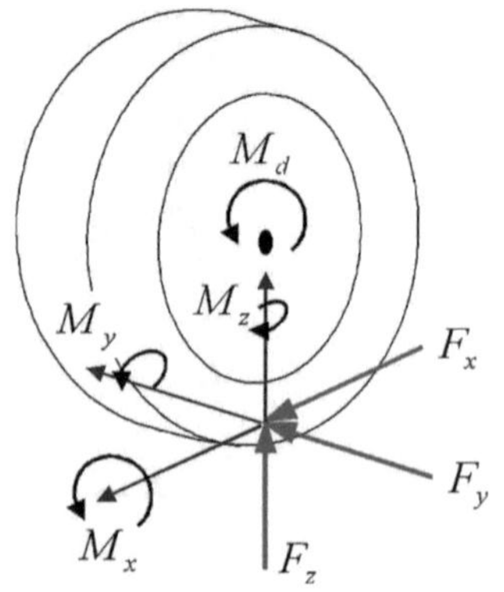

Figure 2.8 Forces and moments acting on the tires of vehicles

ing moment M_x). For a four-wheel-drive (4WD) vehicle, an additional moment of the driving torque M_d also acts on the turning wheel. The vertical force F_z is the reaction of the ground to the axle load, and this plays a critical role in vehicle steering. Rubber tires are the most common type of wheels used on modern off-road vehicles. In essence, tires provide three basic functions: first, they offer stable support for the vertical load; second, tires allow the existence of the longitudinal force needed for acceleration or braking; and third, they also permit the development of the cornering lateral force that makes a vehicle maneuverable. A typical rubber tire is an elastic structure consisting of a flexible carcass of high tensile strength cords fastened to steel cable beads (Figure 2.14). The inflation pressure stresses the structure of the tire in such a way that any external force causing carcass deformation results in a reaction force from the tire. The behavioral characteristics of rubber tires are normally very complex and highly nonlinear.

The power management of conventional off-road vehicles requires the presence of a combustion engine that produces the proper torque and a mechanical transmission system that delivers the torque to the driven wheels (overcoming the rolling resistance), which is technically called the *traction* of the wheels. It is important to emphasize that the amount of torque available is determined by the amount of traction in the powered wheels, not by the maximum torque of the diesel engine. *Traction* is defined as the maximum amount of force that a wheel can apply against the ground to push the vehicle and originate its motion. The acting point of the force on the tire is called the *wheel–ground contact point*, often abbreviated to WGCP. Equations 2.34 and 2.35 are often used to estimate the location of the WGCP (n_L, n_S) in relation to the geometrical center of the wheel–ground contact surface of the rubber tire, where l_0 and l_1 are wheel parameters, F_z is the force acting vertically at the WGCP, F_{z0} is the nominal vertical force at the wheel contact point, C_{press} is a coefficient to correct for the distribution of tire pressure, α is the slip angle of the wheel, and F_y is the lateral force. Figure 2.9 shows the migration of the WGCP from the middle of the tireprint area. This misalignment creates a torque with a longitudinal force that increases the self-aligning torque while accelerating and decreases the self-aligning torque while braking. F_{WS} and F_{WL} are the wheel frictional forces

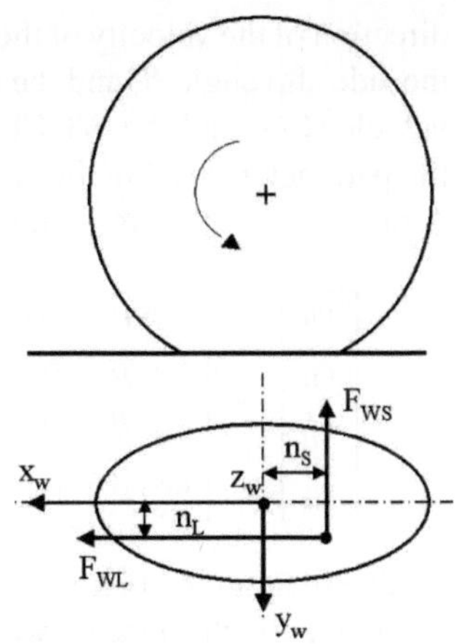

Figure 2.9 Model for situating the wheel–ground contact point (WGCP) within the tireprint area

applied at the WGCP along the lateral and longitudinal directions, respectively, as indicated in the scheme of Figure 2.9.

$$n_{\mathrm{L}} = \frac{1}{2} \cdot \left(l_0 + l_1 \cdot \frac{F_Z}{F_{Z0}} \right) \tag{2.34}$$

$$n_{\mathrm{S}} = 3 \cdot n_{\mathrm{L}} \cdot \tan \alpha + \frac{F_y}{C_{\mathrm{press}}} \tag{2.35}$$

The approximate determination of the linear velocity at the position of the contact point between each tire and the road requires a knowledge of the magnitude and

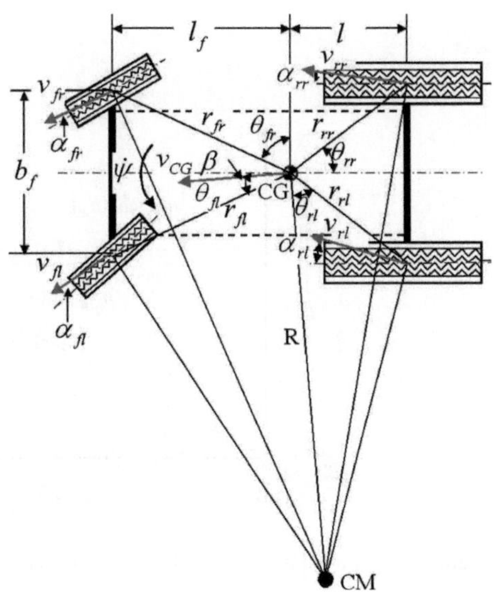

Figure 2.10 Approximate determination of the wheel–ground contact point velocities

direction of the velocity of the vehicle at its center of mass v_{CG}, the yaw rate $(d\psi/dt)$, the side slip angle β, and the distances and angles between the center of mass of the vehicle (CG) and the WGCP of each wheel (r_{ij} and θ_{ij}). Figure 2.10 shows all of the parameters used in the calculation of the wheel–ground contact point velocities. A mathematical expression for them is given in matrix form in Equation 2.36.

$$\begin{bmatrix} \overrightarrow{v_{fl}} \\ \overrightarrow{v_{fr}} \\ \overrightarrow{v_{rl}} \\ \overrightarrow{v_{rr}} \end{bmatrix} = \begin{bmatrix} \cos\beta & -r_{fl}\cdot\sin\theta_{fl} & \sin\beta & r_{fl}\cdot\cos\theta_{fl} \\ \cos\beta & r_{fr}\cdot\sin\theta_{fr} & \sin\beta & r_{fr}\cdot\cos\theta_{fr} \\ \cos\beta & -r_{rl}\cdot\cos\theta_{rl} & \sin\beta & -r_{rl}\cdot\sin\theta_{rl} \\ \cos\beta & r_{rr}\cdot\sin\theta_{rr} & \sin\beta & -r_{rr}\cdot\cos\theta_{rr} \end{bmatrix} \cdot \begin{bmatrix} v_{CG}\cdot\overrightarrow{e_x} \\ \dot{\psi}\cdot\overrightarrow{e_x} \\ v_{CG}\cdot\overrightarrow{e_y} \\ \dot{\psi}\cdot\overrightarrow{e_y} \end{bmatrix} \tag{2.36}$$

The diagram of Figure 2.11 provides greater detail regarding the calculation of the geometric distance between the center of mass of the vehicle chassis CG and the WGCP of the front right tire. Exact calculations of all of the distances r_{ij} and their corresponding angles θ_{ij} are given in Equations 2.37 to 2.44, where $n_{L,ij}$ and $n_{S,ij}$ represent the coordinates for each WGCP considered, and δ_W is the wheel's turning angle. The schematic of Figure 2.11 also serves as a model for the rest of the tires in the vehicle: the procedure is exactly the same for each but the subscript is changed appropriately.

$$r_{fr} = \sqrt{(l_f - n_{L,fr}\cos\delta_W + n_{S,fr}\sin\delta_W)^2 + \left(\frac{b_f}{2} + n_{S,fr}\cos\delta_W + n_{L,fr}\sin\delta_W\right)^2} \tag{2.37}$$

$$r_{fl} = \sqrt{(l_f - n_{L,fl}\cos\delta_W + n_{S,fl}\sin\delta_W)^2 + \left(\frac{b_f}{2} + n_{S,fl}\cos\delta_W - n_{L,fl}\sin\delta_W\right)^2} \tag{2.38}$$

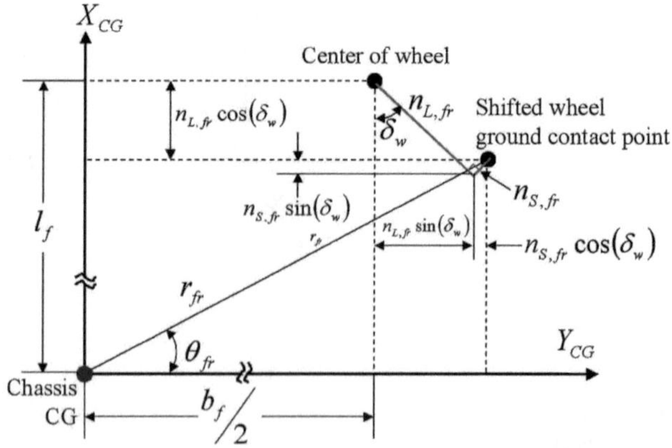

Figure 2.11 Calculation of the distance between the CG and the WGCP for the front right wheel

$$r_{\mathrm{rr}} = \sqrt{(l_{\mathrm{f}} + n_{\mathrm{L,rr}})^2 + \left(\frac{b_{\mathrm{r}}}{2} + n_{\mathrm{S,rr}}\right)^2} \tag{2.39}$$

$$r_{\mathrm{rl}} = \sqrt{(l_{\mathrm{r}} + n_{\mathrm{L,rl}})^2 + \left(\frac{b_{\mathrm{r}}}{2} - n_{\mathrm{S,rl}}\right)^2} \tag{2.40}$$

$$\theta_{\mathrm{fr}} = \arctan\left(\frac{l_{\mathrm{f}} - n_{\mathrm{L,fr}} \cos\delta_{\mathrm{W}} + n_{\mathrm{S,fr}} \sin\delta_{\mathrm{W}}}{\frac{b_{\mathrm{f}}}{2} + n_{\mathrm{S,fr}} \cos\delta_{\mathrm{W}} + n_{\mathrm{L,fr}} \sin\delta_{\mathrm{W}}}\right) \tag{2.41}$$

$$\theta_{\mathrm{fl}} = \arctan\left(\frac{\frac{b_{\mathrm{f}}}{2} - n_{\mathrm{S,fl}} \cos\delta_{\mathrm{W}} - n_{\mathrm{L,fl}} \sin\delta_{\mathrm{W}}}{l_{\mathrm{f}} + n_{\mathrm{L,fl}} \cos\delta_{\mathrm{W}} + n_{\mathrm{S,fl}} \sin\delta_{\mathrm{W}}}\right) \tag{2.42}$$

$$\theta_{\mathrm{rr}} = \arctan\left(\frac{\frac{b_{\mathrm{r}}}{2} - n_{\mathrm{S,rr}}}{l_{\mathrm{r}} + n_{\mathrm{L,rr}}}\right) \tag{2.43}$$

$$\theta_{\mathrm{rl}} = \arctan\left(\frac{l_{\mathrm{r}} + n_{\mathrm{L,rl}}}{\frac{b_{\mathrm{r}}}{2} - n_{\mathrm{S,rl}}}\right) \tag{2.44}$$

The velocity of a vehicle is actually determined by the velocities of the four wheel–ground contact points (previously represented as the WGCP velocities in Figure 2.10). It is important to bear in mind that the actual velocity of the vehicle is often different from the linear velocities of the tires at these contact points, due to the slippage between tires and ground. There are two types of contact that tires can make with the ground: static and dynamic. When there is no slippage between tire and ground, both surfaces are said to be in *static contact*, but when the tire slips relative to the ground, they are in *dynamic contact*. The coefficient of friction for dynamic contact is lower than that for static contact, so the latter always provides better traction. Most of the time when we talk about wheel slippage, we are referring to longitudinal slip without considering the side slip. The *longitudinal slip* at breaking is different from the driving conditions, as expressed in Equations 2.45 and 2.46, where v_{t} is the rotational equivalent velocity (often called the theoretical velocity), and v_{CG} is the vehicle's actual velocity at CG.

$$s_{\mathrm{L,braking}} = \frac{v_{\mathrm{t}} \cdot \cos\alpha - v_{\mathrm{CG}}}{v_{\mathrm{CG}}} \tag{2.45}$$

$$s_{\mathrm{L,driving}} = \frac{v_{\mathrm{t}} \cdot \cos\alpha - v_{\mathrm{CG}}}{v_{\mathrm{t}} \cdot \cos\alpha} \tag{2.46}$$

Apart from the longitudinal slip, there is unavoidably some *side slip* – especially during turning maneuvers – between the tire and the ground, as shown in Figure 2.12. Such side slip causes the wheel to slide, and may lead to serious vehicle stability problems if it is not handled properly. Similar to the determination of longitudinal slip, the side slip differs when breaking and when driving, as shown in

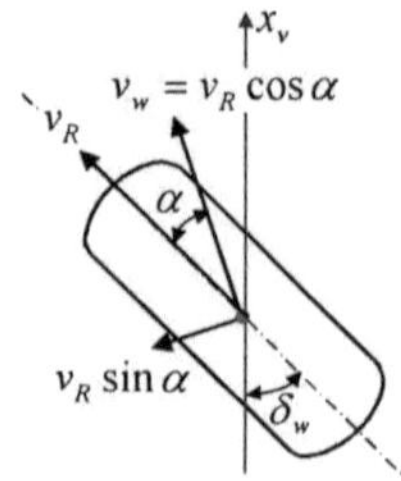

Figure 2.12 Side slippage on tires

Equations 2.47 and 2.48. The *overall wheel slip* s_R is a combination of the longitudinal and the side slip, as shown in Equation 2.49.

$$s_{S,\text{braking}} = \frac{v_t \cdot \sin\alpha}{v_{CG}} \tag{2.47}$$

$$s_{S,\text{driving}} = \tan\alpha \tag{2.48}$$

$$s_R = \sqrt{s_L^2 + s_S^2} \tag{2.49}$$

When a vehicle is moving under driving conditions (*i.e.*, without braking), a driving moment is being applied to the tire axis, and so the tread of the tire will be compressed in the tireprint zone. In this situation, the tire is moving slower than a free tire, implying that $0 < s_R < \infty$. Good traction requires that the overall slippage is kept as close to zero as possible. During breaking conditions, a braking moment is applied to the wheel axis, and the tread of the tire will be stretched in the tireprint zone. This time the tire is moving faster than a free tire, and therefore $s_L < 0$. When the applied braking moment is large enough to lock the tire, the wheel speed is zero, resulting in $s_L = -1$. Both of these conclusions indicate that the longitudinal slip will be between -1 and 0 when braking. The slip is typically quantified by the *slip angle* α_{ij} of each wheel, which can be mathematically defined by a geometric derivation of the wheel velocity vectors defined in Equation 2.36. The particular expression for each slip angle is provided in Equations 2.50–2.53, where the angle γ_{ij} establishes a relationship between the x- and y-components of each velocity, and δ_W is the turning angle of the studied wheel. Note that the overall slip angle of the vehicle β is usually very small, and the simplification $\sin\beta \approx \beta$ and $\cos\beta \approx 1$ has been introduced into the equations for the slip angles α_{ij}. The other parameters are the same as those used and deduced in Equations 2.36–2.44. Also note that there is no δ_W for the rear wheels (subscripts rl and rr), as the vehicle considered in the previous derivations steers by turning the front wheels. A deeper treatment of forces and moments on vehicles can be found in [6, 7].

$$\alpha_{fl} = \delta_W - \arctan\gamma_{fl} = \delta_W - \arctan\left(\frac{v_{CG}\cdot\beta + \dot{\psi}\cdot r_{fl}\cdot\cos\theta_{fl}}{v_{CG} - \dot{\psi}\cdot r_{fl}\cdot\cos\theta_{fl}}\right) \tag{2.50}$$

$$\alpha_{fr} = \delta_W - \arctan\gamma_{fr} = \delta_W - \arctan\left(\frac{v_{CG}\cdot\beta + \dot{\psi}\cdot r_{fr}\cdot\cos\theta_{fr}}{v_{CG} - \dot{\psi}\cdot r_{fr}\cdot\sin\theta_{fr}}\right) \tag{2.51}$$

$$\alpha_{\mathrm{rl}} = -\arctan \gamma_{\mathrm{rl}} = -\arctan\left(\frac{v_{\mathrm{CG}} \cdot \beta - \dot{\psi} \cdot r_{\mathrm{rl}} \cdot \cos\theta_{\mathrm{rl}}}{v_{\mathrm{CG}} - \dot{\psi} \cdot r_{\mathrm{rl}} \cdot \sin\theta_{\mathrm{rl}}}\right) \quad (2.52)$$

$$\alpha_{\mathrm{rr}} = -\arctan \gamma_{\mathrm{rr}} = -\arctan\left(\frac{v_{\mathrm{CG}} \cdot \beta - \dot{\psi} \cdot r_{\mathrm{rr}} \cdot \cos\theta_{\mathrm{rr}}}{v_{\mathrm{CG}} - \dot{\psi} \cdot r_{\mathrm{rr}} \cdot \sin\theta_{\mathrm{rr}}}\right) \quad (2.53)$$

2.4 Vehicle Tires, Traction, and Slippage

Rubber tires began to replace steel wheels in the 1930s, dramatically improving vehicle performance and ride quality. The tire represents the interface between the vehicle and the ground and therefore affects a number of factors, including steering, traction, braking and vibrations caused by uneven ground. It supports the vertical load of the vehicle while cushioning shocks over surface irregularities, and it develops the longitudinal forces needed for traction and braking and the lateral forces required for vehicle cornering (thus ensuring steering control and direction stability). The mechanics of pneumatic tires on both hard and deformable surfaces have been studied in detail, with the latter being of particular interest for off-road vehicles. Key tire dimensions and terminology are shown in Figure 2.13, and are included in

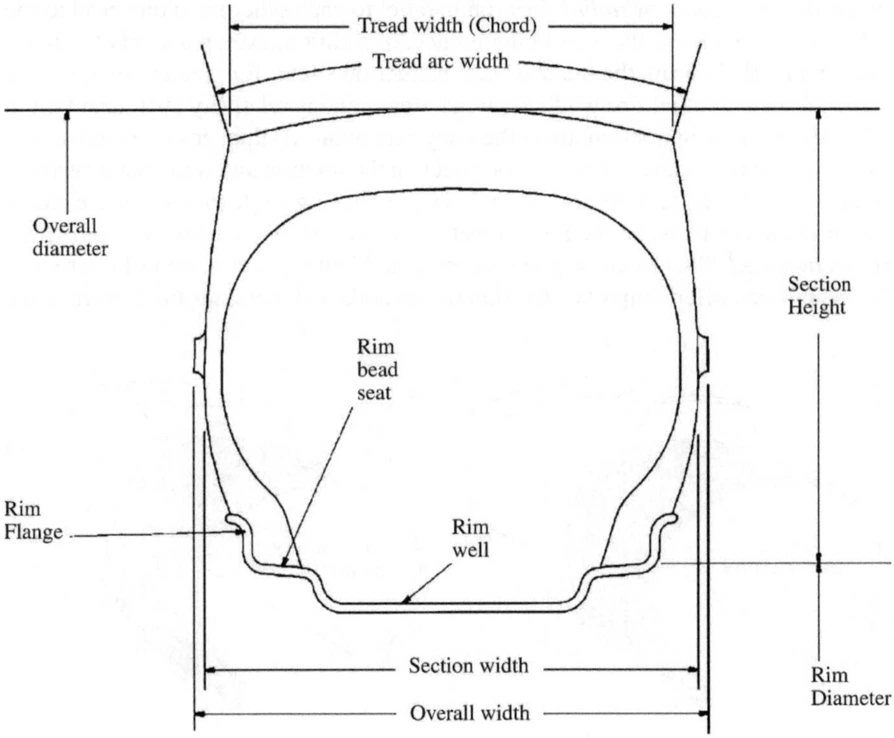

Figure 2.13 Tire dimensions and terminology [8]

a much broader listing of terminology related to traction published in the ASABE Standard S296.5 [8]. The section height and width are undeflected, inflated tire dimensions, and the *aspect ratio* is the ratio of these two measurements expressed as a percentage. The static loaded radius is measured from the axle centerline to the ground when the tire is under load. The rim diameter is the nominal diameter at the intersection of the bead seat and vertical portion of the rim flange [8], and the overall diameter is twice the section height plus the rim diameter. Tire specifications typically include size and load rating as well as tire construction (radial or bias ply). The size of the tire is given by the rim diameter and the tire section width, with the metric size also specifying the aspect ratio. Bias-ply tires used to be rated by the actual number of plies molded into the tire to carry the vertical load [9]. However, radial tires were found to be able to utilize a higher load more effectively than bias-ply tires for better traction. Subsequently, a star rating system was applied to radial tires to indicate the maximum pressure for a given load rating.

Bias-ply and radial tire constructions are shown in Figure 2.14, and comprise layers of fabric that add strength and rigidity to the tire walls. For a *bias-ply tire*, the cords run diagonally from one bead to the other, with the cords of adjacent fabric layers or plies running in opposite directions, thus adding stability and stiffness to the tire walls. The advantage of stiff sidewalls is resistance to cuts and punctures in demanding applications such as forestry. One consequence of the criss-crossing of the body plies is more internal friction, and therefore a greater build-up of heat in the tire. The cords of *radial tires* run parallel to each other from one bead to the other at right angles to the tire circumference. In addition, extra plies or belts are included directly beneath the tread. This construction allows for greater flexing of the sidewalls of the radial tire while the tread area remains relatively stiff, resulting in a 6–14% increase in traction, fuel efficiency, and reduced slippage compared to bias tires [9]. The tread design has a major effect on the traction and wear characteristics of tires. Key design features may include varying the lug angle to improve the cleaning of the space between the lugs in wet soils, and also increasing this void space (to as much as 70%) to allow good cleaning and better penetration in firmer soils. A deeper tread offers improved traction in wet soils and increases tire life when the

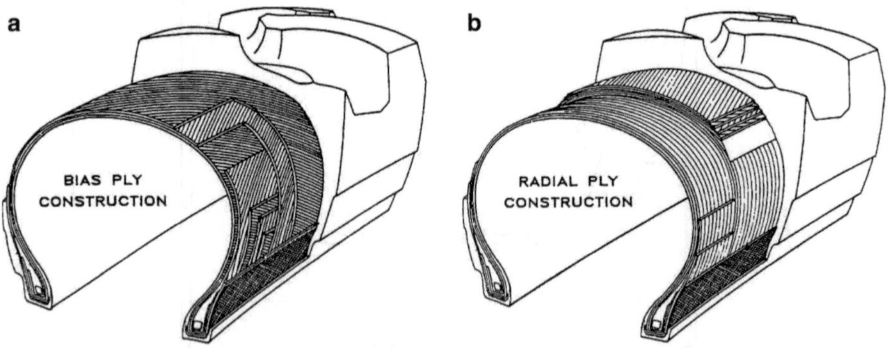

Figure 2.14 Tire constructions: (**a**) bias ply; (**b**) radial ply [9]

vehicle is used on paved road surfaces. For vehicle applications that require minimal ground disturbance, such as airports and golf courses, a non-aggressive, relatively closed tread pattern with a void area of about 30% is available.

A traction analysis can be carried out by first studying a single wheel with all of its related velocities and applied forces, as shown in Figure 2.15. The dynamic load W_d, which acts down on the wheel via the axle, is supported by a vertical surface reaction force R_v. For a driven wheel, the gross tractive force GT is generated by the input torque T applied to the wheel. The motion resistance MR accounts for losses caused by the deformation of the tire and the terrain, and the net traction NT represents the force available to pull an implement. The surface reaction force is represented by R, ω is the angular velocity, and the forward velocity is V.

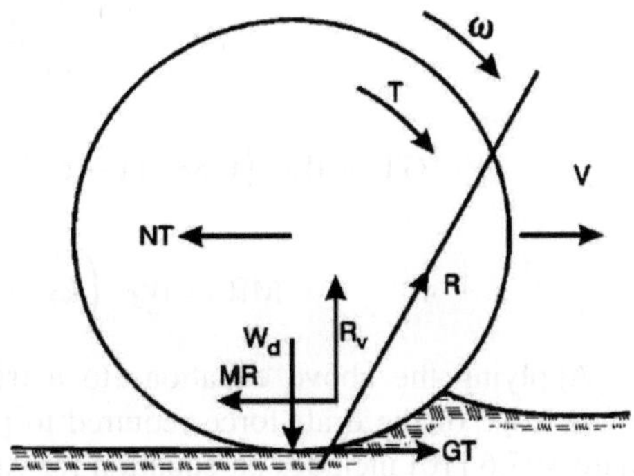

Figure 2.15 Velocities and forces on a single wheel [8]

A tire or track is able to develop a tractive force on a deformable surface such as soil by shearing the soil. This shearing action results in wheel slippage, conventionally determined as in Equation 2.54, where V_t is the theoretical forward travel speed of the vehicle when the rotational motion of the driving axle is perfectly transformed into forward motion, and V_a is the actual forward travel speed (previously denoted as v_{CG}). Longitudinal slip is required to generate a drawbar pull. A slip of 100% means that the actual speed is zero and the traction device is spinning with no forward motion.

$$\text{slip} = s = 100 \cdot \left(\frac{V_t - V_a}{V_t} \right) \tag{2.54}$$

A widely used dimensional analysis approach to predicting off-road traction is specified in the ASABE Standard D497.6 [10], and makes use of a number of ratios and equations developed from extensive soil bin tests with tires. A mobility number B_n accounts for both the shape of the tire and the soil strength, and is a combination of three dimensionless ratios. The soil strength is represented by the cone index, CI, which is the average force per unit area required to force a cone-shaped probe vertically into the soil at a steady rate, as described in ASABE Standard EP542 [11]. The mobility number B_n can be estimated with Equation 2.55, where W_d is the dynamic wheel load normal to the soil surface, CI is the cone index for the soil, b is the unloaded tire section width, d is the unloaded overall tire diameter, h is the tire section height, and d is the tire deflection. Initially, each wheel can be analyzed separately with respect to its contribution to the overall traction performance of the vehicle. The *gross traction* GT developed by a wheel can be determined from Equation 2.56, and the *motion resistance* MR is given by Equation 2.57, where k_1, k_2, and k_3 are constants. The values of the constants k_1, k_2, and k_3 are assigned based on whether the tire is of radial or bias-ply construction. The net traction developed by a single wheel is NT = GT – MR. Equations 2.55–2.57 are applied to each tire for a given soil and slip level, and W_d is the wheel load acting on that specific tire. After the net traction has been determined for all of the wheels supporting the vehicle, the overall net traction can be calculated. In the case of a tractor pulling an implement, this net traction is the horizontal drawbar pull that can be generated.

$$B_n = \frac{\mathrm{CI} \cdot b \cdot d}{W_\mathrm{d}} \cdot \frac{1 + 5 \cdot \frac{\delta}{h}}{1 + 3 \cdot \frac{b}{d}} \tag{2.55}$$

$$\mathrm{GT} = W_\mathrm{d} \cdot \left(0.88 \cdot \left(1 - \mathrm{e}^{-0.1 \cdot B_n}\right) \cdot \left(1 - \mathrm{e}^{-k_1 \cdot s}\right) + k_2\right) \tag{2.56}$$

$$\mathrm{MR} = W_\mathrm{d} \cdot \left(k_2 + \frac{k_3}{B_n} + \frac{0.5 \cdot s}{\sqrt{B_n}}\right) \tag{2.57}$$

Applying the above equations to a tractor–implement combination requires knowledge of the draft force required to pull the implement. The ASABE Standard 497.6 [10] includes an equation for calculating this draft force for a range of different implements for which a set of machine-specific parameters are listed. The draft force is strongly dependent on the field speed of the implement. However, the field speed in turn depends on the tractive performance of the tractor. In addition, the dynamic load on each tire is affected by the transfer of tractor weight from front to rear, as determined by the drawbar pull. Therefore, when analyzing a tractor–implement system with the aid of Equations 2.55–2.57, it is necessary to apply an iterative process to arrive at a situation where the tractor and implement are matched while traveling through the field at a certain speed and wheel slip level. One approach is to assume a slip level in an outer loop of iteration and then initially also assume zero pull, with the dynamic load on each tire being equal to the static load. The equations are then applied to each tire and the net traction for all of the tires representing the drawbar pull is determined. This drawbar pull is then used to calculate a new dynamic load on each tire, and Equations 2.55–2.57 are once again applied until the difference between the previous drawbar pull calculation and the new one is less than a selected amount. For the implement, the assumed slip value is used to determine the actual travel speed, with the theoretical speed being determined from our knowledge of the wheel's rotational speed and the loaded radius of the tire. The draft force for the implement for this actual travel speed can then be calculated and the slip level can be adjusted according to the ratio of the implement draft force and the drawbar pull generated by the tractor. This new slip value is then used in the next iteration of calculations for the drawbar pull as well as for the implement draft force, from which the slip is once again adjusted until there is sufficient convergence with little change in the slip value. These calculations can be used to analyze different tractor–implement configurations for optimal matching.

The tractive performance of a wheel can be represented by the *tractive efficiency*, which is the ratio of the drawbar power to the input power to the wheel. The *drawbar power* is calculated from the product of the drawbar pull and the travel speed. The input power to the wheel requires knowledge of the torque input to the wheel axle and its rotational speed. While the tractive efficiency is useful for assessing the performances of single tires or tractive devices, it does not account for other factors affecting the overall tractor performance. *Power delivery efficiency* is an alternative measure of tractor performance that is determined from the ratio of the drawbar power to the engine power. This efficiency value accounts for differences in drivetrain systems resulting from the use of tracks instead of tires for example.

Tire deformation depends primarily on the inflation pressure of the tire, and to a lesser extent on the tire's construction. Typically, the average surface contact pressure is slightly higher than the tire's inflation pressure, with the difference being caused by tire stiffness [12]. The mean contact pressure of the tire multiplied by the ground contact area must be equal to the applied dynamic load. Therefore, reducing the tire pressure can be expected to increase the contact area through tire deformation for the same dynamic load. Similarly, for a given inflation pressure, an increase in vertical load will result in increases in tire deflection and contact area. The maximum inflation pressure is limited by the tire construction and the number of plies in the tire. Maximum load ratings and tire dimensions are published by the tire manufacturer and are associated with a corresponding maximum inflation pressure at a given maximum travel speed. Larger tires allow for operation at lower pressures, thereby improving tractive capability and reducing surface soil pressures and hence soil compaction. Off-road vehicles frequently operate under different conditions and at different travel speeds. For example, a tractor with an implement may initially travel to a field or between fields on firm road surfaces before entering the field and carrying out an operation under potentially softer soil conditions. In each case, the tractor could benefit from a change in tire inflation pressure. *Central tire inflation* (CTI) systems have been developed and are fitted to vehicles to allow the tire inflation to be adjusted on-the-go so as to optimize traction. CTI systems have been used in Europe for a number of years. Such systems increase the inflation for the relatively high transport speeds permitted and then reduce the inflation for slower field speeds. This inflation pressure adjustment leads to extended tire durability on roads and less soil compaction in the fields.

Rubber tracks are commonly fitted to many tractors and other off-road vehicles as an alternative to tires because they are able to improve traction, especially on tilled soil surfaces. They have been shown to outperform tires in terms of tractive efficiency. However, as mentioned above, differences in the powertrain requirements needed to drive the tracks may cause this improvement in tractive efficiency to be offset by increased losses in the vehicle powertrain.

References

1. Manigal J, Leonhard L (1992) Vehicle control by computer vision. IEEE Trans Ind Electron 39(3):181–188
2. Kiencke U, Nielsen L (2000) Automotive control systems. SAE Int., Warrendale
3. Wong JY (1993) Theory of ground vehicles, 2nd edn. Wiley, New York
4. Zhang Q, Qiu H (2004) A dynamic path search algorithm for tractor automatic navigation. Trans ASAE 47(2):639–646
5. Julian AP (1971) Design and performance of a steering control system for agricultural tractors. J Agric Eng Res 16(3):324–336
6. Centa G (1997) Motor vehicle dynamics: modeling and simulation. World Scientific, Singapore
7. Pacejka HB (2002) Tire and vehicle dynamics. Society of Automotive Engineers/Butterworth–Heinemann, Oxford
8. ASABE (2009) General terminology for traction of agricultural traction and transport devices and vehicles (ASABE Standard S296.5). ASABE, St. Joseph
9. Brodbeck KN (2004) Choosing the right tire. (ASABE Distinguished Lecture No. 28, ASABE Pub. 913C0204). ASABE, St. Joseph
10. ASABE (2009) Agricultural machinery management data (ASABE Standard D497.6). ASABE, St. Joseph
11. ASABE (2006) Agricultural machinery management. (ASABE Standard EP496.3). ASABE, St. Joseph
12. Goering CE, Stone ML, Smith DW, Turnquist PK (2003) Off-road vehicle engineering principles. Publ. ASABE, St. Joseph, Mich.

Chapter 3
Global Navigation Systems

3.1 Introduction to Global Navigation Satellite Systems (GPS, Galileo and GLONASS): the Popularization of GPS for Navigation

Global navigation satellite systems, also known as GNSS, are the set of localization systems that use artificial satellites to determine the position of a receiver in a global (Earth-referenced) system of coordinates. The principle on which GNSS are based is often referred to as *triangulation*, and it involves estimating the distances from at least three satellites orbiting the Earth along different and sufficiently separated trajectories. Given that distances are estimated rather than angles, this calculation method is more accurately termed *trilateration*. The distances between the receiver and the satellites are calculated by measuring the time that an electromagnetic wave takes to travel between the emitter of the satellite and the receptor in the receiver antenna. In theory, three satellites suffice to determine the three coordinates that define the exact position of a point in space, but in practice a fourth satellite is necessary to compensate for uncertainties in the measurement of the time elapsed. However, it is always recommended that the signals from five satellites or more should be detected for redundancy purposes, as atmospheric, multipath, or mask configuration errors are likely to occur. Figure 3.1 illustrates the basic principle of trilateration for satellite localization.

The design and implementation of a satellite positioning system is an ambitious project that only the most powerful nations, or a consortium of them, can afford. The generic term GNSS includes all of the satellite positioning systems that are already functional or being developed at present, such as GPS (USA), Galileo (European Union), GLONASS (Russia), and Beidou-Compass (China). *GLONASS* (*Glo*bal *Na*vigation *S*atellite *S*ystem) was developed by the former Soviet Union in the 1980s and is now operated by the Russian Space Forces. This system suffered from maintenance problems after the collapse of the Soviet Union, but Russia, in collaboration with India, committed to making the system fully operational (24 satellites) by 2009; it was almost fully restored by July 2010, with 21 opera-

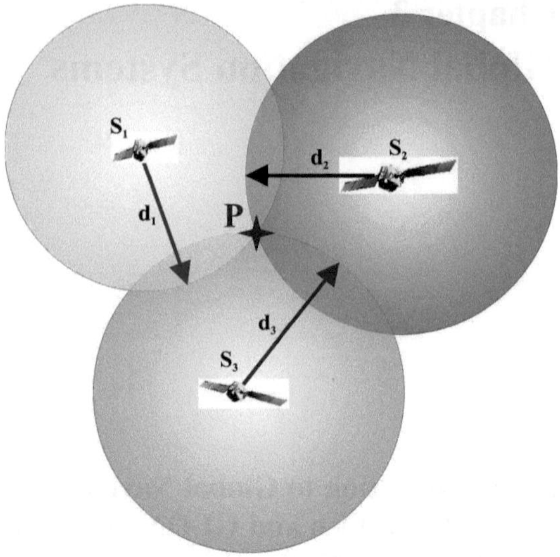

Figure 3.1 Conceptual representation of satellite trilateration

tional satellites and two spares. The *Galileo* positioning system is currently being built by the European Union (27 countries; 2007) and the European Space Agency (ESA). The goal is for it to be completely functional by 2013. The *Beidou Satellite Navigation and Positioning System* is being developed by China. At present, it has a few experimental satellites that provide limited coverage, but the objective is to expand the constellation to 35 satellites under the denomination *Compass*. Unlike other GNSS that use medium Earth orbits (MEOs) that are typically located at altitudes of 19,000–23,000 km, Beidou places its satellites in geostationary orbit, approximately 36,000 km above sea level in the plane of the equator. Geostationary orbits are also employed by *satellite-based augmentation systems* such as StarFire and OmniSTAR, which are of significant importance for agricultural GPS-based applications. *StarFire* is a wide-area differential GPS developed by John Deere for precision agriculture (PA) operations. *OmniSTAR* is also a wide-area differential GPS service provider. Both commercial systems supply proprietary correction signals that require subscription from the user.

Despite the fact that several global positioning satellite systems are currently being developed, the only operational system available today is the NAVSTAR GPS, and so all of the applications described henceforth will refer to this system in particular. However, the rest of the GNSS are expected to be operational in the near future, and positioning corrections from several systems will be available soon, which will probably be captured with a unique generic receiver. The *Global Positioning System* (GPS) provides full coverage of the entire Earth except for the poles via 36 satellites distributed in six orbits. GPS receivers process the signals sent by the satellites to estimate the time and the three geodesic coordinates *latitude*, *longitude*, and *altitude*. There are other parameters available too, such as precision indices or velocities, as specified in Section 6.5 about the NMEA code. A significant milestone

Figure 3.2 GPS receivers typically used in agricultural applications

in the application of satellite navigation to agriculture, where high accuracy is required to navigate between crop or tree rows, has been the deactivation of *selective availability* on 2 May 2000 [1]. Selective availability was an on-purpose GPS signal degradation introduced by the US Department of Defense to restrict accuracy when the GPS was used for civil purposes. Figure 3.2 shows several GPS receivers that find frequent use in agricultural applications.

Documentation on GNSS technology is abundant, and there are many monographs that provide a complete description of these navigation systems [2, 3]. However, this chapter focuses in particular on the applications of GPS in intelligent off-road vehicles. It is important to emphasize that the accuracy requirements for an agricultural machine operating between crop or tree rows is very high, probably the highest of all current commercial applications, as the vehicle's wheels often need to maneuver within tolerances of a few centimeters. The first applications of GPS centered on field mapping and guidance assistance. The dream of automatic steering in the field has recently become a reality thanks to the popularization and accessibility of global real-time positioning via the GPS. Nevertheless, early GPS-based solutions did not actually control vehicle steering; rather, they provided the driver with steering indications so that he or she could guide the machine more accurately. This method, which is still in use, is typically realized through *lightbar displays*. Figure 3.3a shows a lightbar display mounted on a medium-sized tractor used for specialty crops. Lateral displacements of the vehicle are signaled by turning on an equivalent number of lights on the corresponding half (left or right correction) of an

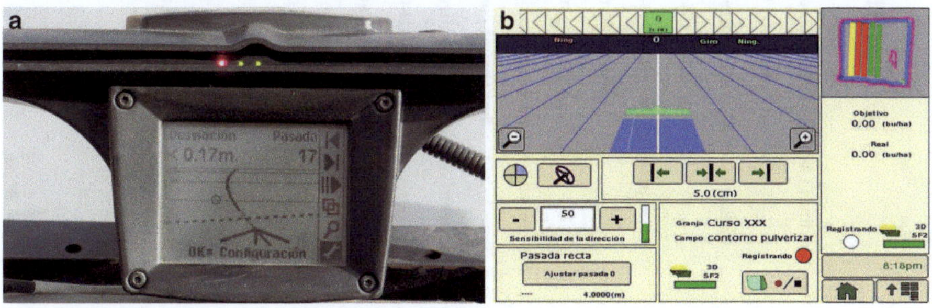

Figure 3.3 GPS-based steering assistance: (**a**) lightbar system; (**b**) GUI screen

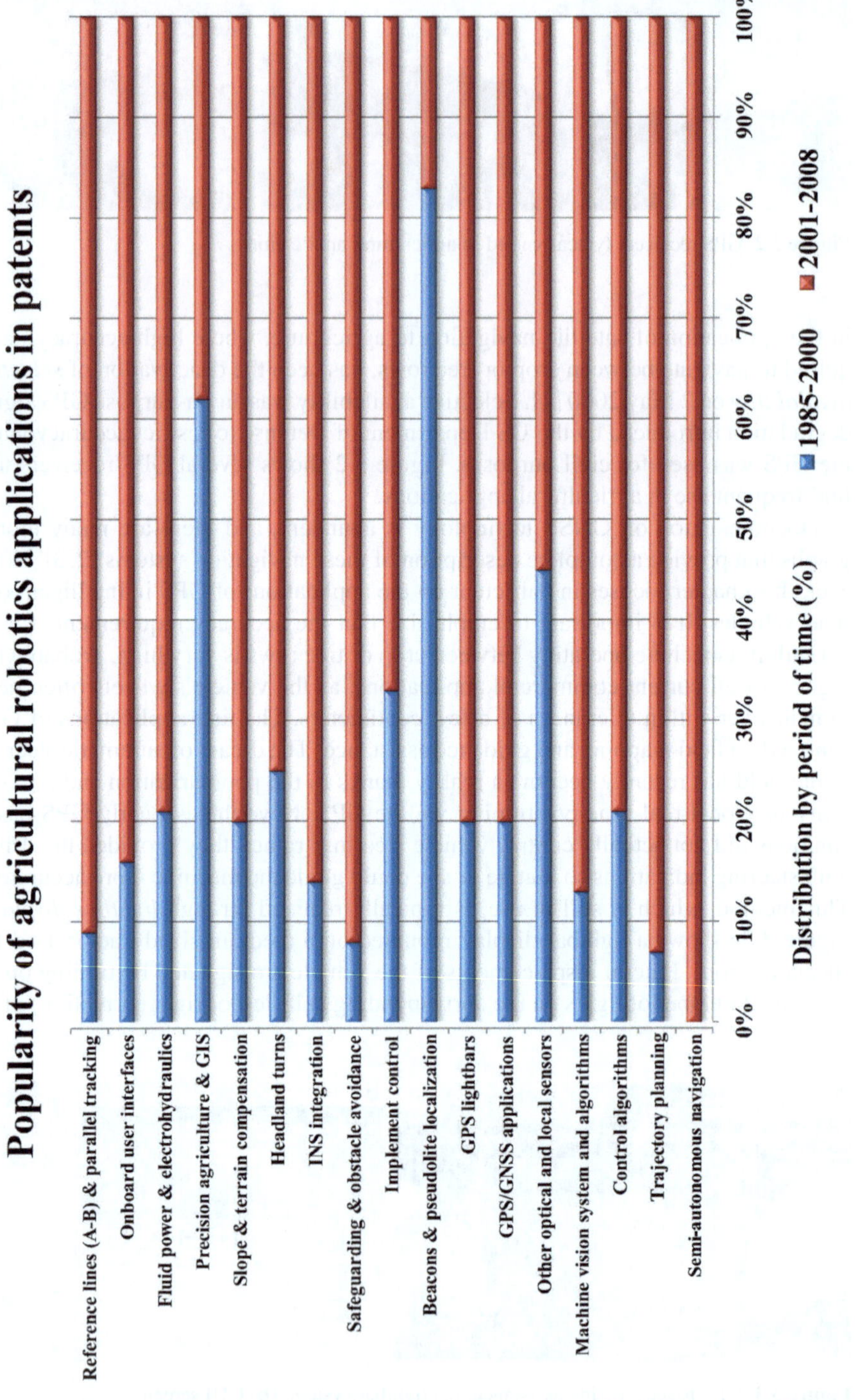

Figure 3.4 Popularity of autoguidance technologies for the periods 1985–2000 and 2001–2008 [1]

LED bar. As technology continues to evolve, lightbars are being replaced by more sophisticated monitors where a graphical user interface (GUI) provides the driver with more friendly guidance commands. Figure 3.3b includes one of these screens used for steering assistance.

The unprecedented success of lightbar systems for guidance assistance soon led to automatic vehicle steering along crop rows, a technique termed *parallel tracking*. The popularity of GNSS-based guidance systems has grown since the turn of the century due to the ever-growing number of commercial systems available, a constant drop in cost, and the suppression of selective availability. Figure 3.4 compares the popularity of autoguidance technologies before and after the year 2000, estimated as the percentage of patents granted according to the study reported in [1]. The graph indicates that 80% of the patents on GPS lightbars and other GNSS applications examined were issued after 2000.

Parallel tracking requires the definition of a reference line, also called the *A-B line*, which sets the initial path from which the rest of the passes will be delineated by simply offsetting the implement working width. The A-B line is typically traced by the driver and recorded with the GPS receiver. Subsequent passes will be calculated by a vehicle-embedded algorithm based upon the inter-pass spacing introduced by the user. The reference line does not need to be straight, and curved segments are often inserted into the user-driven path. The calculated trajectory followed by the guided vehicle commonly consists of a series of points named *waypoints*. Unlike guidance along parallel paths, headland turning is (generally speaking) not commercial and is currently the source of intense developmental activity. Many research institutions, universities and vehicle manufacturers are investing a great deal of resources into deriving a reliable solution. Nonetheless, most farming time is spent driving along the rows, and so any system that is able to successfully navigate within the rows should be considered a triumph. Furthermore, it would appear to be more important to assure robustness and safety while navigating along the rows than to execute vulnerable turns at the ends of rows automatically. After all, the commercial autopilots that are extensively used in aviation do not perform landing and take-off operations but are still considered indispensable.

3.2 Positioning Needs of Agricultural Autosteered Machines: Differential GPS and Real-time Kinematic GPS

The accuracy requirements for the horizontal positioning of aircraft, on-road automobiles, and maritime vessels can be on the order of decameters. This is not the case for off-road vehicles, which often need to navigate within narrow lanes, tight structures, or dense crop alignments. The tractor of Figure 3.5a was autosteered along corn rows separated by 75 cm. Notice the space available for sideways corrections, as the front wheels travel along strips of ground between adjacent rows. The robotic tractor shown in Figure 3.5b needs to travel between orange trees, which barely leaves enough space for the vehicle to move freely. These physical needs result in

Figure 3.5 Positioning needs of agricultural vehicles: (**a**) bulk crops; (**b**) specialty crops

the highest demands for applications based on GPS localization, typically requiring specialized high-accuracy solutions such as differential and real-time kinematic GPS.

Unlike aircraft and on-road vehicles, which tend to travel over large distances, off-road vehicles usually cover smaller areas, which allows some important GPS errors to be removed. One efficient technique used for such error mitigation consists of correcting the original satellite signal with the signal generated by a GPS reference receiver whose location is precisely known. This method is known as *differential GPS*, shortened to DGPS. The worst sources of error are similar for all receivers located in the same local neighborhood, and can therefore be effectively eliminated using data from an additional receiver located at a well-known location. This allows satellite ephemeris and clock errors to be canceled out completely, whereas atmospheric error cancelation degrades with distance [3]. It should always be kept in mind that not all sources of errors can be eliminated using a differential correction signal; multipath and receiver errors will still be present when using DGPS receivers.

There are many practical ways of producing differential corrections for GPS receivers. One is to create land-based reference stations in the vicinity of the area where the vehicle operates, an approach denoted *local-area DGPS*. Widening the area covered, in order to provide corrections beyond oceans and across continents, leads to the *wide-area DGPS*, which is a centralized system that incorporates a network of reference stations that are strategically distributed over wide areas of the globe. The dimensions of the terrain covered often require the use of geostationary satellites emitting real-time corrections, although networks of ground-based transmitters are also feasible. At present, several wide-area correction systems are available, such as the US-operated *Wide Area Augmentation System* (WAAS), the *Euro-*

pean Geostationary Navigation Overlay System (EGNOS), or the Japanese *Multi-functional Satellite Augmentation System* (MSAS).

The ultimate objective of implementing a DGPS is to improve the standard positioning accuracy achievable with GPS; therefore, it is essential to quantify the enhancement of the vehicle localization after incorporating differential-based estimates. Since the deactivation of selective availability in 2000, the average error has been reduced to less than ten meters [2]. Wide-area DGPS can offer a positioning accuracy of less than two meters. While this progress represents a significant milestone in GNSS applications, agricultural vehicles involved in typical farming tasks, such as the demanding situations of Figure 3.5, call for real-time accuracy at the decimeter level. This precision is usually available with commercial carrier-phase DGPS providers, whose private geostationary satellites assure the range of precision needed in many field operations. OmniSTAR and John Deere StarFire are the most widely used augmentation systems currently seen in agricultural fields. The carrier phase is a particular signal used to improve GPS timing measurements in conjunction with code phase signals.

Not all of the tasks that are or can be performed by an intelligent off-road vehicle necessitate the same level of accuracy. Most of them, such as yield monitoring or variable rate applications, are satisfactorily conducted with carrier-phase DGPS and errors of around 15 cm. Other tasks, however, demand sub-inch precision, such as precise planting or autonomous navigation within tight rows. The latter case requires better performance than that offered by conventional augmentation systems such as WAAS or EGNOS. One solution to this demanding request can be found in *real-time kinematic GPS* (RTK-GPS), a method that is used to enhance GPS accuracy by applying corrections from a particular base station located in the field where the vehicle operates. The vehicle, generally called a *rover*, carries an antenna that allows it to receive data sent by the field station through a radio transmission. An RTK kit usually comprises two receivers, a radio link, and computer software to integrate the correcting data and come up with the final position of the rover [2]. RTK systems constitute the most accurate solution for GNSS applications, yield-

Figure 3.6 RTK-GPS receiver antenna mounted on a wheel-type agricultural tractor (**a**); base station (**b**) (courtesy of Deere & Co)

ing typical errors of less than 2 cm. An increasing number of farms in the North American Midwest are using RTK systems to improve their productivity. The main disadvantages of RTK solutions are, on the one hand, higher acquisition costs than alternative options, although once such a system has been purchased there is no need to pay for a subscription. On the other hand, the base station cannot provide corrections for distances of over 10 km (6 miles), which limits its use when farms are large or spread out. GPS receivers, both augmented DGPS and RTK, require an initialization time of at least 15 min before they provide reliable data, as it takes time for the system to search for and lock onto the signals from the minimum number of satellites. Figure 3.6 shows an RTK-GPS solution implemented in an agricultural tractor. Note the base station in the foreground (b) and the receiver antenna on the rover (tractor; Figure 3.6b).

3.3 Basic Geometry of GPS Guidance: Offset and Heading

The characteristics of a particular GPS-based navigation system are determined by the performance specifications demanded by its users in the field. A list of requirements for automatic row guidance of agricultural vehicles could be as follows:

1. capable of following contour and straight rows;
2. can adapt to "guess row" conditions;
3. works at forward speeds of 0.5–12 km/h;
4. can be guided within standing and lodged crops;
5. immune to weeds or other potential disturbances;
6. not affected by the sizes and positions of implements;
7. smooth transition to higher signal accuracy or alternative guidance systems.

Once the key features desired for the system have been identified, the trajectory planner must adapt to them. For instance, according to the previous list, the trajectory generated must consist of segments with lengths that permit curved rows to be faithfully traced, and the vehicle should be able to follow them at speeds of up to 12 km/h regardless of the implement engaged. The trajectory generated by the path planner needs to link the current vehicle location (determined by local tangent plane coordinates) with the desired target point. The position of the vehicle is readily available using the GPS receiver mounted onboard. The coordinates of the target point are calculated by the navigation planner; so, for example, if a parallel track system has generated a series of waypoints parallel to the user-defined reference line (the A-B line), the navigation engine of the vehicle will compare its current position with the location of the subsequent point of the trajectory, and deduce the best steering command required to reach the targeted waypoint. This comparison between real position and desired point is expressed in terms of two fundamental parameters: offset error and heading error. The *offset* δ of the vehicle is its lateral displacement with respect to a hypothetical line that is parallel to the center plane of the vehicle and crosses the target point. The *heading error* ϕ_e of the vehicle quanti-

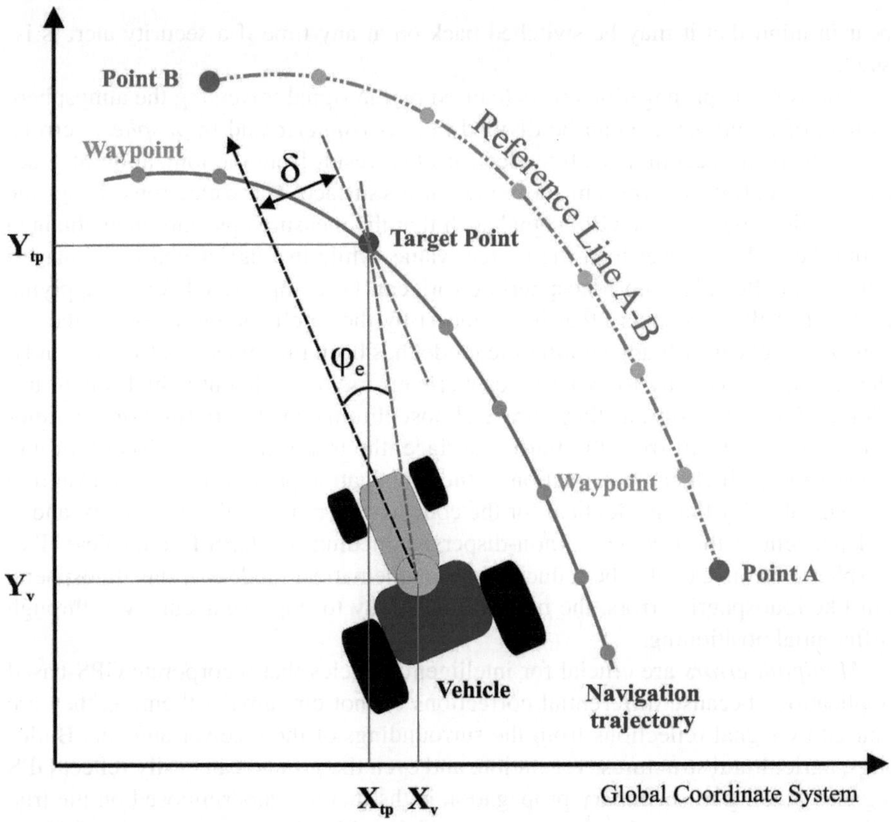

Figure 3.7 Geometry of the GPS guidance of an intelligent off-road

fies its misalignment with respect to the target point, measured as the angle between the moving direction of the vehicle and the desired position. Both variables (offset and heading) are usually combined in a mathematical expression that ultimately yields the steering command sent to the navigation controller. The controller, which is also designed to meet the previous guidance requirements, executes the turns that best track the planned path. Figure 3.7 schematizes the geometry of GPS guidance via the key parameters *offset* δ and *heading error* ϕ_e.

3.4 Significant Errors in GPS Guidance: Drift, Multipath and Atmospheric Errors, and Precision Estimations

The errors caused by selective availability are not considered regular GPS errors here because the US Department of Defense canceled this signal degradation in May 2000. Nevertheless, frequent users of GPS-assisted applications should always bear in mind that it may be switched back on at any time if a security alert is issued.

Atmospheric propagation errors (caused by the signal traversing the atmosphere instead of empty space) can be divided into *ionospheric* and *tropospheric* errors. The ionosphere accumulates free electrons that result from the ionization of gases by solar radiation 50–1000 km above the Earth's surface. These electrons change the propagation speed of the GPS signal such that the measured pseudorange obtained using the code is larger than the correct value, while the carrier phase estimate is equally smaller [3]. The ionospheric errors can be compensated for by applying a model of the ionosphere that is broadcast by the satellites, but errors at the decameter level can still appear after the model has been implemented. Consequently, the recommended way to avoid ionospheric errors in intelligent vehicles is to use *differential corrections*, as they can be almost eliminated. The portion of the atmosphere below 50 km from the Earth's surface, the troposphere, consists of air and water vapor, which cause refraction of the propagation path of the signal, resulting in a signal delay that is identical for the code and carrier signal components, and is independent of the frequency (non-dispersive medium for GPS frequencies). Tropospheric errors can also be reduced with mathematical models of the atmosphere, but like ionospheric errors, the most efficient way to improve accuracy is through differential positioning.

Multipath errors are crucial for intelligent vehicles that incorporate GPS-based applications because differential corrections cannot circumvent them, as they are caused by signal reflections from the surroundings of the receiver antenna. Buildings, agricultural structures, vegetation, and even the ground can easily reflect GPS signals, leading to secondary propagation paths that are superimposed on the true (direct) path signal. Trying to compensate for multipath signals when the GPS receiver is attached to a moving vehicle is not an easy task. Several processing techniques to mitigate their negative effects have been developed, although most of them have been created by receiver manufacturers who, obviously, keep the filtering algorithms as proprietary information. The simplest way to reduce the impact of multipath errors is by placing the antenna in the optimal position where reflected signals are less prone to hit it. In any case, some sort of consistency check must be implemented in the navigation engine to detect severe GPS errors when they appear, as the consequences of such errors for an autonomous vehicle could be catastrophic.

Selective availability and atmospheric errors are independent of the code structure and the receiver–antenna design. Multipath errors, on the contrary, are related to the particular application developed; a simple change of antenna location can make a difference to the signal quality. In addition to multipath effects, every receiver can be affected by random measurement noise, known as *receiver noise*, which is a general term for the noise generated by the antenna, amplifiers, non-GPS radiation, and signal quantization [2]. Measurement inaccuracies derived from receiver noise depend on the signal strength, which varies with satellite elevation angles. Multipath errors also increase as the satellites get lower in the sky and signal power decreases.

These multiple sources of errors will affect the final positioning estimates, and it is not possible for the user to decouple satellite clocks and ephemeris errors, atmospheric delays, multipath errors, and receiver noise. It is practical, however, to classify errors from the user perception standpoint. A first step in the mitigation of control segment errors and atmospheric delays is to implement differential corrections, as these errors change slowly and are similar for receivers that operate locally. Once DGPS is available, accuracy improves significantly but receiver noise and multipath are still present. What are the *noticeable effects on the vehicle*? A favorable antenna location can reduce multipath but receiver noise will still remain. Figure 3.8 represents the variations in the positioning estimates over time of two receivers mounted on a stationary vehicle: an RTK-GPS and a satellite-based DGPS. The true position was determined by the RTK-GPS, and the data shows a small amount of dispersion around the actual location. The RTK-GPS was first initialized with the coordinates given by the DGPS. After that, both receivers were turned off for a period of time and turned on again for data collection. When the vehicle position was registered by the (geostationary) satellite-based DGPS, two phenomena were observed: first, the initial points determined by the DGPS were several meters from the true position; and second, there was a constant loss of positioning accuracy over time [4]. The first error can be seen as an initiation jump, and represents a primary source of inaccuracies that we can consider *initial bias*. The second, conversely, evolves with time and gets worse after several minutes of use. It results in a continuous signal degradation that is generally denoted the *drift error*. At this point, we are not concerned about how much of the error is attributable to multipath, receiver noise, and so on; what matters is the fact that a DGPS can place the vehicle several meters away from its true location, and that this initial positioning will drift with time, reaching significant offsets after several minutes of work in the field.

Given the accuracy requirements for GPS applications in agricultural fields, the assessment of localization errors is vital to assure acceptable robotic performance in intelligent off-road vehicles. In particular, a study of the potential *bias* and a quantitative evaluation of the *drift* are essential for these machines, and there is no better place to investigate these parameters than actual fields. Figure 3.9 shows the trajectories traced by an agricultural tractor following several corn and soybean rows with approximate lengths of 250 m and a row spacing of 0.75 m. The vehicle carried two receivers: an RTK-GPS and a satellite-based DGPS (left side of Figure 3.6). Passes A and E were recorded with the RTK receiver while the others (B, C, D, F, G, and H) were acquired with the DGPS. The RTK receiver is typically used to mark reference lines, as it is accurate to the centimeter level, and so it tends to provide a reliable source of validation data. Given that the separation between two adjacent rows is 75 cm, the spacing between passes must be a multiple of this quantity. As a matter of fact, the distance between passes A and E, L_R, is 6 m, which corresponds exactly to eight rows. This result is not surprising, as RTK measurements are precise and free of drift and bias. However, a DGPS typically suffers from these two errors, and an analysis of runs B, C, D, F, G, and H should illustrate this fact. Actually, a quick inspection of Figure 3.9 reveals that passes B and F are biased by approximately 50 cm, as they represent the same runs used for passes A and E but recorded with a different receiver. Nevertheless, this bias is not the only effect seen in runs B and F; their average separation L is around 5.75 m, which implies an error

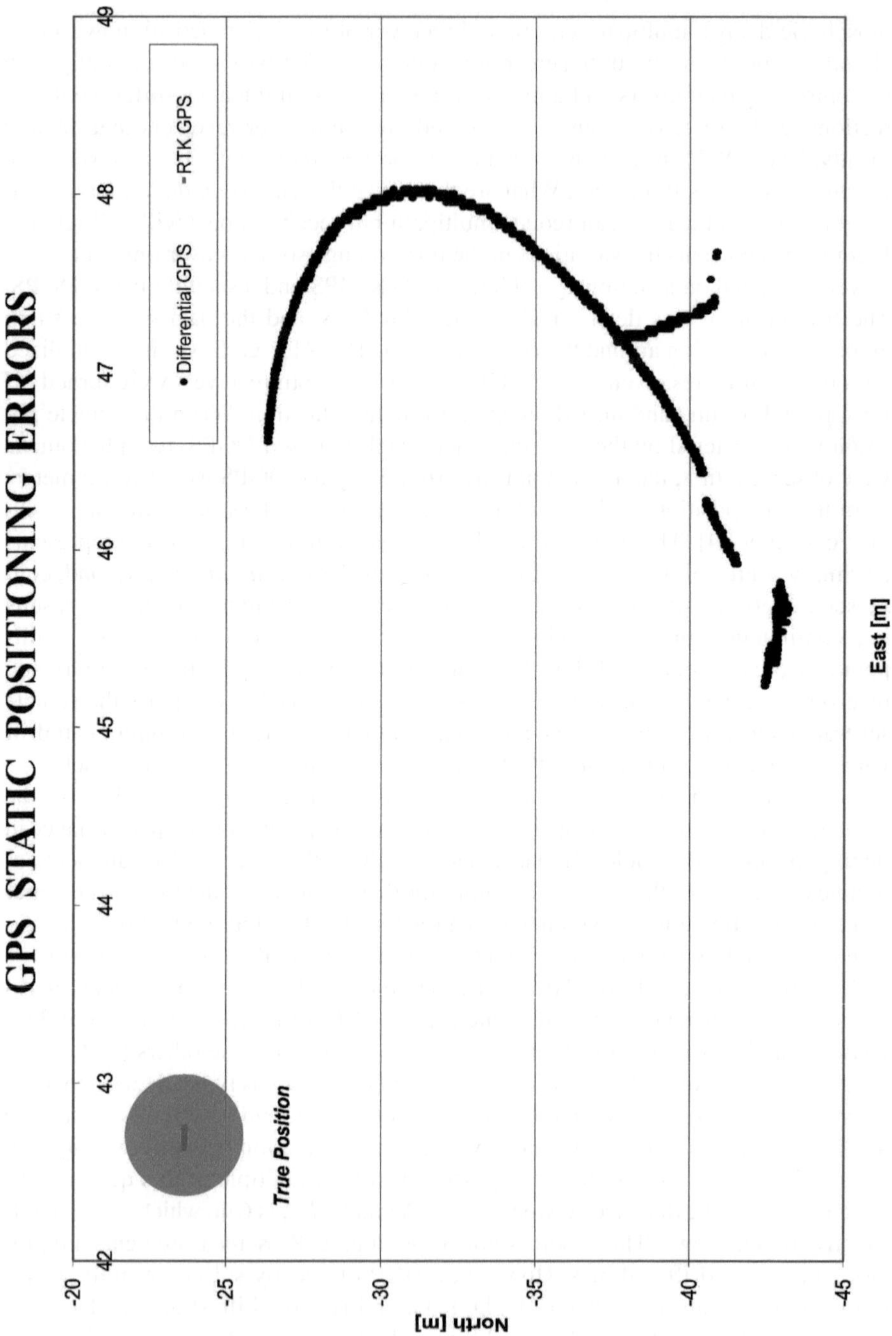

Figure 3.8 Observable GPS-based positioning errors in a static vehicle

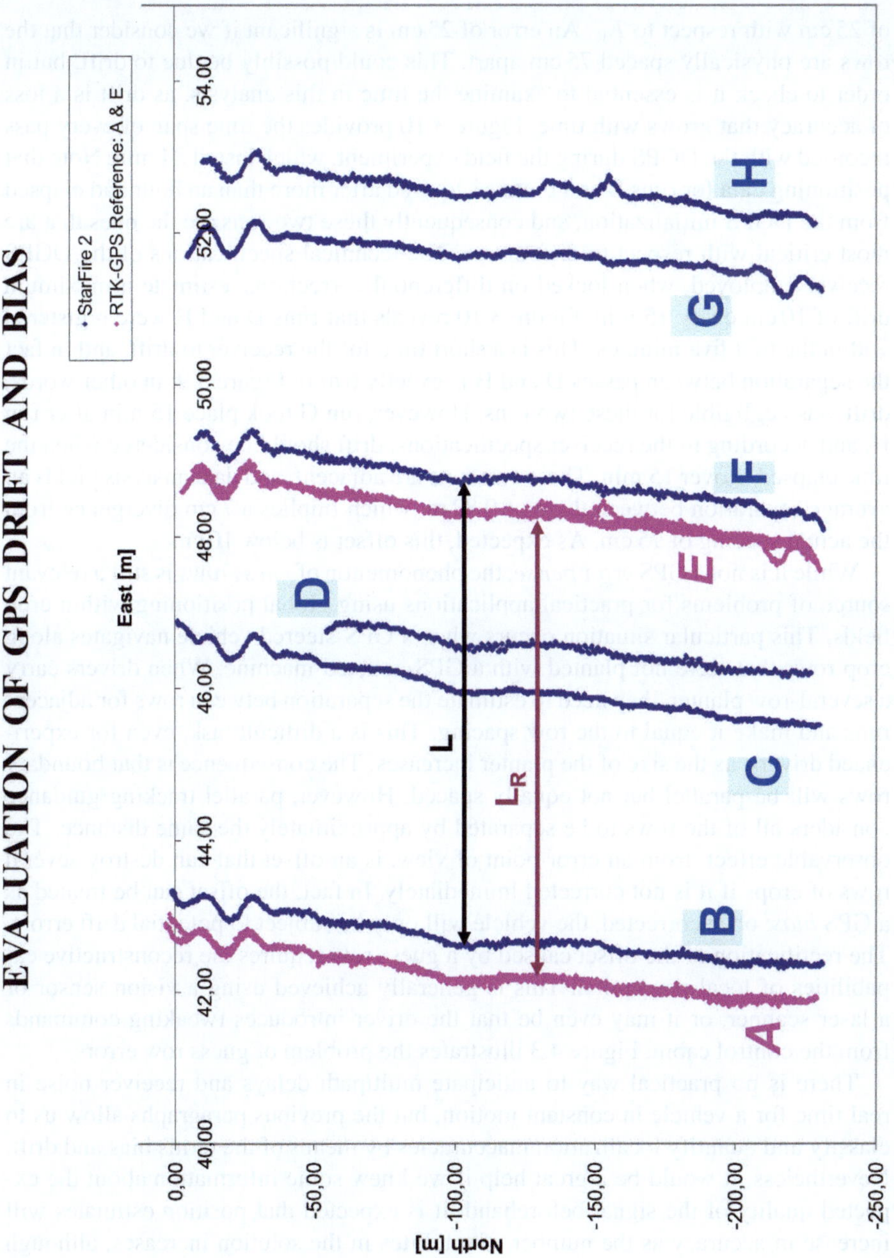

Figure 3.9 In-field evaluation of GPS drift and bias

of 25 cm with respect to L_R. An error of 25 cm is significant if we consider that the rows are physically spaced 75 cm apart. This could possibly be due to drift, but in order to check it is essential to examine the time in this analysis, as drift is a loss of accuracy that grows with time. Figure 3.10 provides the time span of every pass recorded with the DGPS during the field experiment, which lasted 84 min. Note that positioning data for runs B and F were acquired after more than an hour had elapsed from the DGPS initialization, and consequently these two runs are the ones that are most critical with respect to drift errors. The technical specifications of the DGPS receiver employed, when locked on differential corrections, estimate a maximum drift of 10 cm every 15 min. Figure 3.10 reveals that runs D and H were registered within the first five minutes. This is a short time for the receiver to drift, and in fact the separation between passes D and H is exactly 6 m in Figure 3.9; in other words, drift was negligible for these two runs. However, run G took place 15 min after run H, and according to the receiver specifications, drift should be considered when the time elapsed is over 15 min. These two runs are adjacent, and data analysis yields an average separation between them of 0.82 m, which implies a 7 cm divergence from the actual spacing of 75 cm. As expected, this offset is below 10 cm.

While it is not a GPS error *per se*, the phenomenon of *guess rows* is still a relevant source of problems for practical applications using global positioning within crop fields. This particular situation occurs when a GPS-steered vehicle navigates along crop rows that were not planted with a GPS-assisted machine. When drivers carry a several-row planter, they need to estimate the separation between rows for adjacent runs and make it equal to the row spacing. This is a difficult task, even for experienced drivers, as the size of the planter increases. The consequence is that boundary rows will be parallel but not equally spaced. However, parallel tracking guidance considers all of the rows to be separated by approximately the same distance. The observable effect, from an error point of view, is an offset that can destroy several rows of crops if it is not corrected immediately. In fact, the offset can be treated as a GPS *bias*; once corrected, the vehicle will only be subject to potential drift errors. The rectification of the offset caused by a guess row requires the reconstructive capabilities of local perception. This is generally achieved using a vision sensor or a laser scanner, or it may even be that the driver introduces tweaking commands from the control cabin. Figure 4.3 illustrates the problem of guess row errors.

There is no practical way to anticipate multipath delays and receiver noise in real time for a vehicle in constant motion, but the previous paragraphs allow us to classify and quantify localization inaccuracies by means of the terms bias and drift. Nevertheless, it would be a great help if we knew some information about the expected quality of the signal beforehand. It is expected that position estimates will increase in accuracy as the number of satellites in the solution increases, although the relative positions of the satellites also have a notable influence on the final results too. The control segment facilitates this task by transmitting the number of locked satellites and a GPS quality index: the dilution of precision. The *dilution of precision* (DOP) is, in reality, a set of parameters that quantify how favorable the distribution of satellites in the sky is in relation to a specific user location. The dilution of precision can refer to the clock bias estimation error, resulting in the *time*

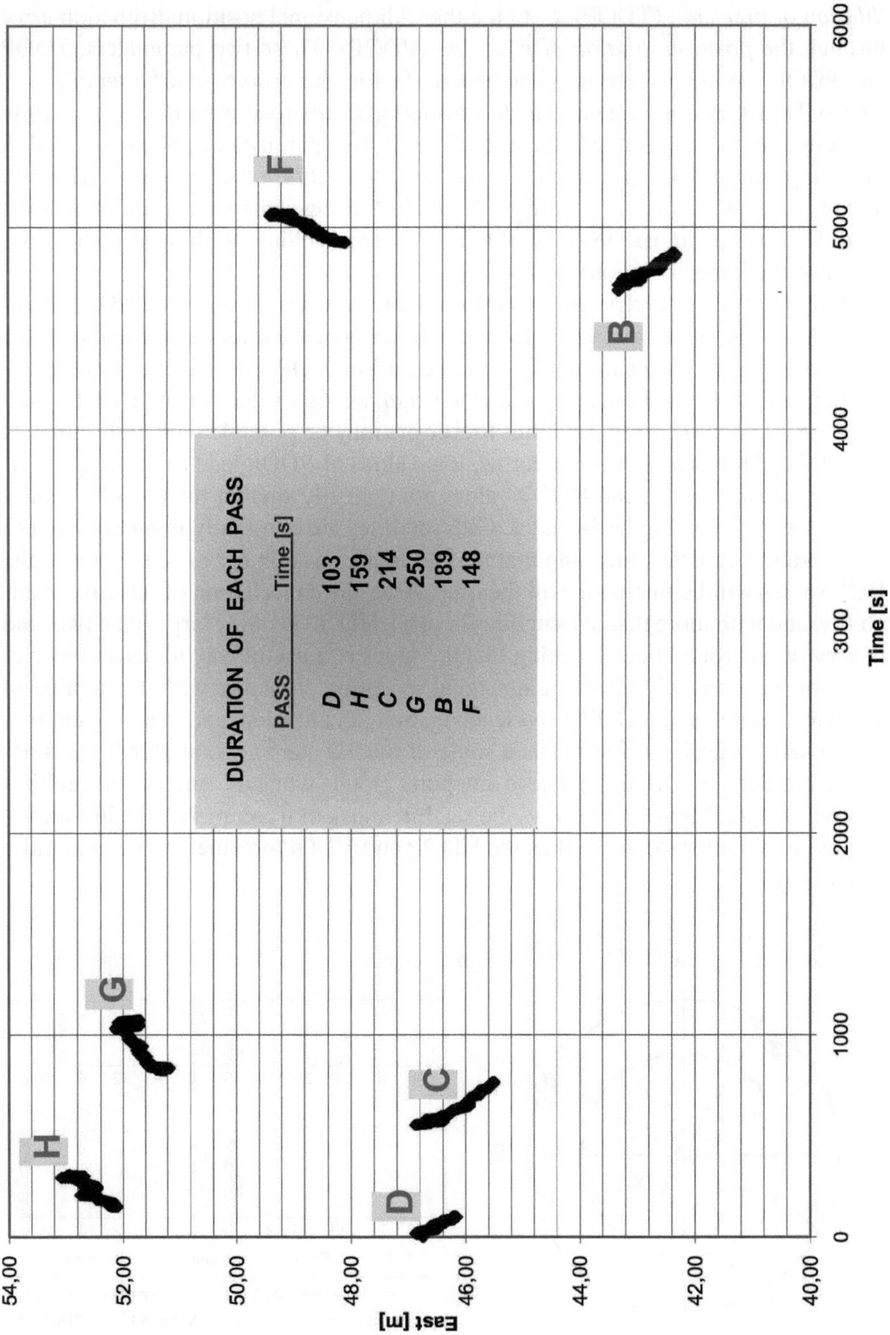

Figure 3.10 Time frame for tests evaluated in Figure 3.9

dilution of precision (TDOP), or to the three-dimensional position estimation error through the *position dilution of precision* (PDOP). These two parameters, TDOP and PDOP, can be merged into one general factor: the *geometric dilution of precision* (GDOP). An intelligent vehicle traversing an off-road terrain will be mainly interested in real-time knowledge of the PDOP. This parameter can be expressed as the composite of two specific terms: the *horizontal dilution of precision* (HDOP), and the *vertical dilution of precision* (VDOP). A forage harvester working on a flat field will pay attention to the HDOP, but a forestry machine climbing steep hillsides will also track the evolution of VDOP.

It is important to understand the information conveyed through the DOP parameters. As a rule, the more favorable the distribution of the satellites tracked over the GPS antenna, the lower the DOP. However, a low DOP value does not automatically mean a low position error, and *vice versa*, as high values of the HDOP result in scattered position errors, and thus low errors may be possible with high values of HDOP [2]. However, in general terms, low values of PDOP lead to favorable positioning measurements, and HDOP values are typically smaller than VDOP values, which are bigger at higher latitudes. GPS satellites are constantly in motion, and so are operating vehicles, meaning that the instantaneous number of satellites and the DOP values will be fluctuating all the time. After the cancellation of selective availability, and with more than 24 satellites in orbit, HDOP is rarely larger than two, and is very often around one. Tracking PDOP values is a useful way to detect receiver faults or locations with challenging signal reception. Together with the number of satellites, HDOP and VDOP provide navigation, localization, and mapping engines with a safety check, so for instance some of the 3D stereo vision global maps described in Section 5.10 did not save any point clouds when the number of satellites was below four. Figure 3.11 shows the satellite mask of a commercial DGPS implemented in a tractor (a), as well as the HDOP and VDOP obtained at that particular instant (b).

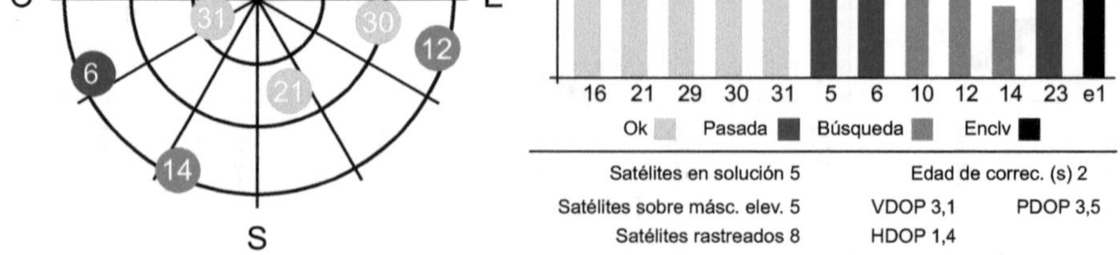

Figure 3.11 Satellites in view for a vehicle-mounted DGPS (**a**); HDOP and VDOP (**b**) (courtesy of Deere & Co.)

3.5 Inertial Sensor Compensation for GPS Signal Degradation: the Kalman Filter

Most bulk crops are produced in vast regions located far from urban areas, where the absence of tall buildings and dense structures facilitates the favorable reception of GPS signals. Specialty crops, in contrast, are arranged in rows of trees, and the heights of these trees very often induce multipath reflections and even drastic signal drops. Fifteen minutes may be required to recover fixes to an acceptable number of satellites at appropriate positions in the sky following a complete loss of GPS signal. During this recovery time, the vehicle is totally blind to global positioning, and if it needs real-time localization data to navigate or map, it must either cease its motion or estimate its current position based upon the last reliable data loaded from the GPS and inertial information to get dead-reckoning estimates. The supplementary information that can assist in a sudden short-term GPS signal loss is typically provided by a compass or inertial measurement unit, the data from which are fused with GPS noise estimates using the Kalman filter technique. An alternative solution to the use of inertial systems is to estimate pitch and roll by mounting a plurality of GPS antennas on the vehicle [5]. This solution has been implemented in a tractor to create a *terrain compensation unit* capable of minimizing the lateral position error of the tractor.

When GPS data start to degrade, it is essential to estimate the vehicle's attitude, which is determined by the three Euler angles, usually called the *yaw*, *pitch*, and *roll*. The yaw is also known as the *heading* and provides the orientation of the vehicle in the horizontal plane defined by east and north coordinates. For an off-road vehicle, the heading is usually the most important angle, as the pitch and roll tend to be small when operating on flat terrain, whereas the heading is crucial to orientation, path planning, mapping, and navigation. Recall, for instance, the definition of heading error illustrated in Figure 3.7; without a reliable way of measuring headings in real time, this error cannot be tracked in a timely manner. When the quality of the GPS signal is favorable (*i.e.*, we have a differential fix with more than seven satellites and low values of PDOP), the heading can easily be calculated from a sequence of coordinates that provide a real-time estimate of the orientation of the vehicle. However, the idea is to get the heading independently from the GPS as an alternative navigation aid when the GPS data degrade. The GPS heading requires the vehicle to be in motion, and can be treated as redundant information to enhance reliability. However, the vehicle heading should be determined by an independent source, and two options are generally accepted: a compass and an inertial measurement unit. *Compasses* are sensitive to magnetic fields and can be inadequate for vehicles with a large amount of electronics kept close together (a situation commonly termed by some experts an "onboard magnetic mess"). *Inertial measurement units*, or IMUs, are the most extensively used attitude sensors. They comprise gyroscopes and accelerometers that measure angular rate and linear acceleration along three perpendicular (Cartesian) axes. The advantage of IMUs is their capacity to estimate the three angles yaw, pitch and roll with the same sensor, unlike compasses, which only supply the yaw. Their main disadvantage, though, is the drift they introduce, which makes them unreliable for extended periods of time, as errors quickly accumulate. Nevertheless, they work well for short-term corrections.

The fact that gyroscopes accumulate errors over time, and the poor results obtained with dead-reckoning navigation in off-road environments where wheel slip is frequent, mean that inertial navigation can only provide very temporary support for GPS, but global positioning needs to be regained with certain degree of assiduity in order to reinitialize the gyro and prevent slippage. One successful method to fuse the information from such different sources of data is to use the Kalman filter. The *Kalman filter* (KF) is a powerful technique for estimating the states of a system (vehicle) with uncertain dynamics by updating those states with noisy observations provided by sensors such as GPS receivers or inertial units. Uncertain vehicle dynamics may be provoked by a defective response of the steering mechanism, a change in soil properties, or wheel slippage, among other reasons. The *vehicle state vector* $\boldsymbol{x}_k$ usually consists of the variables of interest and those measured by the available sensors, but may also include variables that are helpful for the estimation process, such as signal quality parameters. Typical components of the vehicle state vector are position, velocity, acceleration, attitude, and attitude rate. The state vector is propagated from one time step (k) to the next $(k+1)$ by means of the *state transition matrix* A_k, which takes into account the vehicle dynamics involved in the physical process of navigation. This step is regarded as the *state prediction* and it projects the motion of the vehicle forward in time considering the movement dynamics and the process noise. Equation 3.1 gives an example of a vehicle state prediction model where $\boldsymbol{\varepsilon}_k$ is the *process noise vector*, whose covariance matrix Q is known or can be estimated, X and Y give the position of the vehicle in a north-east frame, U and V are the linear velocities in the same frame, and φ is the heading angle.

$$\boldsymbol{x}_{k+1} = A_k \cdot \boldsymbol{x}_k + \boldsymbol{\varepsilon}_k \quad \rightarrow \quad \begin{bmatrix} X_{k+1} \\ Y_{k+1} \\ U_{k+1} \\ V_{k+1} \\ \varphi_{k+1} \end{bmatrix} = \begin{bmatrix} 1 & 0 & \Delta t & 0 & 0 \\ 0 & 1 & 0 & \Delta t & 0 \\ 0 & 0 & 1 & 0 & 0 \\ 0 & 0 & 0 & 1 & 0 \\ 0 & 0 & 0 & 0 & 1 \end{bmatrix} \cdot \begin{bmatrix} X_k \\ Y_k \\ U_k \\ V_k \\ \varphi_k \end{bmatrix} + \begin{bmatrix} \varepsilon_{x,k} \\ \varepsilon_{y,k} \\ \varepsilon_{u,k} \\ \varepsilon_{v,k} \\ \varepsilon_{\varphi,k} \end{bmatrix} \tag{3.1}$$

In addition to the state prediction, the other fundamental step in the Kalman filter is represented by the *measurement update*. Its general expression is given in Equation 3.2, where $\boldsymbol{z}_k$ is the *measurement vector* at time t_k, H_k is the *measurement sensitivity matrix*, and $\boldsymbol{v}_k$ is the *measurement noise vector*. The vector $\boldsymbol{v}_k$ corresponds to the electronic noise generated during sensor output. This is assumed to be additive [3] and zero-mean Gaussian noise with a known covariance expressed by the matrix R. The measurement vector $\boldsymbol{z}_k$ is formed from the variables that can be measured in the process: east and north coordinates, heading, and forward speed for the DGPS; attitude and attitude rate for the IMU. According to Equation 3.2, the measurement vector $\boldsymbol{z}_k$ and the vehicle state vector $\boldsymbol{x}_k$ are related by the measurement sensitivity matrix H_k, which expresses in matrix form the mathematical correspondence between the states observed and the variables measured:

$$z_k = H_k \cdot x_k + v_k \tag{3.2}$$

The main difficulties when applying the Kalman filter do not arise from its mathematical formulation, but from determining the noise covariance matrices Q and R. The process noise covariance matrix Q embodies a model for estimating the uncertainty in the process. This uncertainty is represented by zero-mean white Gaussian noise whose covariance constitutes the main diagonal of matrix Q; the rest of the elements in Q must be zero for sampling to be considered independent. It is hard to deduce process noise from direct observations of the process, but positive results can be achieved when the components of Q are properly tuned. Obviously, these components do not need to be constant throughout the process, so, for instance, if noise in the process is caused by wheel slippage, components may depend on the vehicle ballast or soil conditions. An iterative process for tuning Q, complemented by field tests, has been the usual procedure applied to determine Q. The composition of the measurement noise covariance matrix R is favored by the fact that sensor manufacturers include their expectations of the sensor's performance in its technical specifications. Like Q, only the elements of the principal diagonal of R are non-zero, and the electronic noise at the sensor output is assumed to be zero-mean Gaussian noise of known covariance. If, for example, a vehicle is equipped with an RTK-GPS that provides horizontal accuracy with a standard deviation of 2 cm and velocity precision of 0.1 m s^{-1}, and an INS can estimate headings with an accuracy of half a degree, the covariance matrix R would be given by Equation 3.3. Needless to say, the components of R do not have to be constant, and they may vary as the number of satellites, HDOP, or INS drift decreases. A practical implementation of the use of the Kalman filter to fuse GPS and machine vision can be found in [6]. Because the integration of GPS and INS through the Kalman filter has several limitations, mainly due to the difficulty involved in establishing an accurate dynamic model, and the need to rely on well-defined stochastic models of sensor errors, several different artificial intelligence (AI) techniques have been proposed to either replace or augment KF-based GPS-inertial systems. Many of these techniques apply artificial neural networks, fuzzy logic, and hybrid neuro-fuzzy algorithms. Overall, during medium-to-long GPS outages, an AI module will perform better than a KF module; in contrast, KF modules tend to give better results during short GPS outages [7].

$$R = \begin{bmatrix} \left(\sigma_{\text{East}}^{\text{GPS}}\right)^2 & 0 & 0 & 0 \\ 0 & \left(\sigma_{\text{North}}^{\text{GPS}}\right)^2 & 0 & 0 \\ 0 & 0 & \left(\sigma_{\text{V}}^{\text{GPS}}\right)^2 & 0 \\ 0 & 0 & 0 & \left(\sigma_{\varphi}^{\text{INS}}\right)^2 \end{bmatrix} = \begin{bmatrix} 0.02^2 & 0 & 0 & 0 \\ 0 & 0.02^2 & 0 & 0 \\ 0 & 0 & 0.1^2 & 0 \\ 0 & 0 & 0 & 0.5^2 \end{bmatrix} \tag{3.3}$$

3.6 Evaluation of GPS-based Autoguidance: Error Definition and Standards

All of the errors described in Section 3.4 will affect the vehicle in different ways according to the particular GNSS system implemented; that is, not all of the systems will provide the same level of performance. This may be confusing for system integrators, as well as users, if there is no practical means of evaluating the behavior of a positioning system, and therefore of establishing a comparison among possible solutions. To date, there are no generally applied dynamic test standards for satellite-based localization due to the novelty of these systems, but efforts are being made in this direction. In particular, the American Society of Agricultural and Biological Engineers (ASABE) is currently elaborating Standard X587 [8] as the technical basis for the Standard ISO/TC23/SC19/FDIS 12188-1, which is in the approval state (September 2009). Document X587 provides a means for evaluating the position and velocity of a GNSS device mounted on a vehicle performing ground-based agricultural tasks. The specifications for such dynamic tests include minimum requirements for the testing course (two straight segments that are each 90 m long and linked by a turn of radius 5–10 m), traveling speeds (0.1, 2.5, and 5 m/s), test durations of under an hour, and four repetitions per combination of speed and direction. Signal reacquisition capabilities are also considered in the test by simulating signal losses in the field with an attenuator that is specially adapted to the antenna. These specific norms for testing GPS receivers are very useful for manufacturers who need to certify products according to standardized protocols. However, a particular user who is trying to assess the performance of an already implemented system will probably prefer to sacrifice a certain degree of rigor in favor of a more practical evaluation. The conventional approach to determining how close an automated vehicle can track a desired trajectory has been to fit a regression line to the reference path and then calculate the Euclidean distance between the actual position recorded in the field with the receiver and the optimum regressed line. These distances can be considered tracking errors whose root mean square (RMS) provides the mathematical basis for a quantitative evaluation. While this procedure has been extensively used to study in-field guidance, it cannot be regarded as a general solution because it only works for straight rows, and many fields are made of curved sections where linear regression cannot be applied. Furthermore, even straight rows present a certain degree of curvature when they are long, as it is quite difficult for human operators to drive along an imaginary perfect line. Hilly fields, in addition, complicate the problem even more. The following paragraphs propose a methodology to evaluate the quality of autoguidance for agricultural vehicles, and it is the result of a need originating in the field to compare the performances of several automatic solutions.

The proposed *trajectory matching algorithm* for evaluating autoguided vehicles requires the definition of a reference path and the real-time acquisition of the guided vehicle's trajectory. Both sets of coordinates are determined with GNSS receivers and, for convenience, represented in the local tangent plane (LTP) system of coordinates (Section 3.8). The final objective is the quantitative evaluation of how well

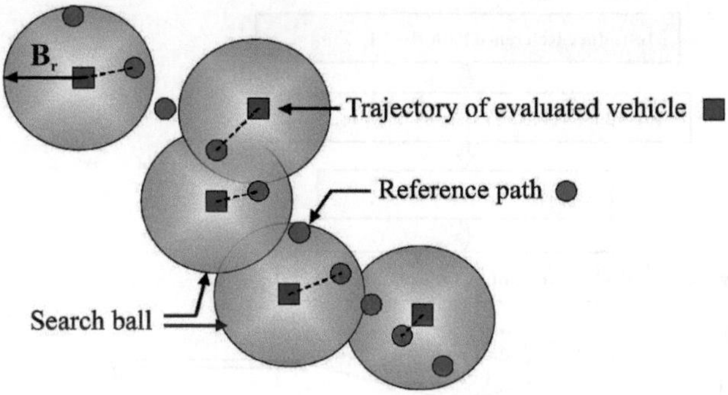

Figure 3.12 Conceptual representation of the trajectory matching algorithm

both trajectories match; that is, how close the evaluated solution is to the optimum. In practice, for each point on the autosteered course a search is conducted to find the nearest point on the reference path. Once determined, the distance between these points represents the deviation, or error, of that individual point. The process continues for all of the components of the examined trajectory, resulting in the error vector. The search is spatially bounded by the user-defined *search ball*, as depicted in Figure 3.12.

The radius of the search ball, B_r, allows a certain level of flexibility on the user side; however, a well-adjusted magnitude makes the algorithm more computationally efficient. Although the trajectory followed by the vehicle has to be recorded in real time, the performance evaluation is usually analyzed *a posteriori*. For that reason, processing time is not a critical issue for the trajectory matching algorithm. The search radius can be seen as the maximum error permitted under the particular circumstances evaluated. The statistical analysis of the *error vector* comprising the vehicle deviations characterizes its performance, and therefore provides a means to compare different automatic guidance systems. A flowchart of the trajectory matching algorithm is shown in Figure 3.13, and this can be mathematically formulated as follows. Let $T = \{1, \ldots, m\}$ be the set of indices of points in the trajectory being evaluated, and let $P = \{1, \ldots, n\}$ be the set of indices of reference path points. If E and N represent the east and north coordinates respectively, for each $j \in T$, a new set $B_j = \{i \in P / d_{ij} < B_r\}$ can be defined, where distances d_{ij} are calculated with Equation 3.4.

$$d_{ij} = \sqrt{[E(i) - E(j)]^2 + [N(i) - N(j)]^2} \tag{3.4}$$

The error, or deviation, for all $j \in T$ is computed according to Equation 3.5. Notice that any point j such that $B_j = \{\emptyset\}$ will not be considered in the evaluation.

$$\text{error}\,(j) = \min\,(d_{ij} \in B_j) \tag{3.5}$$

Figure 3.14 depicts the reference path with star dots and a simulated trajectory for a vehicle under evaluation with squares. In this case, both of the sets T and P consist of the same number of elements ($m = 15$; $n = 15$), but in general they

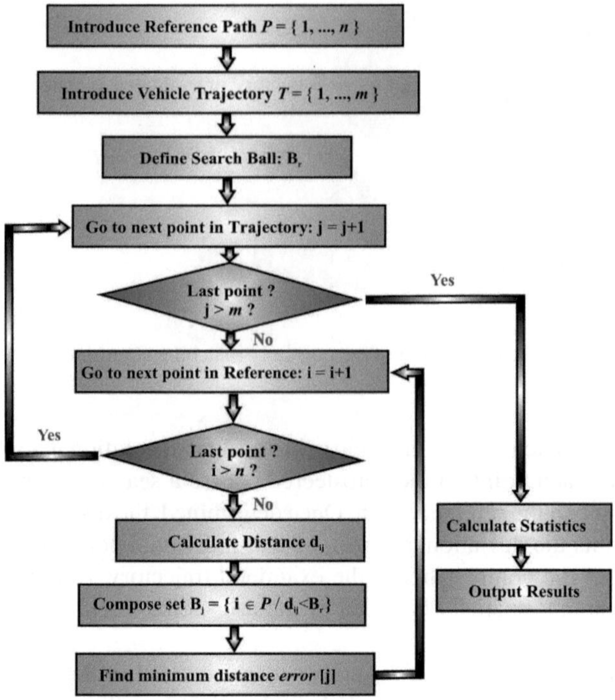

Figure 3.13 Flowchart of the trajectory matching algorithm

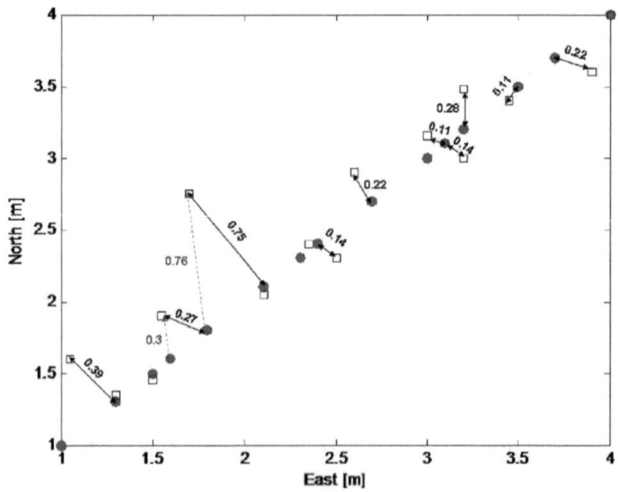

Figure 3.14 Simulated example for the trajectory matching algorithm

will have different dimensions. Given that T is composed of 15 elements ($m = 15$), there will be m sets $B_j(B_1, B_2, B_3, \ldots, B_{15})$ with a different number of elements depending on the ball radius selected. The first point ($j = 1$) on the generated path in Figure 3.14 is located at coordinates (1.05, 1.60). The direct application of Equation 3.4 to point $j = 1$ would yield 15 distances d_{i1}; however, only the points encircled by the defined search ball will constitute the set of distances B_1. If, for instance, the ball radius is set at $B_r = 0.6$ m, the set B_1 will consist of $\{0.55, 0.39, 0.46\}$. The calculation of the error for $j = 1$, and therefore the first element of the vector error VE, is realized by applying Equation 3.5, as detailed in Equation 3.6.

$$\mathrm{VE}_1 = \mathrm{error}\,(1) = \min\,(d_{i1} \in B_1) = 0.39\ \mathrm{m} \tag{3.6}$$

Repeating the same procedure for the rest of the points in $T(j = 2, \ldots, 15)$ leads to the completion of the vector error. Equation 3.5 keeps the results somewhat independent of the choice of B_r, and following the calculations set by Equation 3.6, if B_r had been 0.5 m or 2 m, the final results would have been identical. This approach relies on the availability and appropriateness of the reference path. Excessive spacing between consecutive points in the reference path could impede the correct evaluation of a vehicle. To avoid this inconvenience, a limiting function can be defined that sets a boundary on the interval between referencing points as a function of search radius, receiver accuracy, and sampling rate. In the case of excessively separated points, linear interpolation helps to bridge over-spaced points and thus bring the reference path to completion. A reliable reference course is a guarantee of a successful evaluation, and consequently should be verified before collecting data with the evaluated vehicle.

The trajectory matching algorithm was first developed to assess the autoguidance performance of a self-propelled forage harvester (SPFH) at different traveling speeds. The plot of Figure 3.15 represents the reference path, consisting of two straight rows and the headland turn linking them, with the trajectory followed by the SPFH while being automatically guided at 6 km/h shown over the north row. The reference course was recorded with an RTK-DGPS receiver when the harvester was driven slowly and very carefully over the geometrical center of hay windrows. The same RTK receiver was also used to register the trajectory followed by the vehicle.

The application of Equations 3.4 and 3.5 to the set of points depicted in Figure 3.15 leads to a vector error of 2033 elements, one for each point of the evaluated trajectory. The average error is 33 cm, whereas the median is 31 cm. The standard deviation is helpful in order to compare the stability of the vehicle during various tests. Thus, for instance, when the harvester forward speed increased from 6 to 12 km/h, the standard deviation also increased from 17 to 19 cm. Large oscillations due to unstable behavior tend to increase the magnitude of the standard deviation. Figure 3.16 summarizes the error occurrence for several intervals, and this indicates that 82% of the errors were 0.1–0.5 m. Similarly, deviations of over half a meter were below 15%. The main advantages of the *trajectory matching algorithm* over conventional *regression analysis* can be summarized in three ideas: first, there is

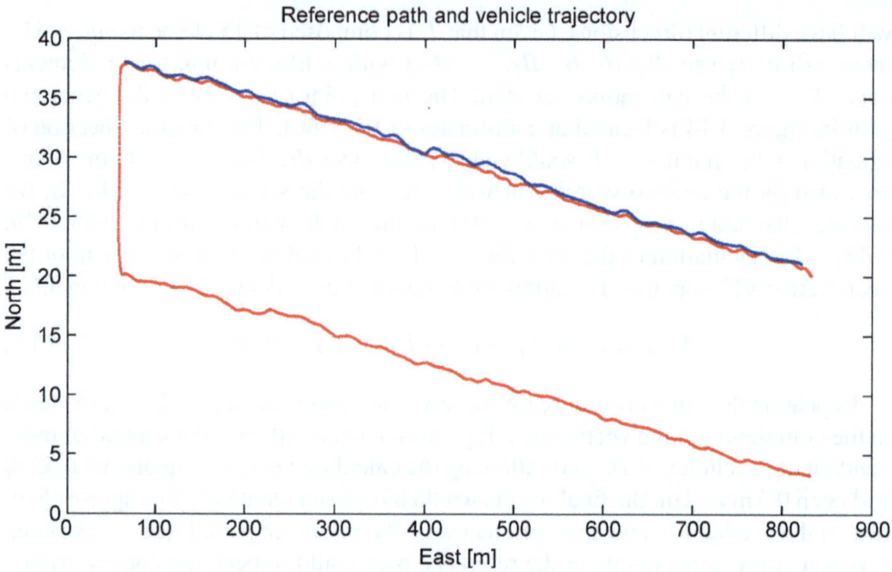

Figure 3.15 Reference path and trajectory followed by an autoguided harvester at 6 km/h

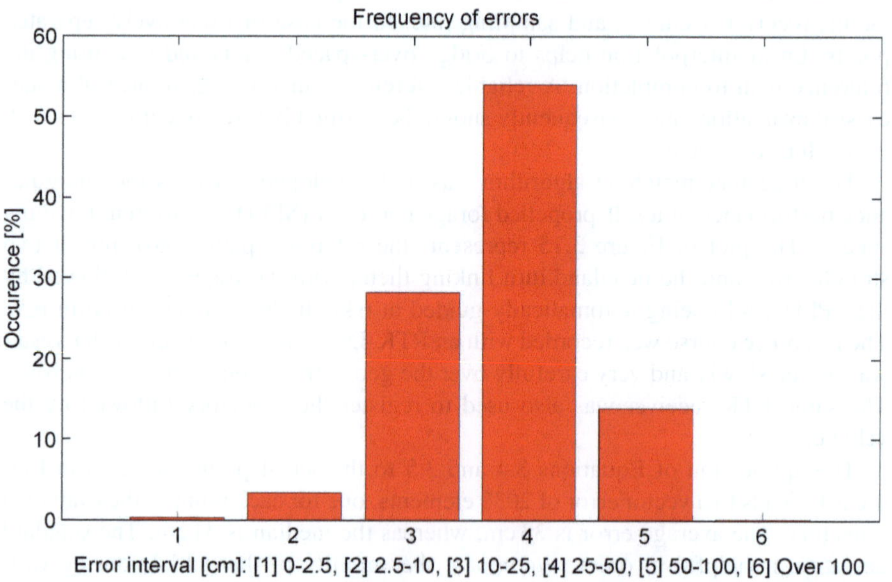

Figure 3.16 Error occurrence for a harvester automatically guided at 6 km/h

no need to pre-process the reference path by removing headlands and non-straight portions, which would complicate the calculation of the regression line; second, the algorithm can be applied to rows of any shape and size; and third, this method is immune to outliers, as points outside the search ball have no influence on the final results. These benefits make the trajectory matching algorithm more versatile, efficient, and easier to apply than conventional regression. A more detailed description of this method can be found in [9].

3.7 GPS Guidance Safety

The tremendous popularity of global positioning systems has reached multiple aspects of engineering and daily life: automobiles, cell phones, aviation, and of course agriculture. There are many different GPS-based applications that are used in agriculture, and not all of them demand the same level of reliability. A yield map, for example, can accept accuracies at the meter level, but guiding a powerful and heavy machine is quite a delicate task. Those who have spent many hours in the field dealing with GPS receivers will be able to recall severe signal drops and the not too infrequent receipt of spurious data. An error in just one parameter of the NMEA code can change the latitude from north to south, pointing at a random location thousands of kilometers away from the actual position of the receiver. While post-processing applications can easily get rid of these obviously wrong messages, vehicles that demand real-time actuation can create dangerous situations that may result in fatal accidents. Agricultural applications require special treatment in terms of safety because the majority of activities involved possess certain risks. We should always bear in mind that the rate of occupational fatalities in agriculture is notably elevated, and the introduction of automation cannot afford to increase these already high figures; indeed, the availability of complementary "intelligence" should instead improve the current situation. Data reported by the United States Bureau of Labor Statistics for the year 2007 indicate that the agricultural sector has the highest rate of occupational fatalities among high-rate sectors (agriculture and forestry, mining, transportation, warehousing, and construction), at 28 deaths per 100,000 workers, which is more than seven times the average rate for all occupations [10]. Some manufacturers of GPS-based commercial solutions for automatic steering incorporate safety measurements to avoid the misuse of this technology; for example, immediate disengagement if the operator leaves the driver's seat. Nevertheless, there are no systems, at present, which can detect if the driver falls asleep, and the possibility of an object suddenly entering and interfering with the vehicle's trajectory is always likely. For this reason, a good robotic design needs to start by ensuring that basic safety principles are upheld, especially in agricultural robots, which can easily weigh several tonnes. Local perception is an invaluable tool to add robustness to GPS-based navigation systems, since it can assist in the execution of slight adjustments or the detection of unexpected obstacles. It seems to be generally assumed that mistakes that would be accepted from humans are not normally allowed from machines; they typically need to outperform the average human driver.

3.8 Systems of Coordinates for Field Applications

GPS receivers provide the real-time position of the on-vehicle antenna in *geodetic coordinates*; namely, *latitude* (λ), *longitude* (ϕ), and *altitude* (h). The *World Geodetic System 1984* (WGS 84), developed by the US Department of Defense, defines an *ellipsoid of revolution* that models the shape of the Earth, upon which the geodetic coordinates are defined. The WGS 84 also defines the *geoid*, another surface that models the mean sea level as the reference of altitudes. According to the schematic representation of Figure 3.17, *latitude* is the angle measured in the meridian plane containing point P between the normal N_o (line perpendicular to the surface of the ellipsoid at P) and the equatorial plane of the ellipsoid, considered positive north from the equator; *longitude* is the angle measured in the equatorial plane between the prime meridian (statistically determined mean Greenwich meridian [2]) and the projection of the point of interest P on the equatorial plane, measured positive east from the reference meridian; while *altitude* (or *height*) is the normal distance between the surface of the ellipsoid and P. Note that the geodetic altitude has no physical correspondence as it is referenced to the ellipsoid, which is a mathematical model. This difficulty is overcome by defining the geoid as an idealized representation of the mean sea level. The most accurate geoid model currently available is the WGS 84 geoid. The circles that connect all points of the ellipsoid with constant

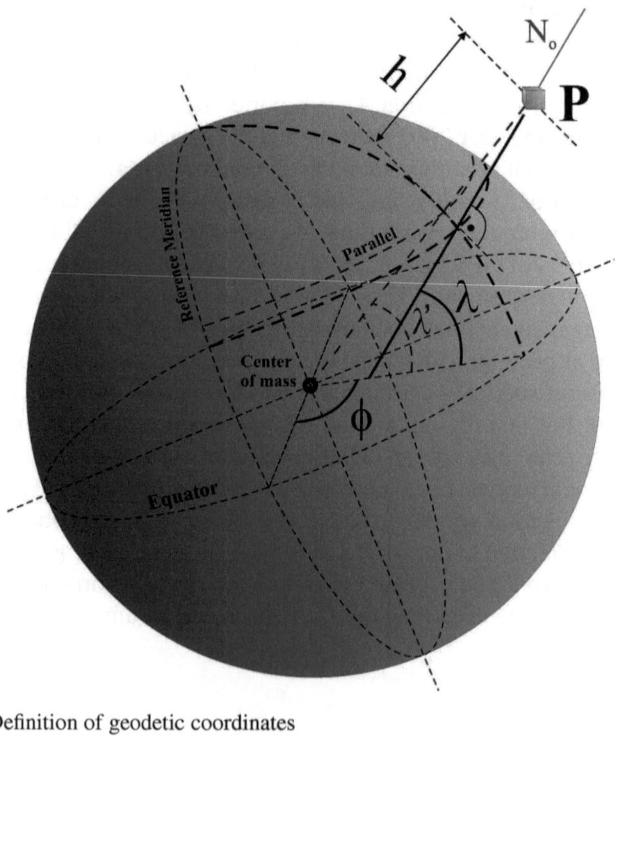

Figure 3.17 Definition of geodetic coordinates

latitude are called *parallels*, and the ellipses that connect all points with the same longitude are called *meridians*. Note that in Figure 3.17 latitude can be defined according to the center of mass of the ellipsoid (*geocentric latitude* λ') or referred to the perpendicular line to the ellipsoid at P (*geodetic latitude* λ), which does not cross the center of the Earth as the ellipsoid is slightly flattened at the poles.

Apart from facilitating geodetic coordinates, the WGS 84 also defines a Cartesian coordinate system that is fixed to the Earth and has its origin at the center of mass of the Earth. This system is called the *Earth-centered Earth-fixed* (ECEF) coordinate system, and provides an alternative way to locate a point on the Earth's surface with the conventional three Cartesian coordinates X, Y, and Z. The Z-axis coincides with the Earth's axis of rotation and therefore crosses the Earth's poles. In reality, the axis of rotation is not fixed; it wanders around, following a circular path, moving several meters a year in what is called polar motion. A change in the position of the poles alters the definition of latitude and longitude, as this means that the equatorial plane is not fixed either. Nevertheless, these slight variations do not have a strong influence on the results for the applications pursued by intelligent vehicles operating in the field. In any case, given that the Z-axis of the ECEF frame needs to be defined unambiguously, geodesists have estimated the average position of the Earth's pole of rotation, which is called the conventional terrestrial pole (CTP) and crosses the Z-axis by definition. The X-axis is defined by the intersection of the equatorial plane (0° latitude) with the prime meridian (0° longitude). The Y-axis is also contained in the equatorial plane and completes the right-handed coordinate system. Figure 3.18 represents the ECEF coordinate system.

The majority of the applications developed for off-road vehicles do not require the coverage of large surfaces in a short period of time with the same vehicle; rather,

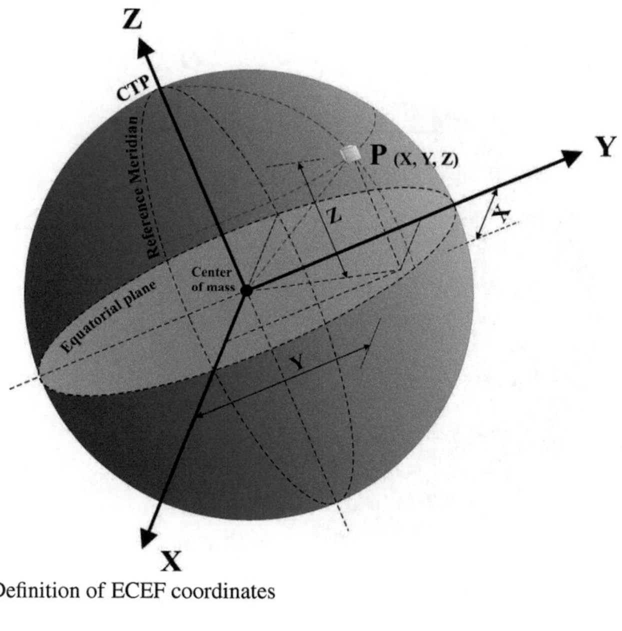

Figure 3.18 Definition of ECEF coordinates

given that off-road equipment cannot travel at high speeds, these machines tend to remain within areas of limited size. In these circumstances, the curvature of the Earth has a negligible effect, and fields can be considered flat. Both geodetic and ECEF coordinates are defined in such a general manner that figures and calculations are cumbersome, and origins are quite abstract, as nobody has seen the center of mass of the Earth. It would be much more practical to define a local system of coordinates where the origin is close to the operating field. Furthermore, modest changes in the position of the vehicle would result in very tiny changes in latitude, longitude, X or Y. However, a local origin and more intuitive coordinates such as east or north would simplify the practical realization of GNSS-based tasks. This need has been fulfilled by the development of the *local tangent plane* (LTP) system of coordinates. The local tangent plane coordinates, also known as NED coordinates, are north (N), east (E), and down (D). These coordinates are measured along three orthogonal axes in a Cartesian configuration generated by fitting a tangent plane to the surface of the Earth at an arbitrary point selected by the user and set as the LTP origin. A simplified sketch of the local tangent plane coordinate system is depicted in Figure 3.19. Notice that the scale of the plane has been greatly exaggerated to illustrate the definitions of the axes; the tangent plane should be small enough to be able to disregard the Earth's curvature.

For practical field applications we would definitely use the LTP, but the GPS receiver typically provides geodetic coordinates through the NMEA code. In other words, the receiver acquires the latitude, longitude, and altitude of the vehicle's position, but it would be much more helpful to know the NED coordinates for a real-time locally based operation. It is therefore necessary to *transform geodetic coordinates to LTP coordinates*, and this can be carried out using the ECEF coordinates, as detailed in the following paragraphs. The first step in the transformation process is to

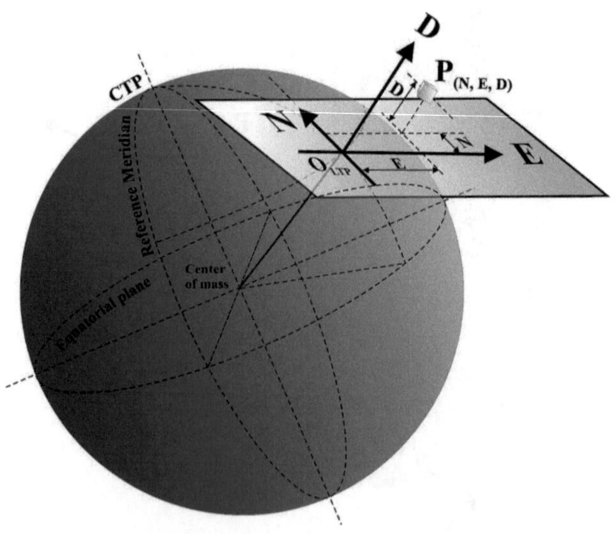

Figure 3.19 Definition of LTP coordinates

select an ellipsoid as a reference to model the shape of the Earth. The widespread use of GPS is turning WGS 84 from a global datum to an international datum, a *de facto* world standard [2] that provides excellent data that can be used to perform the transformation from geodetic coordinates to the local tangent plane. Equations 3.7–3.11 provide the fundamental parameters of WGS 84, revised in 1997. These parameters are the *semi-major axis* (a), the *semi-minor axis* (b), the *ellipsoid flatness* (f), the *eccentricity* (e), and the *length of the normal* (N_o).

$$a = 6\,378\,137.0\ \text{m} \tag{3.7}$$

$$b = a \cdot (1 - f) = 6\,356\,752.3\ \text{m} \tag{3.8}$$

$$f = \frac{a - b}{a} = 0.00335281 \tag{3.9}$$

$$e = \sqrt{f \cdot (2 - f)} = 0.0818 \tag{3.10}$$

$$N_o(\lambda) = \frac{a}{\sqrt{1 - e^2 \cdot \sin^2 \lambda}} \tag{3.11}$$

The length of the normal N_o, defined in Equation 3.11 and depicted in Figure 3.17 for point P, is the distance from the surface of the ellipsoid to its intersection with the axis of rotation Z. The relationship between the geodetic (λ, ϕ, h) and the ECEF (X, Y, Z) coordinates is given by Equations 3.12–3.14.

$$X = (N_o + h) \cdot \cos\lambda \cdot \cos\phi \tag{3.12}$$

$$Y = (N_o + h) \cdot \cos\lambda \cdot \sin\phi \tag{3.13}$$

$$Z = \left[N_o \cdot \left(1 - e^2\right) + h\right] \cdot \sin\lambda \tag{3.14}$$

The transformation from ECEF to LTP (N, E, D) coordinates requires the selection by the user of the origin of the coordinates on the tangent plane (X_0, Y_0, Z_0). Note that the origin belongs to the plane and is a point tangent to the surface of the Earth. The last step in the coordinate transformation leading to the LTP coordinates can be executed with Equation 3.15. Note that Equation 3.15 requires the coordinates of the arbitrary origin of the tangent plane in ECEF coordinates. Once a favorable point in the field has been chosen as the origin, its geodetic coordinates are easily established with a GPS receiver, but they need to be transformed into the ECEF frame with Equations 3.12–3.14 before they can be introduced into Equation 3.15.

$$\begin{bmatrix} N \\ E \\ D \end{bmatrix} = \begin{bmatrix} -\sin\lambda \cdot \cos\phi & -\sin\lambda \cdot \sin\phi & \cos\lambda \\ -\sin\phi & \cos\phi & 0 \\ -\cos\lambda \cdot \cos\phi & -\cos\lambda \cdot \sin\phi & -\sin\lambda \end{bmatrix} \cdot \begin{bmatrix} X - X_0 \\ Y - Y_0 \\ Z - Z_0 \end{bmatrix} \tag{3.15}$$

3.9 GPS in Precision Agriculture Operations

The birth of *precision agriculture* (also known as *precision farming*) and the advent of *GPS* are intimately related. It was the availability and the expansion of the latter

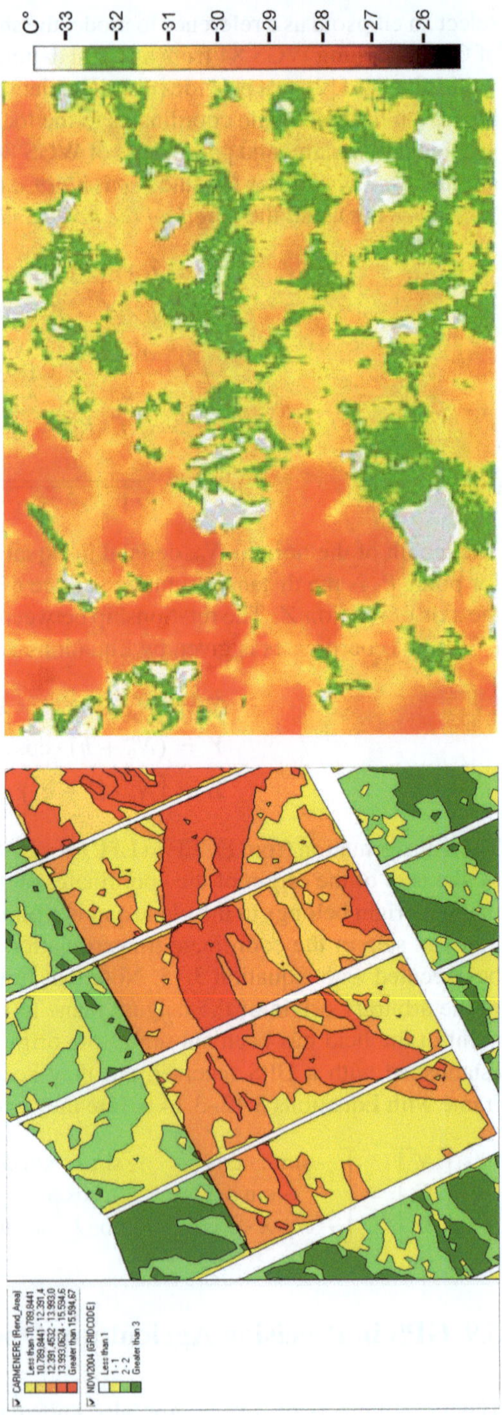

Figure 3.20 Precision farming maps: (**a**) yield; (**b**) thermal gradient for water stress (courtesy of Stanley Best, Chile)

that motivated the creation of the former. Real-time knowledge of the position of a vehicle, person, structure, margin, or even a tree in the field allows the practical application of *information technology* (IT) to the field, in order to take advantage of an intrinsic property of agricultural production: *spatial variability*. The central idea is to apply only *what* is needed *where* it is needed and *when* it is needed, and this can only be achieved if large amounts of reliable data are available in a timely fashion. Thus, for instance, producers greatly benefit from mapping areas with low yields and comparing them with other maps of soil properties, pest or weed distributions, or climatic differences over large fields. Furthermore, handling historic data has proven to be essential for planning future farming tasks through the use of *prescription maps*, which, for example, may reduce pesticides in clean portions of the field and increase the amount of fertilizer added to depressed patches, thus executing what is known as *variable rate application*. An efficient way, which is already in use and highly appreciated by farmers, to increase production turnover is to reduce overlaps in applications with large machines, such as oversize sprayer booms. All of these applications require many different sensors to acquire critical data from the field, but these data can only be linked together through a map by means of the global positioning coordinates provided by the GPS or any other GNSS receiver. This has both economical and environmental benefits; making more rational use of resources such as fertilizers, pesticides, water, and soil will surely result in an improvement in rural area welfare. GPS has been shown to be a unique tool for enhancing the efficiency of agricultural and forestry production since it makes large amounts of meaningful site-referenced data available. Figure 3.20 shows a yield map (a) and a thermal image for water stress detection (b) of Chilean vineyards.

References

1. Rovira-Más F (2009) Recent innovations in off-road intelligent vehicles: in-field automatic navigation. Recent Pat Mech Eng 2:169–178
2. Misra P, Enge P (2006) Global positioning system. Ganga–Jamuna, Lincoln
3. Grewal MS, Weill LR, Andrews AP (2001) Global positioning systems, inertial navigation, and integration. Wiley, New York
4. Rovira-Más F, Han S, Wei J, Reid JF (2005) Fuzzy logic model for sensor fusion of machine vision and GPS in autonomous navigation (ASABE Publ. 051156). ASABE, St. Joseph
5. Parkinson BW, O'Connor ML, Elkaim GH, Bell T (2000) Method and system for automatic control of vehicles based on carrier phase differential GPS. US Patent 6052647
6. Rovira-Más F, Han S (2006) Kalman filter for sensor fusion of GPS and machine vision (ASABE Publ. 063034). ASABE, St. Joseph
7. Chanthalansy L, Noureldin A (2009) AI for INS/GPS integration (Inside GNSS, Sept/Oct 4(5)). Gibbons Media & Research LLC, Eugene
8. ASABE (2007) X57 dynamic testing of satellite-based positioning devices used in agriculture (PM-54, Draft 10). ASABE, St. Joseph
9. Rovira-Más F, Han S, Reid JF (2008) Trajectory-based algorithm for the quantitative evaluation of automatically steered vehicles. Trans ASABE 51(5):1545–1552
10. Myers ML (2009) The agricultural safety and health challenge (Resource, Oct/Nov). ASABE, St. Joseph

Chapter 4
Local Perception Systems

4.1 Real-time Awareness Needs for Autonomous Equipment

Before we can proceed with an account of the needs of autonomous – or more appropriately semiautonomous – vehicles in terms of real-time perception and local awareness, it would be useful to describe the potential environments around which intelligent off-road vehicles are going to rove. Agricultural production sites are very diverse, which means that the potential surroundings of a vehicle that performs mechanized tasks are equally as varied. They range from vast regions in the North American Midwest, Australia, or Central Europe, the intensive farming and horticulture of Japan and Mediterranean countries, the highly mechanized specialty crops of South America, the west coast of the United States, and Florida, to the abundant paddy fields of Southeast Asia. All of these settings might appear to demand very special needs, but from a robotic vehicle standpoint, open environments (excluding greenhouses) can be classified into three main categories. Of course, each application will require specific tuning, but, generally speaking, the core systems will need to cope with the following three situations:

- rows of equally-spaced plants (Scene I);
- rows of equally-spaced trees (Scene II);
- rows of plants or trees growing within limits imposed by man-made structures (Scene III).

Figure 4.1 shows a typical example of each of these environments. Image (a) represents a field of potatoes in Hokkaido (Japan), image (b) is an orange grove in Valencia (Spain), and image (c) portrays apple trees guided by a trellis in Washington State (USA). The differences between these main categories are quite obvious: the vehicle can travel over the plants in Scene I, but it will probably need to move between the trees in Scene II, not over them; the structures of Scene III may limit the vehicle's clearance when traveling, but they never will in Scene II. The consequences of an accident in each scenario will also be different: running over soybeans will not hurt the driver, but crashing against a big tree or a concrete structure can have disastrous consequences for the operator and the machine. On the other hand,

F. Rovira Más, Q. Zhang, A.C. Hansen, *Mechatronics and Intelligent Systems for Off-road Vehicles*. © Springer 2010

Figure 4.1 Scenarios considered when defining the perception needs of intelligent off-road vehicles: (**a**) field of potatoes in Hokkaido (Japan); (**b**) orange grove in Valencia (Spain); (**c**) apple trees guided by a trellis in Washington State (USA)

variations within the same category will result in small adjustments and operation-specific tuning. For example, corn spacing in Illinois is 30 inches (75 cm), but in Texas it is 40 inches (1 m). The same perception system will work for both situations after precise tuning has been performed to adapt the machine to each case, but a different philosophy will have to be implemented to traverse an orchard whose tree rows are spaced several meters apart.

Once the possible rural environments have been classified, the perception needs for the vehicles will be determined by specific farming tasks and the potential hazards derived from them. The situations portrayed in Figure 4.2 illustrate four important scenarios where local awareness is vital for reliable and safe operation of an automated vehicle for agricultural applications. Figure 4.2a is a barren field that is prepared for the production of hops, a vine of the hemp family. These twining plants require a network of wires through which leaves can climb up. The wires are supported by rows of wooden posts whose presence must be detected by the perception engine of a vehicle navigation unit. Global navigation systems do not always locate

Figure 4.2 Potential hazards for intelligent vehicles traveling in agricultural fields: (**a**) wooden posts in a barren field; (**b**) large wooden box left inside an apple orchard; (**c**) unexpected appearance of person in path of vehicle; (**d**) cattle meandering around a field

the posts accurately, and new structures may have appeared since positioning data were last recorded. Therefore, given that for safety reasons every single post needs to be identified in real time, local perception is essential in a situation like this. Figure 4.2b shows a large wooden box left by apple pickers inside an apple orchard. The box is about one cubic meter in size, and its presence in the middle of a lane can compromise the completion of a mission tackled by an intelligent vehicle; proper estimation of the box's position and dimensions will determine if the vehicle, given its turning geometry and size, will need to go around the obstacle or turn back and find an alternative route. Figure 4.2c provides the most critical situation for autonomous vehicles: a person shows up, unexpectedly, in front of the vehicle, thus interfering with its trajectory. Only local perception can deal with this situation, and the more reliably the system behaves the better. People need to be precisely detected in the vicinity of a vehicle that is moving with any degree of autonomy, and if a person is present ahead of the vehicle, it must cease its motion or accurately dodge the object. A less compromising case, but one that is important anyway, is given in Figure 4.2d, where cattle meander around a field, presenting danger to any vehicle traversing it. These are typical situations encountered in agricultural milieu, but there are many others that also justify the need for accurate and reliable local awareness.

The diversity of the agricultural activities that can benefit from automation, together with the seriousness of potential hazards associated with it (such as those shown in Figure 4.2), result in many important applications for local perception sensors. Navigation takes advantage of two aspects: on the one hand, a local sensor helps to locate the vehicle accurately according to local features whose coordinates are not known exactly or have changed with time; on the other, it provides a safeguarding awareness that can detect unexpected obstacles such as machines, people, or animals. The first aspect is fairly common when planting is done manually, caus-

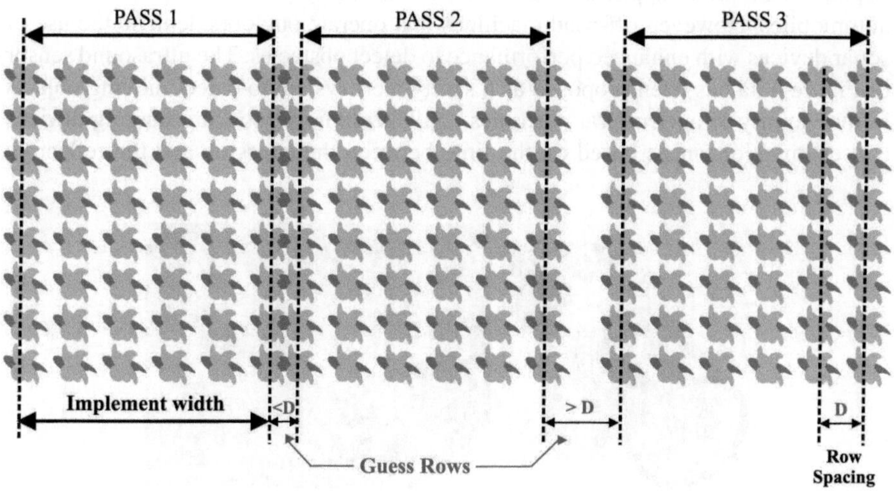

Figure 4.3 The problem of guess rows for parallel tracking autoguidance

ing the problem of *guess rows*. This happens when the operator cannot accurately estimate the distance between passes. Since GNSS-based parallel tracking requires all of the rows to be parallel, heterogeneous spacing between passes will result in the vehicle running over crops unless slight real-time in-field corrections are provided by a local sensor. Figure 4.3 illustrates the problem of guess rows. Local perception devices are also adequate for high-resolution mapping, since they can attain a degree of detail that cannot be reached by satellite imagery. These devices are also ideal for monitoring tasks, such as assisting with reversing maneuvers and helping to locate the grain tank of a loading truck.

4.2 Ultrasonics, Lidar, and Laser Rangefinders

Before the satellite positioning became popular, localization was achieved through beacons that were strategically placed at fixed landmarks in the working area [1]. While the popularity of this approach has been greatly overtaken by GNSS technology nowadays, this procedure is still used for some specific applications, in particular in areas where satellite signals cannot penetrate, such as in underground mines, or in open areas where fixed landmarks offer a reliable solution that is independent of atmospheric errors. This is the case for autonomous stadium mowers, which often need to trace elaborate patterns on the lawn. Aside from these atypical examples, ranging devices based on *time-of-flight* calculations are mainly employed for safeguarding and obstacle avoidance, mapping and SLAM (simultaneous localization and mapping), and for precise positioning during navigation.

Ultrasonic sensors have been used for mobile robot guidance for a long time, mainly following the commercialization of the low-cost Polaroid ranging sonar. One popular widespread application is an ultrasonic device used as a parking aid for automobiles. However, off-road machines that operate outdoors demand the use of sonar devices with enhanced performance to detect obstacles. The ultrasound sensor of Figure 4.4a has been proposed as a safety alert system to detect moving objects in the vicinity of agricultural machines [2]. Ultrasonic sensors are ranging devices that estimate distances based on the time between the emission and the reflection

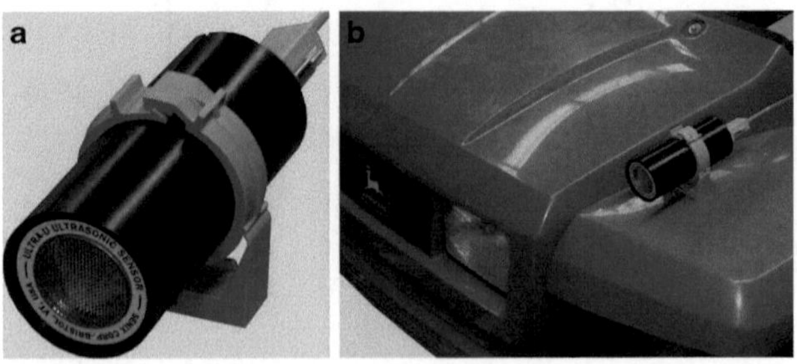

Figure 4.4 Ultrasonic sensor (**a**) used in the safety alert system of an agricultural vehicle (**b**)

(echo) of an ultrasound wave. The wave is transmitted in a cone shape, and the cone diameter determines the detectable region. Obviously, the diameter of the cone grows with the distance from the emitter, but typical devices for off-road vehicles possess a useful range of 0.05–11 m, and a cone angle of 15°. Ultrasound sensors need to be calibrated; that is to say, the relationship between output voltage and estimated distance must be determined. Manageable ratios of one meter per volt are common for sensors with output signals of 0–10 V. However, sensor outputs tend to be noisy, and moving averages or other filtering techniques are frequently employed to smooth the estimates obtained from the sensor. Precision can be on the order of 2–5% when detecting moving objects, especially within the first five meters [2]. Ultrasonic devices are sensitive to strong winds due to a loss of reflected signal.

The scouting robot of Figure 1.5a has a laser rangefinder at the front to locate the center of the lane between corn rows. The off-road robotized vehicle that won the Grand Challenge competition in 2005 (Figure 1.3) features five lidars on the roof. What is the reason for the popularity of this device in field robotics? *Lidar* stands for *light detection and ranging*, and it is an optical sensor that is used to estimate ranges based on the time of flight of a light beam. The most frequent light source employed is a *laser*, due to its coherency, which makes lidars accurate industrial laser scanners. The laser emitter sends out a pulse of light and starts the timer; the timer stops when the reflection of the pulse is captured by the receptor. Lidars usually provide polar range data by means of a rotating mirror. Typical scanning fields vary between 180 and 270°. The field range depends on the particular application, and a number of solutions are available, but the majority of the sensors fall in the range 2–20 m. The object detection resolution is distance dependent, but it can be as high as 10 mm. While it provides accurate estimates, the laser beam is very narrow, and in order to cover the entire area in front of a vehicle, lidars need to rotate back and forth. Mechanical rotation introduces several disadvantages compared with imaging sensors: first, a mechanical rotating device on a moving vehicle traveling off-road and subjected to vibrations and rough terrain can break or lose accuracy; second, it will need a significant period of time to cover the whole area to be monitored, whereas a camera will acquire all of the information needed in just one snapshot that can be grabbed at 30 frames per second. Further complications arise if three-dimensional perception is desired, as the head of the lidar needs to rotate in two perpendicular planes simultaneously, with obvious complications for data synchronization. The standard output of a nodding laser is a *range map*, which yields the profile of the obstacles in the detectable zone ahead of the sensor. An accepted method of processing the information presented by range maps is through occupancy grids, where the environment ahead of the vehicle is represented in terms of the probability of finding an object in a set of evenly spaced cells (Section 5.5).

4.3 Monocular Machine Vision

Since the popularization of machine vision for robotic applications in the 1980s,, mainly due to improvements in computer processing power, monocular cameras have been present in many mobile robots, such as *Shakey* and *Stanley* in Figure 1.3. Machine vision offers significant advantages over other sensors that provide similar information. High-rate images provide rich information about the scene ahead of an intelligent vehicle. In addition, it provides instantaneous updates on what is occurring in the vicinity of the vehicle, allowing the perception of a variety of object properties such as position, dimensions, texture, and color. The greatest challenges when implementing monocular cameras in off-road vehicles are the computational loads of the processing algorithms needed and coping with outdoor illumination patterns.

4.3.1 Calibration of Monocular Cameras

The final output of a monocular camera is a *digital image* of resolution $H \times V$, where V represents the height of the image in *pixels* (picture elements) and H is the horizontal dimension of the image, also in pixels. The information conveyed through the image belongs to the *image space*, and can be monochrome (grayscale) or color (typically RGB). The image space is a two-dimensional representation of a three-dimensional scene, usually referred to as the real world or *world space*; therefore, more than one point in the real world can be represented by the same pixel in image space, establishing a relationship that is not univocal. In fact, when the sensitive array of the camera, also called the *imager*, is not parallel to a flat surface, image perspective will lead to inaccurate correlations between the image space and the real world. The mathematical function that establishes a relationship between the image space (pixels) and the world space (meters) is called the *calibration function*.

Monocular cameras used for vehicle navigation purposes tend to be mounted on the front of the vehicle and to look ahead. They usually point downwards somewhat, forming an inclination angle with respect to an imaginary vertical line crossing the optical center of the lens. In this camera configuration, perspective can be sensed in the images, and parallel lines in the real scene converge at a vanishing point in image space. Every image acquired with the camera needs to be transformed to physical units before critical information about the real scene can be extracted. This operation is performed through calibration equations, which depend on the camera's resolution, the focal length of the lenses, the pixel size, lens aberration, *etc.*, making it difficult to develop an exact equation that relates all of these parameters. However, a calibration routine based on establishing a correlation between pixel position and real position, when both are accurately known, is feasible, accurate, and efficient for the applications described here. This relationship is calculated using optimization functions such as least squares techniques, given that the calibration of monocular cameras involves overdetermined systems. A practical way to conduct a calibration

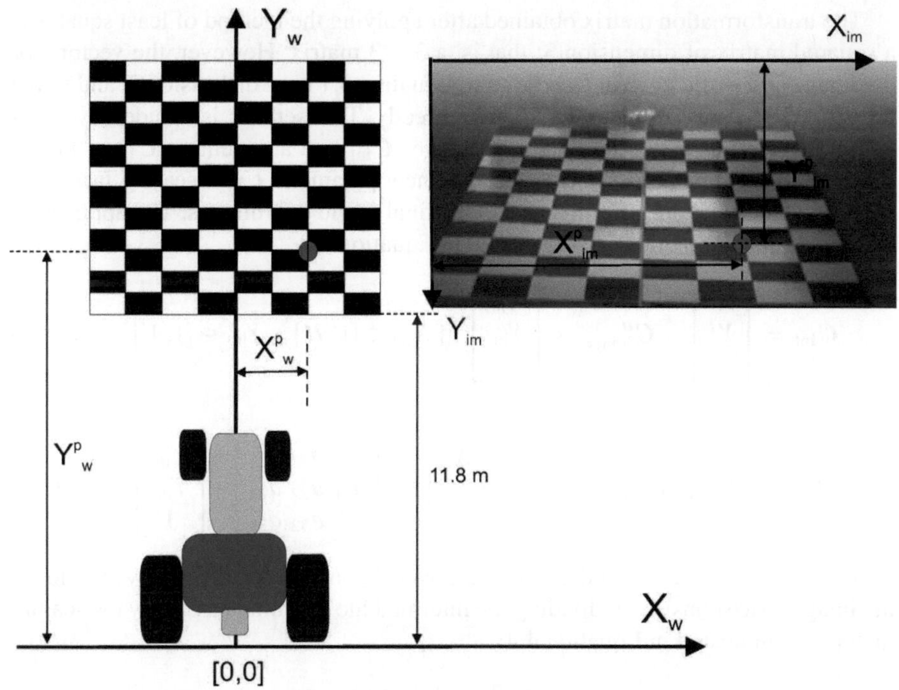

Figure 4.5 Calibration procedure for a monocular camera mounted on an agricultural tractor

test is to place a board tiled in a chessboard arrangement in front of the vehicle, as shown in Figure 4.5. The rectangular board displayed in the figure has an area of $2.4 \times 3.6\,\text{m}^2$, as every tile is square and one foot long. The coordinates must be centered and fixed on the vehicle. In the particular case of Figure 4.5, the origin was set at the drawbar to facilitate its access, rather than placing it at the center of mass, as represented in the diagram. The testing board image represented in the figure was employed to apply the calibration routine, and a detailed calculation of this routine is given in Equations 4.1–4.4. The intersections between the tiles (black and white squares in the image) yield a total of 115 points for which both image-space and world-space coordinates can easily be easily calculated in pixels and meters, respectively. The final *image to world* transformation matrix, after applying the least squares method to the 115 data points, is given in Equation 4.5.

Once the calibration has been performed and the transformation matrix between image space and world space is known ($T_{\text{i}\to\text{w}}$), we can relate any point in the image to a real position in the field of view. The first step after the camera has been calibrated is to define the input and output vectors for the transformation of coordinates. The image coordinates will be bounded by the image resolution $H \times V$.

$$\boldsymbol{C}_{\textbf{world}} = \begin{bmatrix} X_{\text{w}} \\ Y_{\text{w}} \end{bmatrix}; \quad \boldsymbol{C}_{\textbf{image}} = \begin{bmatrix} X_{\text{im}} \\ Y_{\text{im}} \end{bmatrix} \quad \{X_{\text{im}} \in [1, H]\ ;\ Y_{\text{im}} \in [1, V]\} \tag{4.1}$$

The transformation matrix obtained after applying the method of least squares is a squared matrix of dimension 3; that is, a 3×3 matrix. However, the vectors for the image and world coordinates given in Equation 4.1 have dimension 2, and so the transformation matrix cannot be applied directly. This setback is avoided by introducing the vector of intermediate coordinates $\boldsymbol{C}_{\mathbf{int}}$ and augmenting $\boldsymbol{C}_{\mathbf{image}}$ in one dimension, as shown in Equation 4.2. The new parameter t is a scaling factor between the intermediate coordinates and the final world coordinates. The application of the transformation matrix is provided in Equation 4.3.

$$\boldsymbol{C}_{\mathbf{int}} = \begin{bmatrix} X' \\ Y' \\ t \end{bmatrix} \quad \boldsymbol{C}^{a}_{\mathbf{image}} = \begin{bmatrix} X_{\mathrm{im}} \\ Y_{\mathrm{im}} \\ 1 \end{bmatrix} \quad \{ X_{\mathrm{im}} \in [1, H] \; ; \; Y_{\mathrm{im}} \in [1, V] \} \tag{4.2}$$

$$\boldsymbol{C}_{\mathbf{int}} = T_{\mathrm{i}\to\mathrm{w}} \cdot \boldsymbol{C}^{a}_{\mathbf{image}} \quad \Rightarrow \quad \begin{bmatrix} X' \\ Y' \\ t \end{bmatrix} = \begin{bmatrix} a_{11} & a_{12} & a_{13} \\ a_{21} & a_{22} & a_{23} \\ a_{31} & a_{32} & a_{33} \end{bmatrix} \cdot \begin{bmatrix} X_{\mathrm{im}} \\ Y_{\mathrm{im}} \\ 1 \end{bmatrix} \tag{4.3}$$

The final step needed to determine the world coordinates of any given point in the image space consists of dividing the intermediate coordinates $\boldsymbol{C}_{\mathbf{int}}$ by the scaling factor t, as indicated in Equation 4.4.

$$\begin{bmatrix} X_{\mathrm{w}} \\ Y_{\mathrm{w}} \end{bmatrix} = \frac{1}{t} \cdot \begin{bmatrix} X' \\ Y' \end{bmatrix} \tag{4.4}$$

The particular values of the components of the transformation matrix $T_{\mathrm{i}\to\mathrm{w}}$ will depend on the intrinsic parameters of the camera as well as its position and orientation. The resulting matrix for the calibration test shown in Figure 4.5, which utilized an analog monochrome monocular camera (Cohu 2622, San Diego, CA, USA) with a 16 mm lens, is provided in Equation 4.5. The digitized images had a resolution of 320×240 pixels, and the world coordinates are measured in meters. Note, for example, that the center of the image ($X_{\mathrm{im}} = 160$; $Y_{\mathrm{im}} = 120$) will correspond, according to the transformation matrix of Equation 4.5, to a distance of approximately 17 m from the origin of the coordinates in the Y_w direction.

$$T_{\mathrm{i}\to\mathrm{w}} = \begin{bmatrix} -57.616 & -1.369 & 8560.109 \\ 2.284 & -154.528 & -52465.137 \\ 0.444 & -28.722 & -848.23 \end{bmatrix} \tag{4.5}$$

4.3.2 Hardware and System Architecture

Machine vision sensors can take many forms. The cameras described in this section are *monocular* and sensitive to the *visible spectrum*. Other spectra are discussed in Section 4.4, and binocular stereo cameras are explained in depth in Chapter 5.

Generally speaking, monocular cameras for visible light can be classified according to two concepts:

1. *Type of image*: monochromatic (grayscale) or color (red-green-blue);
2. *Type of output signal*: analog output or digital output.

When the camera is analog, given that the computer algorithm can only deal with digital images, a *frame grabber* is required to convert the analog image to a digital one before it can be sent to the computer. Overall, the camera can be considered an optoelectric device that captures images of scenes and stores the information as digital images for processing later. Visual information is acquired by means of an array of light-sensitive electronic cells. Figure 4.6 shows several monocular cameras used for the automatic guidance of off-road agricultural vehicles and a set of typical lenses.

Figure 4.6 Monocular cameras for visual perception in intelligent vehicles

The selection of the camera lenses is crucial, as images are formed by light, and light beams must traverse lenses to hit the sensitive cells of the imagers. As a result, special care must be taken when choosing lenses for a specific camera and application. Various aspects need to be considered before making a decision about the optics for the camera:

- *Lens quality*. Just as in recreational and artistic photography, the higher the quality of the lens, the sharper the image. High-quality optics usually make a difference, and using them often circumvents the need for post-processing operations and filtering routines, which tend to overload the perception processor. Any improvement that can be achieved through hardware enhancement should not be performed through software, as it will make the system slower and more error-prone.
- *Image sensor size*. The photon-sensitive arrays of video cameras have different sizes and resolutions. The size of the electronic image sensor needs to match with the chosen lens in order to avoid reduced intensity and saturation at the periphery of the image, a phenomenon called *vignetting*. Common sensor sizes for professional machine-vision cameras are (diagonal length) ¼″, ⅓″, ½″, ⅔″, and 1″.

- *Focal length*. Usually referred to as f, the focal length is determined by the particular application; that is to say, the focal length will depend on the scene being sensed. It establishes the shape and dimensions of the field of view ahead of the camera and determines how close objects will appear in the image. Small values of focal length lead to *wide-angle lenses*, which give fields of view that include large portions of the scene. Large values of f, on the other hand, yield *telephoto lenses* with narrow fields of view but which are favorable for capturing distant objects. Standard machine-vision cameras with $^1/_3''$ imagers, for example, can implement wide-angle lenses of 2.8, 3.6, or 4.6 mm; normal lenses of 6 mm or 8 mm; and telephoto lenses of 12 and 16 mm.
- *Type of mount*. This is one of the most important, and consequently most frustrating, sources of problems when assembling a machine-vision system. There are two standard types of mount: *C mount* and *CS mount*. Both camera mounts consist of a male thread that couples with the female thread of the camera, so both types of mounts are screwed into the camera. The difference between them is the flange focal distance: 17.526 mm for C-mount lenses, and 12.52 mm for CS-mount lenses. Most current cameras feature a CS mount, which allows the use of both kinds of lenses; however, C-mount lenses require a 5 mm ring adapter that allows them to be properly focused on a CS camera. However, with C-mount cameras it is not possible to use CS-mount lenses, as the image will never be focused properly. Although C-mount lenses can be screwed into CS-mount cameras by inserting a 5 mm ring between the camera and the lens, this is not a robust solution as the vibration of the vehicle very often ends up unscrewing the lens, since the use of the adapter leads to a looser fit. The best solution is therefore to implement cameras and lenses with CS mounts.
- *Autoexposure capabilities*. Some lenses can automatically adjust the aperture of the iris according to the ambient light, so that sudden changes in illumination can immediately be compensated for by the hardware instead of needing to apply image processing techniques to correct the exposure and enhance the images. This feature is very advantageous for vehicles working outdoors, where the natural light is always changing from dawn to dusk. As mentioned before, any improvement achieved through the hardware reduces computational demands from other image processes.

Even though the cameras described in this section integrate image sensors that are sensitive to *visible light*, which approximately comprises the part of the electromagnetic spectrum (400–750 nm) that an average human eye can perceive, machine perception can easily be extended to the neighboring *near-infrared* band (NIR). The NIR window is adjacent to the human response window, although it lies outside it (it is usually 700–1400 nm). The main advantage of amplifying the sensing spectrum of a monocular visible camera with NIR is that this enhances images of crops. The reflectance of crop plants in the IR window helps to discriminate these plants from the soil, even when there is an intense weed infestation. Figure 4.7 provides a monochrome image of soybean rows (a) and a NIR-filtered image of the same crop (b). Notice how the NIR image reinforces the differences between the crop rows and

Figure 4.7 Image filtering: (**a**) without filters; (**b**) with a NIR filter

the non-crop areas, thus facilitating crop tracking via imaging without the need to consume extra computational resources.

The procedure used to sense in the near-infrared band with a standard monocular camera is to physically attach an IR filter to the camera, positioning it just over the electronic image sensor. NIR filters are available in specialized stores, and two properties need to be considered when choosing them: *center wavelength* and *bandwidth*. The diameter of the filter must also be carefully chosen to ensure that the filter will fit into the camera without interfering with the lens mount and focusing operations. The filter used to acquire Figure 4.7b has a center wavelength of 800 nm and a bandwidth of 100 nm. Figure 4.8 shows filters with various diameters and a camera with a NIR filter mounted on it.

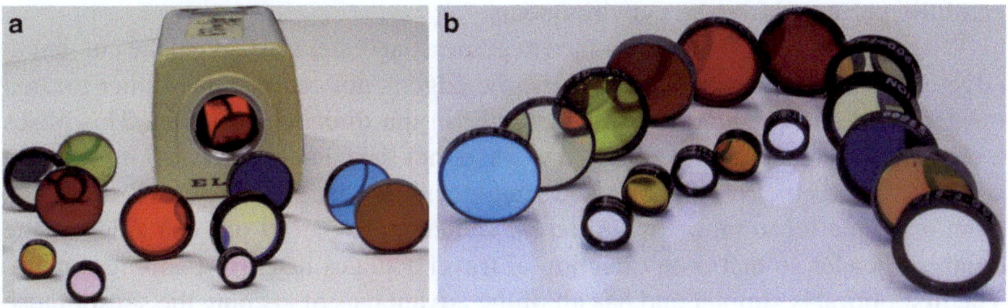

Figure 4.8 IR filters for enhancing agricultural images

After the vision sensor has been chosen and appropriately furnished with the optimum lens, and a filter if needed, it needs to be mounted on the vehicle. Several locations are eligible, and an advantageous placement of the camera can be crucial to successful perception. There are many potential scenarios that an intelligent vehicle might need to cope with, but two fundamental applications should be considered in terms of *camera location*: a tractor guided within a field of crop rows or orchard rows, and a harvester autosteered by detecting the cut–uncut crop edge. When a tractor is set up to track crop rows or move between trees in a grove, the best place for the camera is the center plane of the vehicle, so that the camera will move along the middle plane of the tree rows that forms the traversable lane or the set of crop rows covered by the tractor base. However, the camera can be mounted on the

Figure 4.9 Camera locations: (**a**) front end of tractor nose; (**b**) tractor cabin; (**c**) harvester header

front end of the tractor nose (Figure 4.9a) or on the cabin (Figure 4.9b). Although the first option puts the camera closer to the features tracked, the latter position is recommended. Placing the camera right at the front of the vehicle means that the wires to the onboard computer must be placed under the hood, which is typically lacking in space and heated by the engine. In addition, operating the camera (for example changing lenses or focusing) requires two people, and the results cannot be immediately checked on the screen, as it is far away in the cabin. When the camera is attached to the cabin, the cables are shorter and the driver can operate the camera while checking the resulting images, even if the camera is outside the cabin (*i.e.*, by looking through the windshield). If the camera is mounted inside the cabin, not only it is easier to operate but it will also be protected from dust and other environmental hazards. Cameras that are meant to work outside the cabin should be waterproof or carefully protected from wet environments.

When the vision system is designed to guide a harvester by tracking the cut–uncut edge of the crop, a new problem arises: the edge is not centered but rather located on one side, and the working side depends on the direction of travel. This poses a challenge that can be solved in two ways: either the camera is located in a central section of the vehicle and uses a wide-angle lens so that both edges are detectable from a unique location, or two different cameras are placed on both sides of the combine header so that each direction of travel actuates its corresponding camera. The first option is simpler and has all of the advantages of keeping the sensor close to the processing computer, but it is not practical for large harvester headers (over 8 meters wide). The second possibility assures better framing of the scene, but running cables from the cabin to the extremes of moving headers can result in cut or stripped wires. This option also requires more sophisticated camera control, usually through software, to swap cameras according to the working side. Figure 4.9c illustrates a camera placed on the header of a corn harvester. An alternative option to guide the harvester with a monocular camera would be to place the camera in the center of the vehicle, just on top of the cabin, and detect the crop rows in order to extract guiding features. As will be shown in Section 4.3.4, this alternative is a challenging one for tall crops, such as corn, as plants in late vegetative states are fully developed, which can complicate the process of distinguishing the rows (Figure 4.25b).

4.3.3 Image Processing Algorithms

Selecting the proper hardware for machine vision perception clearly enhances system performance, but the key stage is information extraction and decision-making. Image-processing software represents the "brain" of the perception system and it thus houses the intelligence of the system. As a matter of fact, the appropriateness of a camera–lens arrangement is often judged according to the extent that it simplifies further processing by enhancing raw input images. Two basic qualities are searched for in optimum software design: *robustness* and *processing speed* (frame rate). Depending on the scene and application envisioned, a chain of image analysis operations needs to take place between initial image capture by the camera and obtaining the final and conclusive parameters. There is a rich diversity of image-processing algorithms, but the most popular ones used for local perception in intelligent off-road vehicles are described in the rest of this section.

Selection of a Region of Interest (ROI)

Adequate selection of a lens adjusts the field of view of the camera in such a way that the images acquired capture the critical data from the features tracked. Nevertheless, this adjustment is not perfect in most cases because of image perspective or camera location, and so not every pixel in the image will provide useful information; on the contrary, margin areas may contain confusing and misleading information. Consequently, one of the first operations to perform when programming a vision algorithm is to select a *region of interest* (ROI), which can be defined as the region of the image that is specially chosen for successive image analysis operations; all of the pixels outside of the ROI are discarded. The selection of a suitable ROI is key to successful perception, and the following design ideas must be taken into consideration:

- *Size*. The ROI should include all critical information while maintaining the smallest size possible. An excessively large processing area will overload the algorithm, as calculations are performed pixel by pixel. Off-road vehicles necessitate the processing of several images per second, and high efficiency of the vision algorithm is vital to achieving safe navigation. Top and bottom pixel rows are normally excluded from the ROI.
- *Position*. The vehicle will move laterally to adjust its trajectory to the tracked features. It is important to anticipate the magnitude of these offsets so that the essential visual data is never outside the ROI regardless of the position of the vehicle. The loss of tracking features can have devastating consequences for vision-based navigation.
- *Shape*. Most ROIs tend to be rectangular and thus completely defined by their four corners, because this configuration eases further calculations. However, other shapes can be useful too, such as trapezoidal ROI windows adjusted to the perspective of the scene.

- *Dynamism.* A good trade-off between position and size can be sometimes difficult to achieve if the window is always static and usually in the center of the image. This handicap can be overcome by endowing the ROI selection routine with the ability to move the ROI around the image in search of the location most likely to give informative pixels. A further degree of dynamism can also be incorporated to adapt the size of the ROI to each image, but this approach requires heavier programming and processing during the initial operations.

Figure 4.10a shows a static ROI of dimensions 180 × 100 applied to an original image of resolution 320 × 240, meaning that the ROI represented 23% of the image. The selected window was rectangular and centered, designed to capture the two central rows of soybeans ahead of an automatically guided tractor. An upgrade of the window on the left converted it into a dynamic window (Figure 4.10b), with a center that moved horizontally according to the vision-based trajectory of the vehicle found in the previous image.

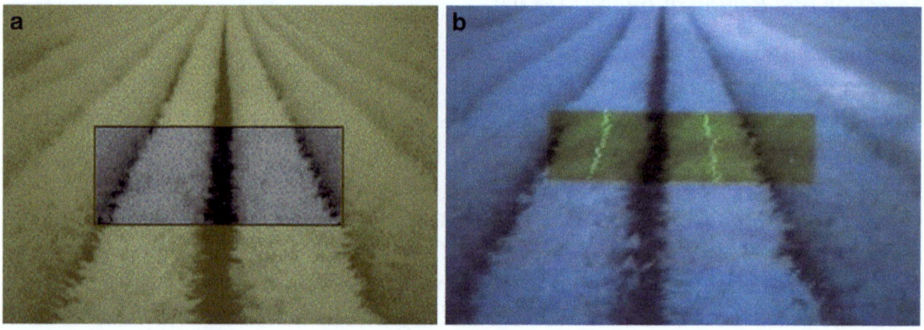

Figure 4.10 Region of interest: static window (**a**), and dynamic window (**b**)

Image Binarization and Thresholding

The ability to perceive color has been successfully used in many machine-vision applications, but most of the algorithms developed for navigation and obstacle avoidance are programmed to work on monochrome images. When a color camera is used to acquire images, different analytical techniques can be utilized:

- Split the image into its three color components: red, green, and blue. Analyze these monochrome images separately and continue processing those that provide useful data.
- Very often, better results are obtained when pixel information is converted into the color space HSI (hue, saturation, and intensity), rather than RGB.
- However, the most frequently used procedure is to convert the original color image to grayscale using one of the conventional formulae that mimic the human eye's perception of luminosity, such as 59% green + 30% red + 11% blue.

Regardless of the methodology followed to get a monochrome image, the starting point is usually the same: a *grayscale image*, typically with a depth of *8 bits*, which provides *256 levels* of intensity (2^8) between black (level 0) and white (level 255). More intensity levels are not usually necessary and result in greater computational cost, which could slow down the algorithm. All of the images considered for the remainder of Section 4.3 are assumed to be 8-bit monochrome.

Binarization is a simple operation that involves mapping all of the pixels in the image, which have values of between 0 (black) and 255 (white), according to their membership of the classes "feature" or "background." Therefore, binarization is a means to separate out interesting features. While the *binarizing function* is fairly simple and straightforward to apply, the main difficulty of this stage is to determine the particular gray level that separates one class from the other. This value is known as the *threshold*, and intensity values below it will be mapped to black whereas the others will be transformed to white. Note that the assignment of binary values to classes is irrelevant beyond its use for discriminating features from background. Complexities arise with threshold selection due to the need to adapt to changes in ambient illumination, which means that using one threshold for all images is unlikely to yield good results. The threshold determination needs to be *dynamic*, meaning that a specific value is calculated for each image acquired. An example of a dynamic threshold was implemented in *Case Study I* of Section 4.5 [3], which followed the mathematical expression of Equation 4.6, where the dynamic threshold is TH, the average gray level within the selected ROI is I_{av}, the maximum gray level within the ROI is I_{max}, and δ_o represents an offset in the interval $[-0.5, 0.5]$ that must be found experimentally after evaluating some sample images.

$$\text{TH} = I_{av} + \delta_o \cdot (I_{max} - I_{av}) \tag{4.6}$$

The left image in Figure 4.11a represents a typical field image used for crop tracking with a binarized region of interest obtained via Equation 4.6. Note that, from the moment a region of interest is selected, all of the computations will take place within its boundaries. The histogram of Figure 4.11b represents the gray level distribution of the entire image; however, Equation 4.6 is exclusively applied to

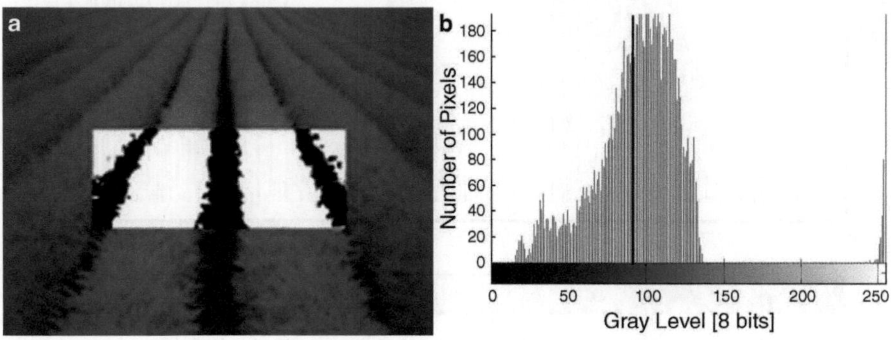

Figure 4.11 Binarization of region of interest (**a**), and histogram of original image (**b**)

the ROI, and so, despite the extreme values shown in the histogram, the average and maximum intensity levels considered are those of the ROI. For this particular example, the average is 102, the maximum gray level in the ROI is 138, and the offset applied is −0.2, giving a final threshold of 94. The division line shown in Figure 4.11b separates the pixels that are mapped to each class. In this particular case, the tracked features are NIR-filtered soybean rows, which were transformed to white pixels after the application of the threshold TH.

When the number of classes is known and they are distinctly represented by somewhat homogeneous gray levels, clustering techniques can be applied to systematically find a dynamic threshold. One such method that has been efficiently applied to threshold agricultural images is the *K-means algorithm*. This algorithm demands *a priori* knowledge of the number of clusters K, and using this, it clusters data points based upon their distances from the centers of the clusters, which are continuously updated. It finally locates the centers of mass for the estimated groups of data in a similar manner to the generic concept represented graphically in the diagram of Figure 4.12. The threshold can, for instance, be the midpoint between the centers of the two clusters. The K-means algorithm is an iterative process, and the number of iterations must be carefully bounded to avoid never-ending loops. Other Bayesian-based methods (*e.g.*, data iteratively updated with newly collected evidence) that are used to discriminate classes can be implemented too as long as they consistently find a threshold in a reasonable time, such as the boundary dashed line of Figure 4.12.

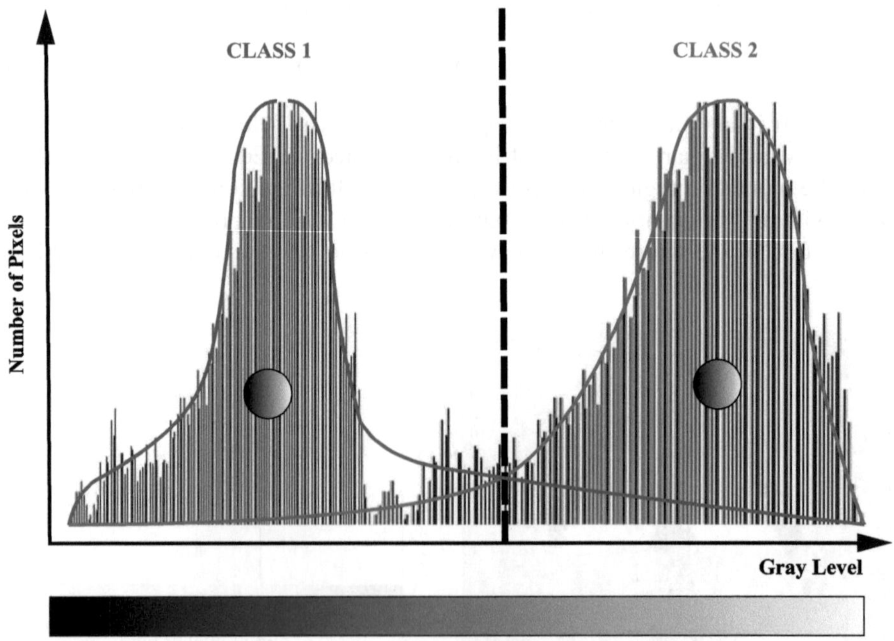

Figure 4.12 Clustering techniques for dynamic thresholding: working principles

Midpoint Encoding

The binarization process leads to a huge reduction in information with regards to gray levels: from 256 levels to 2 levels. This operation results in a great number of pixels holding the same information. Given that the processing speed depends on the number of pixels that take part in the computations, if the same basic information can be carried by a reduced set of pixels, the execution of the algorithm will improve considerably. When the tracked features are crop or tree rows, there is no need to maintain the entire thickness of the row; rather, its skeleton is sufficient to provide navigation commands. This is the purpose of the *midpoint encoding* algorithm [3, 4]. The objective of this operation is to replace the binarized feature with a thin line that holds the critical information. The application of the midpoint encoder to the binarized ROI of Figure 4.11 results in the simplified ROI of Figure 4.13a. Note the tremendous reduction in the number of white pixels, which are the pixels processed in subsequent operations, although the critical information persists. One potential mistake when running the midpoint encoder results from incomplete binarized rows, such as the top corners of the region of interest in Figure 4.11a. In order to avoid false centerlines such as those shown in the training board of Figure 4.13b, the algorithm must skip incomplete rows to achieve robust encoding. This feature is incorporated into the flowchart of the encoding algorithm schematized in Figure 4.14.

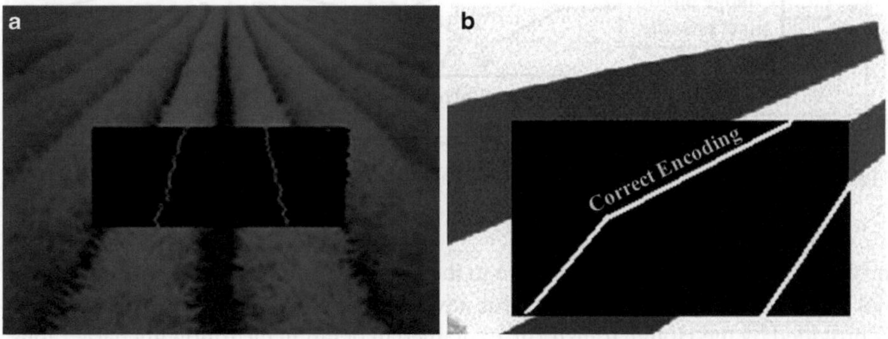

Figure 4.13 Midpoint encoded field image (**a**), and erroneous encoding (**b**)

Line Detection Techniques: Regression Analysis and the Hough Transformation

The three agricultural environments illustrated in Figure 4.1 are different, but they do have something in common: *line structuralization*. Corn, potatoes, and orange trees are planted in equally spaced lines. This fact makes *line detection* a fundamental tool for off-road navigation in agricultural fields. In general, two basic approaches are used to do this: regression analysis and the Hough transform. *Regression analysis* is a popular statistical technique that is employed to estimate the best fit for a line given a scattered set of points. While widely used for other engineering

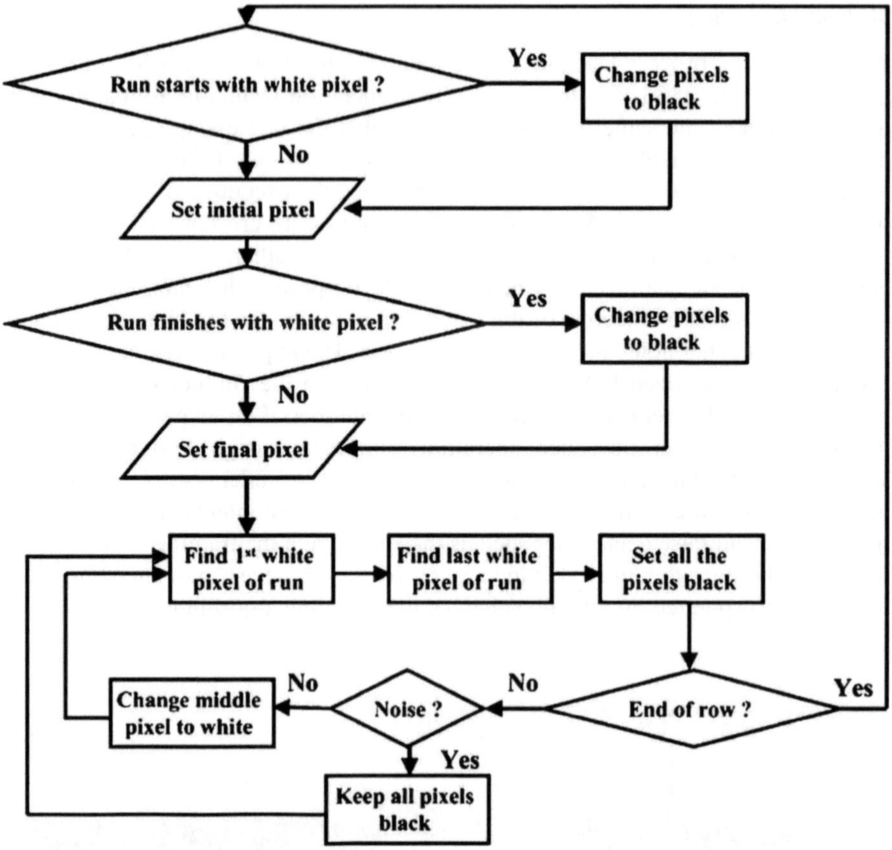

Figure 4.14 Flowchart of the midpoint encoder algorithm

problems, regression is very sensitive to the presence of outliers, which are certainly not rare in digital images acquired for navigational purposes. A superior approach is provided by the Hough transform, explained in detail in the following paragraphs. A deeper study of regression analysis is provided in Chapter 5 (*Case Study II*).

The *Hough Transform* is a mathematical transformation that enhances the alignment of points very efficiently due to its low sensitivity to outliers. The transformation takes place between the *image space* represented by the pixels (x_{im}, y_{im}) of the image processed and the *parameter space* (ρ, θ) in which linear features are detectable. The transformation is carried out by applying Equation 4.7 to every pixel considered in an image of resolution $H \times V$ when the parameter θ varies between $-90°$ and $90°$. The representation of ρ according to Equation 4.7 yields a sinusoid for each image point considered (x_{im}, y_{im}); the resolution of the sinusoids will depend on the number of θ points considered in the interval $[-90°, 90°]$. When the points transformed lie in a straight line, all of the sinusoids in the parameter space will cross at one point defined by (ρ_L, θ_L). The final equation of the estimated line

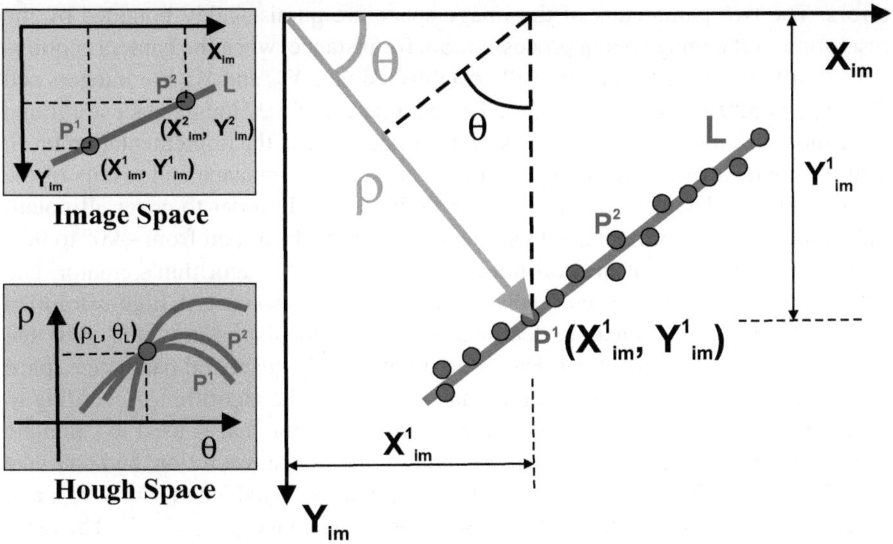

Figure 4.15 Parametric representation of a line with the Hough transformation

must be expressed in image space, so a back transformation must be applied to the results found in the parameter space (ρ_L, θ_L) (also known as *Hough space*). This inverse transformation is given in Equation 4.8, and it represents the explicit equation of a line ($Y = \text{intercept} + \text{slope} \cdot X$) in the image space. A conceptual illustration of the Hough transformation is shown in Figure 4.15. Once the line is defined in image space, it can be further transformed to the real world by following the procedure outlined in Equations 4.1–4.4.

$$\rho = X_{im} \cdot \cos\,\theta \;+\; Y_{im} \cdot \sin\,\theta \qquad \{X_{im} \in [1,\, H];\; Y_{im} \in [1,\, V];\; \theta \in [-90°,\, 90°]\} \tag{4.7}$$

$$Y = \frac{\rho_L}{\sin\,\theta_L} - \frac{\cos\,\theta_L}{\sin\,\theta_L} \cdot X \qquad \{X \in [1,\, H];\; Y \in [1,\, V];\; \theta_L \neq 0\} \tag{4.8}$$

As usually happens with mathematical transformations, the parameter space is helpful for highlighting linear features, but the trigonometric representation of the image points is not intuitive, and so it is important to acquire a profound understanding of how the Hough transformation works before any attempt is made to implement it. The numbers behind the parameters ρ and θ must be meaningful in spite of the fact that they do not represent physical dimensions or distances, unlike the coordinates X_{im} and Y_{im}, which represent the actual positions of pixels in the image. The incorporation of the Hough transform into vision-based perception algorithms is usually realized through computer programs, and consequently the variables involved in the transform calculations should be properly defined to avoid execution errors. The two parameters of the image space are unmistakably bounded by the resolution of the image being processed. So, for instance, when the camera acquires 320×240 images, it is automatically established that X_{im} and Y_{im} are integers and $1 \leq X_{\mathrm{im}} \leq 320$ and $1 \leq Y_{\mathrm{im}} \leq 240$. The parameters of the Hough space are found to be more debatable, as some choices need to be made at the implementation level. These decisions will have significant effects on the performance and results of the transformation. The first parameter to be defined is ρ. In order to cover all potential sinusoidal crosses associated with an image, θ needs to span from $-90°$ to $90°$. The quantization of that interval must be decided by the algorithm's creator, and it leads to a trade-off between precision and computational cost. A high-resolution family of sinusoids will help to deal with potential critical crossings at $\theta_{\mathrm{L}} = 0$ that make Equation 4.8 impractical. However, an extremely quantized parameter space will result in heavy computation that may slow down the algorithm. According to Equation 4.7, which provides the conventional expression that is used to calculate ρ, the operational interval for ρ will depend on the image resolution, and can take negative values. The minimum value of ρ, $\rho_{\min}$, will be found when $\theta = -90°$, and is equal to the negative of the vertical resolution of the image: $\rho_{\min} = -V$. The maximum value of ρ, $\rho_{\max}$, can be determined by finding the value $\theta_{\max}$ that maximizes ρ in Equation 4.7, as shown in Equation 4.9, or from geometrical constraints in the image space, as indicated in Equation 4.10. The use of trigonometric functions in the calculation of ρ implies that this parameter will take non-integer values, and will therefore need to be suitably quantized too. The optimum discretization of the Hough space is essential for the efficient application of the Hough transform. Given a digital image of resolution $H \times V$, the intervals for all of the variables involved in its Hough transformation are given in Equation 4.11.

$$\begin{aligned} &\frac{\partial \rho}{\partial \theta} = \frac{\partial\,(H \cdot \cos\theta + V \cdot \sin\theta)}{\partial \theta} = 0; \\ &\theta_{\max} = \arctan\left(\frac{V}{H}\right); \quad \rho_{\max} = H\cos\theta_{\max} + V\sin\theta_{\max} \end{aligned} \tag{4.9}$$

$$\rho_{\max} = \sqrt{H^2 + V^2} \tag{4.10}$$

$$\text{Image Resolution}\, H \times V : \left\{ \begin{array}{l} X_{\mathrm{im}} \in [1,\ H] \\ Y_{\mathrm{im}} \in [1,\ V] \\ \theta \in [-90°,\ 90°] \\ \rho \in \left[\rho_{\min} = -V,\ \rho_{\max} = \sqrt{H^2 + V^2}\right] \end{array} \right\} \tag{4.11}$$

The direct application of Equation 4.7, under the constraints specified by Equation 4.11, results in a two-dimensional parameter space full of sinusoids, where each sinusoid originates from a pixel belonging to the image space. If the set of points under consideration form a perfect straight line, which is very unlikely in real field images, all of the sinusoids cross at the same point (ρ_{L}, θ_{L}). Figure 4.16 illustrates such a case for a simulated example of five ideal points. Notice the specific sinu-

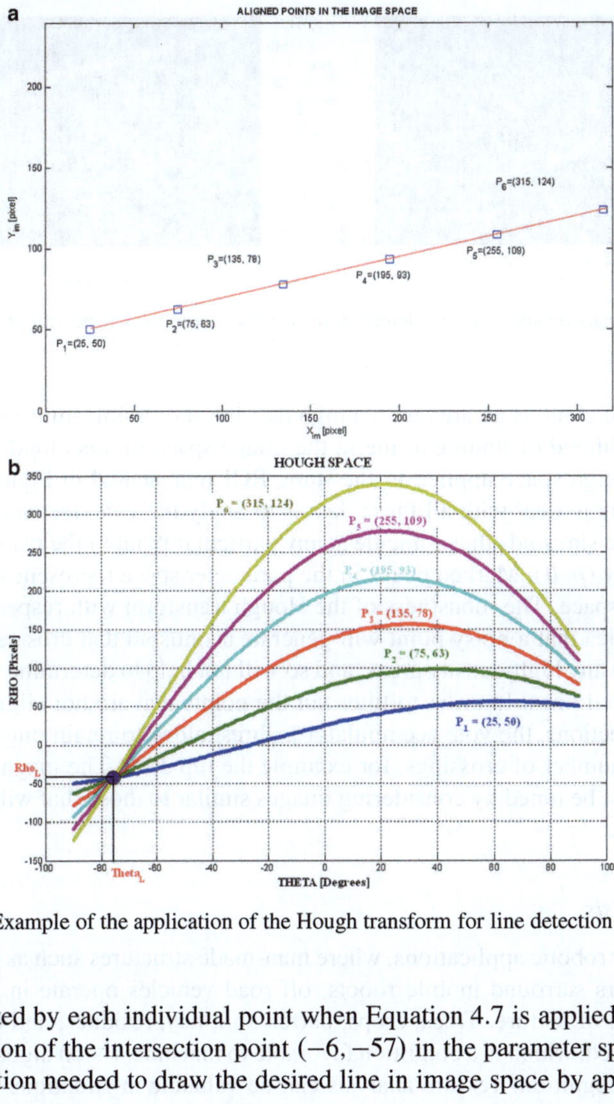

Figure 4.16 Example of the application of the Hough transform for line detection

soid generated by each individual point when Equation 4.7 is applied. The inverse transformation of the intersection point (−6, −57) in the parameter space provides the information needed to draw the desired line in image space by applying Equation 4.8.

The fundamental expression for the Hough transform given in Equation 4.7 requires the generation of a complete sinusoid in the range [−90°, 90°] for every processed point. This procedure requires a massive number of calculations unless the input images are enhanced to retain critical information and remove redundant or unnecessary data. A practical example of this philosophy can be seen in the transformation of the original field image of Figure 4.10a to the elaborated image of Figure 4.13a. The Hough transform would be applied exclusively to the white pixels that outline the crop row. Although the thin line of white pixels that mark the center of the row is approximately straight, it is obvious that the alignment is not perfect. This fact causes the sinusoids to intersect at a cloud of points rather than just

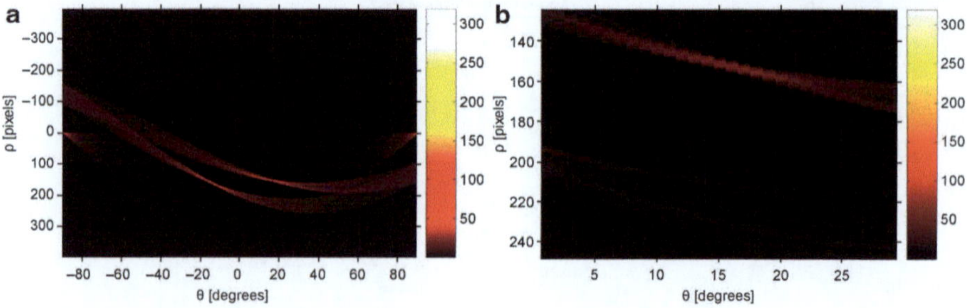

Figure 4.17 Parameter space for the detection of two lines (crop rows inside the ROI)

one point. The cloud is an area with a high density of crossing sinusoids, and there is a high likelihood of finding a line in the image space in this cloud. Figure 4.17 shows the Hough space applied to the static ROI represented in Figure 4.10. The traditional way to deal with all these data is to apply the *vote accumulator*, which sums all of the sinusoids that coincide at any particular point in the parameter space represented by (ρ, θ). Marked peaks in the parameter space represent straight lines in the image space. The robustness of the Hough transform with respect to outliers rests on the idea that a noisy point will generate a sinusoid that crosses other sinusoids far from any high-density areas, and so will not help to determine the equation of the estimated lines. In order to filter out the points that are not aligned with the dominant directions, the vote accumulator is thresholded to retain only those areas with a large number of crossings, for example the top 30%. The magnitude of this threshold must be tuned by considering images similar to those that will be fed into the algorithm.

Texture Analysis

Unlike indoor robotic applications, where man-made structures such as walls, doors and neat floors surround mobile robots, off-road vehicles operate in open fields, which are full of texture. Trees, crops, barren soil, hay, residues, and bales are all richly texturized, which is a condition favorable to machine-vision algorithms based on pattern recognition. Texture analysis can be valuable in situations where color differences are minimal. The cut swath of Figure 4.18a is a potential feature for tracking with machine vision. However, the color of the swath is too similar to the color of the surrounding terrain to attempt segmentation based exclusively on color. Figure 4.18b shows the corresponding monochrome image, where it is even more difficult to distinguish the tracking swath. The resulting histogram (Figure 4.18c) for this scene shows, as expected, a dominant band around medium levels that has a homogeneous distribution with few pixels at the extremes of the spectrum.

A first step in the processing of the input image given in Figure 4.18b can lead to the reduction of gray levels in order to ease the application of texture algorithms. Simplified gray level resolution that preserves the principal texture properties will speed up the detection algorithm and probably enhance the detection of interesting

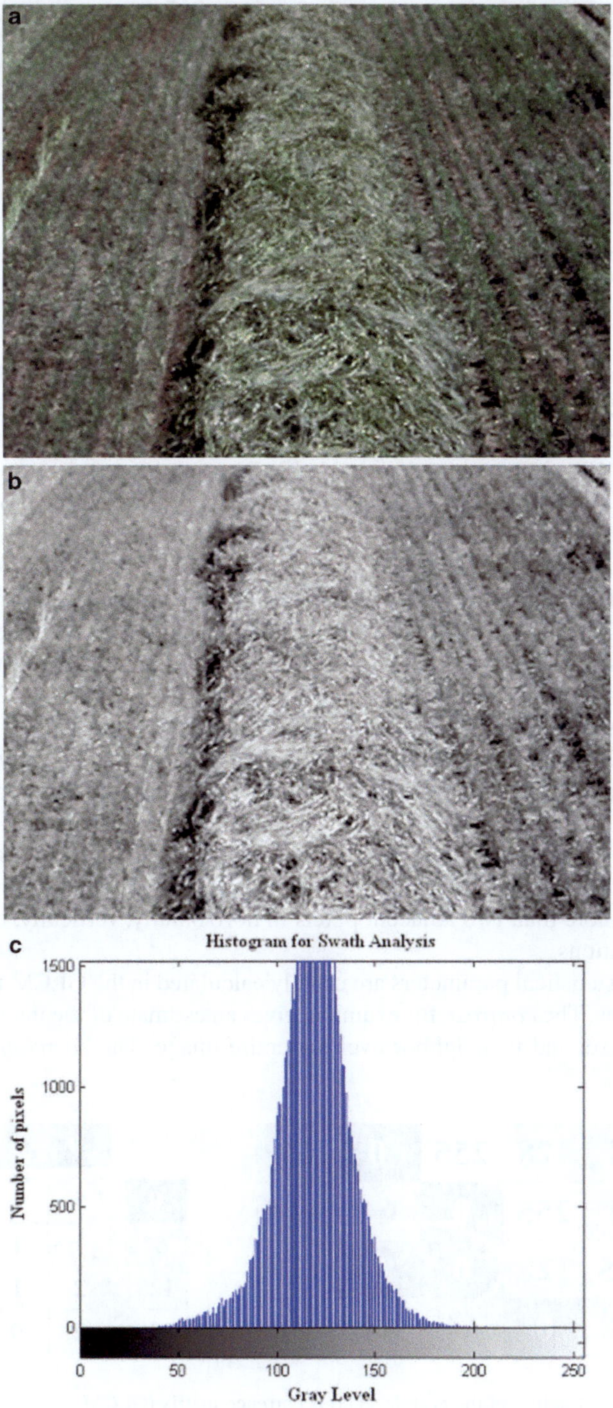

Figure 4.18 Cut swath selected as a tracking feature for machine vision based autoguidance: (**a**) original image; (**b**) monochrome image; (**c**) resulting histogram

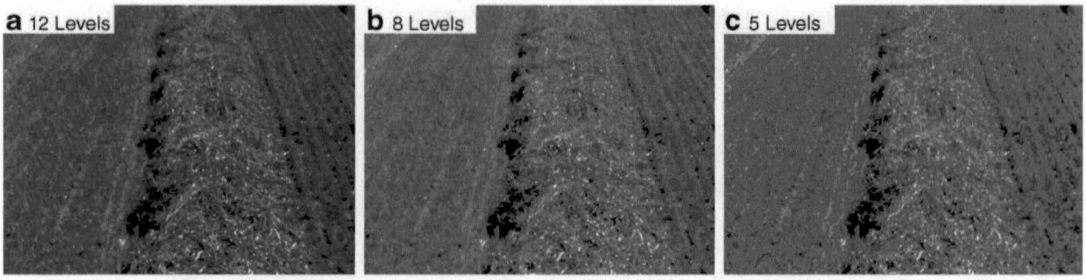

Figure 4.19 Gray level reduction for enhancing texture analysis: (**a**) twelve-level version of image; (**b**) eight-level version; (**c**) five-level version

patterns. Figure 4.19a is the twelve-level (0, 80, 90, 100, 110, 120, 130, 140, 150, 160, 170, 255) version of the original image, Figure 4.19b is the eight-level (0, 110, 125, 140, 155, 170, 185, 255) version of the same image, and finally the five-level (0, 110, 140, 170, 255) version is provided in Figure 4.19c.

The *gray-level co-occurrence matrix* (GLCM) is a statistical method for analyzing the texture content of a group of pixels in a monochrome image. It considers the spatial relationships of the pixels within the examined window, calculating how often pairs of pixels with specific intensity values and keeping a determined spatial pattern (vertical neighbors, diagonal neighbors, 4-connectivity, 8-connectivity, etc.), occur in the studied image. If the co-occurrence of two adjacent pixels is studied for an image of resolution 6×4 and four gray levels, as shown in Figure 4.20, the resulting GLCM will be the four-dimensional square matrix on the right in Figure 4.20. This pictorial demonstrates the benefits of reducing the number of gray levels to get operational matrices. Element (1, 1) in the GLCM is 1 because two horizontally adjacent pixels with the value 0 only occur once, but the second element is 4 because there are four cases of adjacent pixels 0–64. Element (2, 1), on the other hand, is null because there is no pair 64–0 in the entire image. The neighborhood can be extended to more than two adjacent pixels in horizontally, vertically, or diagonally adjacent positions.

Different statistical parameters are usually calculated in the GLCM to assess texture properties. The *contrast*, for example, gives an estimate of the intensity contrast between a pixel and its neighbor over the entire image. The *correlation* measures

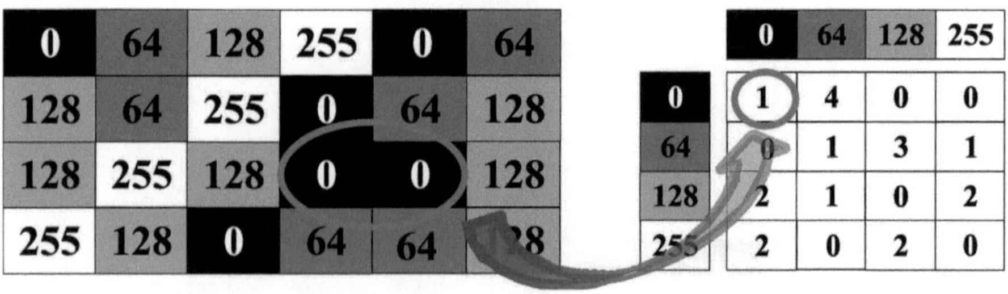

Figure 4.20 Construction of the gray-level co-occurrence matrix (GLCM)

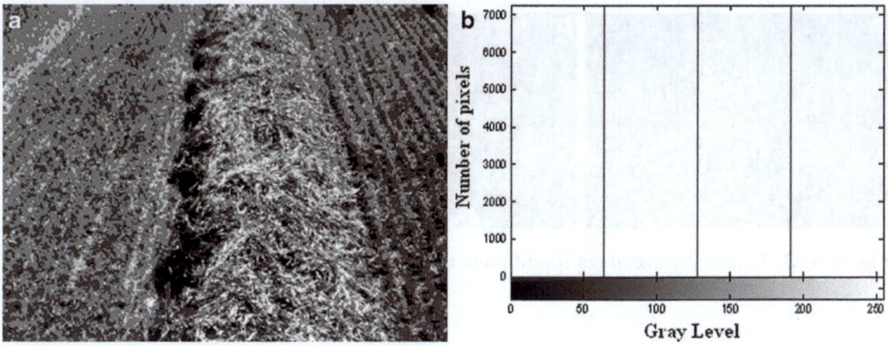

Figure 4.21 Image preparation for GLCM analysis: (**a**) monochrome image; (**b**) histogram

how correlated a pixel is to its neighbor, calculated for the complete image. The *energy* returns the sum of squared elements in the GLCM. The matrix must be normalized so that the sum of its elements is equal to one. The cut swath of Figure 4.18 was reduced to the five levels (0, 64, 128, 191, 255) represented in the histogram of Figure 4.21b, resulting in the monochrome image of Figure 4.21a. For this situation, the GLCM is the 5 × 5 matrix of Equation 4.12, with a contrast of 0.4284, a correlation of 0.6651, and an energy value of 0.8171.

$$\mathrm{GLCM} = \begin{bmatrix} 2434 & 1249 & 139 & 20 & 2 \\ 1286 & 13\,629 & 6541 & 367 & 26 \\ 132 & 6689 & 28\,050 & 4350 & 281 \\ 14 & 345 & 4403 & 4610 & 649 \\ 0 & 29 & 271 & 660 & 384 \end{bmatrix} \tag{4.12}$$

The texture content of an image can also be quantified through the concept of *entropy*. The entropy of an image is a statistical *measure of randomness* that char-

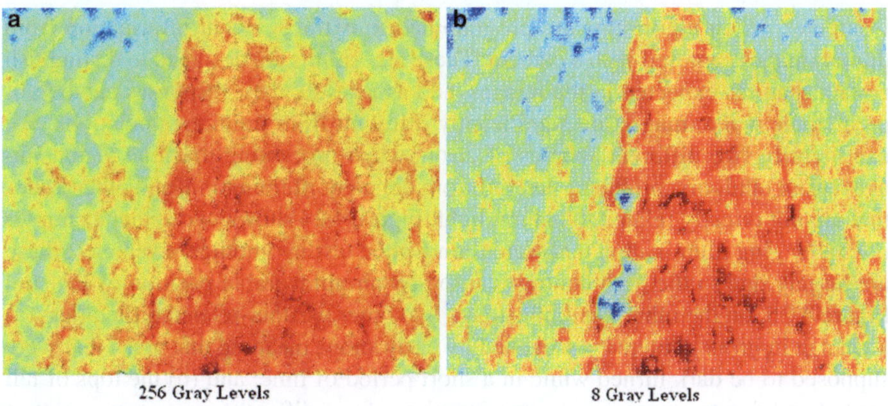

Figure 4.22 Entropy spectra for a cut swath image: (**a**) 256 GL; (**b**) 8 GL

Figure 4.23 Texture enhancement for different neighborhoods: (**a**) 3×3; (**b**) 5×5; (**c**) 7×7

acterizes the texture of an image. It is directly calculated from the histogram of the monochrome version of the original image by applying Equation 4.13, where p is the number of pixels of a given intensity value, and GL is the total number of gray levels. Figure 4.22a shows the entropy spectrum for the cut swath shown in Figure 4.18, considering 256 gray levels and a neighborhood of 9×9. When the number of gray levels is lowered to eight, the resulting entropy image is fairly similar, as shown in Figure 4.22b. This fact demonstrates that entropy yields satisfactory results for images with reduced gray levels. The calculation of entropy allows areas of richer texture content to be distinguished. For instance, the three images of Figure 4.23, which were produced from the entropy spectra of the studied image obtained under different settings, approximately locate the cut swath. These images differ in the size of the neighborhood considered in the calculation, which was 3×3 (a), 5×5 (b), and 7×7 (c).

$$\text{Entropy} = -\sum_{i=1}^{GL} p \cdot \text{Log}_2\, p \tag{4.13}$$

4.3.4 *Difficult Challenges for Monocular Vision*

Computer vision algorithms are programmed to extract critical information from digital images captured with a camera. These algorithms are designed to distinguish certain features expected in the images, and as a result, their responses can be unpredictable if the images are significantly different from training simulations. We can identify two different sources of errors when processing machine vision images taken outdoors: problems derived from deficient illumination, and other issues that are not directly related to natural light. Figure 4.24 shows two actual examples of challenging illumination that were found when testing intelligent vehicles; these are largely due to the high contrast caused by an unfavorable position of the sun. Figure 4.25, on the other hand, illustrates two other weather-related difficulties: (a) a change in ground intensity provoked by early snow, meaning that areas that were supposed to be dark turned white in a short period of time; and (b) the tops of tall crops bend due to strong winds, causing plants from different rows to merge, thus making it difficult to discriminate the rows used as guidance features.

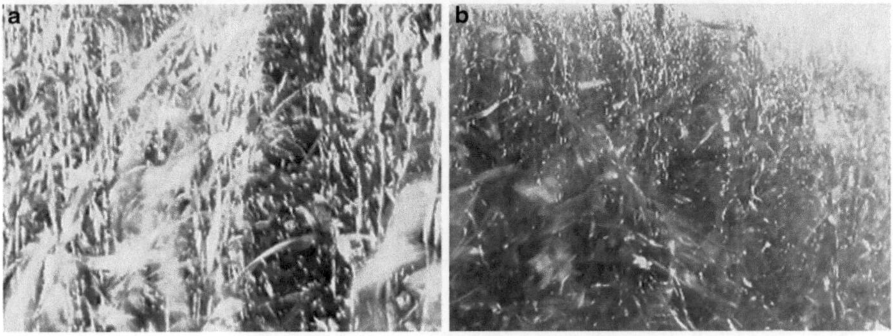

Figure 4.24 Images with difficult illumination for machine vision guidance

Figure 4.25 Challenging images caused by wind (**b**) and snow (**a**)

Other less probable complexities can defy the vision algorithm, and the more flexible the image processing code is, the more robust the navigation commands that will be generated. Figure 4.26a shows a row that has been so severely damaged that significant parts of the rows are missing. Without reliable tracking features, the vehicle could go astray and cause a serious accident. The tractor photographed in

Figure 4.26 Automatic guidance with deteriorated rows (**a**) and at high speeds (**b**)

Figure 4.26b is being automatically steered at speeds of approximately 18 km/h. This vehicle velocity demands high reliability and processing speed for the algorithm. Even when the software works acceptably, if the commands are sent to the navigation engine too late, the system must be considered to be unsatisfactory because it cannot carry out the assigned task optimally. It should not be forgotten that intelligent vehicles are designed to improve farming tasks, and delaying work because of an algorithm is not the right philosophy. They are meant to outperform the average operator.

4.4 Hyperspectral and Multispectral Vision

The visible spectrum, often amplified to the adjacent near-infrared band, fulfills the needs of most applications involving intelligent agricultural vehicles, and mainly those relating to navigation assistance and automation. Most monocular cameras, from basic webcams to sophisticated high-performance imaging sensors, are sensitive to visible light, which means that they see exactly what humans see. However, there are other agricultural applications that demand sensing at different electromagnetic bands. The observation of plant health, for example, benefits from multispectral machine vision as well as remote sensing technologies.

Multispectral cameras can capture light from outside the frequency range of visible light, allowing the detection of special features that are not distinguishable by the red, green, and blue receptors of human eyes. Multispectral imaging sensors often yield different narrow bands separately, some or all of which fall outside the visible range. When numerous bands of fine spectral resolution are pursued, the term *hyperspectral* is more commonly applied. A combination of spectral bands of particular interest in agriculture is obtained by substituting the blue channel for infrared. This arrangement of *green-red-infrared* is particularly effective at detecting features in vegetation. Figure 4.27a shows a multispectral camera used to monitor soybean leaf health. Figure 4.27b is an image of a soybean plant in the visible spectrum,

Figure 4.27 Multispectral imaging: (**a**) camera; (**b**) R-G-B image; (**c**) R-G-IR image (courtesy of Di Cui)

whereas Figure 4.27c is the same image taken with the band combination green-red-infrared. These images were taken for the detection and real-time identification of foliar symptoms of sudden death syndrome (SDS) in soybeans. Figure 4.27b was useful for separating dead areas from healthy ones, whereas Figure 4.27c yielded the best segmentation results during image background removal.

4.5 *Case Study I*: Automatic Guidance of a Tractor with Monocular Machine Vision

The agricultural tractor (John Deere 8200) of Figure 4.28 was successfully autoguided at speeds of between 9 and 11 km/h. The navigation engine of the tractor used a *monocular camera* to find the optimum path for the tractor and an electrohydraulic valve to execute vision-based steering commands. The vision sensor utilized in this study was a charge-coupled device (CCD) camera mounted at the center of the tractor's nose and equipped with a NIR filter and a 16 mm lens, which provided a field of view of approximately ten meters ahead of the vehicle.

The core algorithm used to detect the crop rows and thus obtain the travel directrix was based on the Hough transform. The binarized region of interest of Figure 4.11 and the midpoint encoding of Figure 4.13a prepared the captured images

Figure 4.28 Autoguided tractor: algorithm tuning tests (**a**) and field experiments (**b, c**)

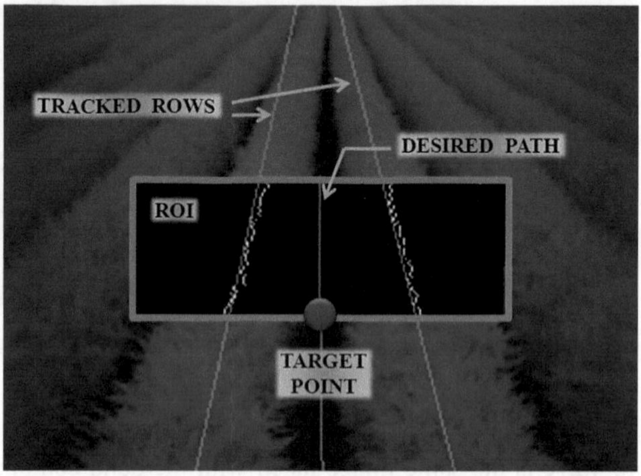

Figure 4.29 Typical outcome from the vision algorithm implemented for *Case Study I*

for line detection analysis with the Hough transform. As a result, a parameter space similar to that shown in Figure 4.17 was obtained for most of the images, yielding the positions of the crop lines represented in Figure 4.29. The *desired trajectory* was calculated as the *center line* of the algorithm-detected central pair of rows. The automated tractor was directed towards the *target point*, defined in this application as the intersection of the desired trajectory and the lower side of the rectangle indicating the region of interest, as illustrated in Figure 4.29.

Even though most of the processed images led to the correct detection of the central tracking rows using just two estimated lines, as in the case of Figure 4.29, sometimes more than two peaks passed the threshold set for the parameter space of the Hough transformation, and so more than two lines indicating the positions of the central pair of tracking rows were obtained. This case is represented in Figure 4.30, where three (a), four (b), and five (c) estimated lines were employed to deduce the desired trajectory. Reliable estimation of the desired trajectory is essential to determine the target point and therefore guide the tractor automatically. When three lines were output by the algorithm as a consequence of three peaks being identified in

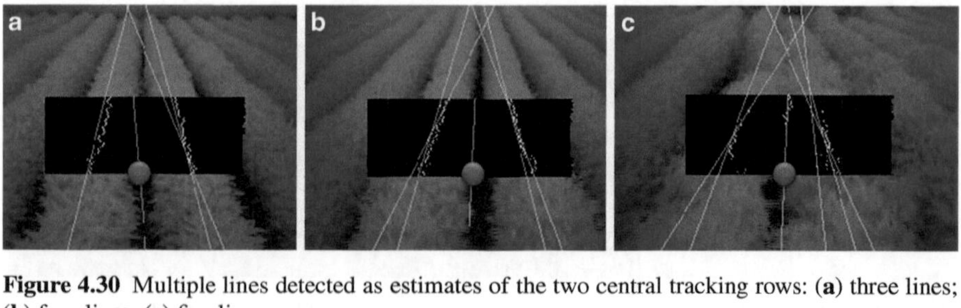

Figure 4.30 Multiple lines detected as estimates of the two central tracking rows: (**a**) three lines; (**b**) four lines; (**c**) five lines

the Hough space, there were two estimated lines associated with the same row, and dropping one of them led to the optimum outcome of Figure 4.29. When four rows were obtained, averaging them gave a good estimate for the desired path, and when five lines were produced, they were reduced to four and then averaged. More than five lines were rarely generated, and when this did occur, only four were actually considered in order to find the vehicle path. Once the target point is satisfactorily calculated, it needs to be converted to world coordinates (X_{tp}, Y_{tp}), as the vision algorithm only works in image space. The transformation matrix applied in this study is given in Equation 4.5, and the calibration procedure utilized is exactly the same one described in Section 4.3.1 and Figure 4.5.

After the position of the target point, expressed in the world coordinates (X_{tp}, Y_{tp}), had been determined by the vision algorithm, a turning angle ϕ had to be sent to the controller in order to reach this point. The position of the target point referred to the system of coordinates represented in Figure 4.5. According to that Cartesian frame, the larger the offset from the tracking rows, the wider the steering angle required to keep the tractor on track. A direct way to acknowledge this relationship is by applying Equation 4.14.

$$\phi = \arctan\left(\frac{X_{tp}}{Y_{tp}}\right) \tag{4.14}$$

The expression used to estimate ϕ implicitly takes into account both the offset and the heading of the tractor. Note that positive values for X will lead to right turns and *vice versa*. This expression can be further elaborated to add, for instance, the effect of the velocity of travel on the selection of the target point. In this form, the algorithm was capable of guiding the tractor across the field at up to 11 km/h.

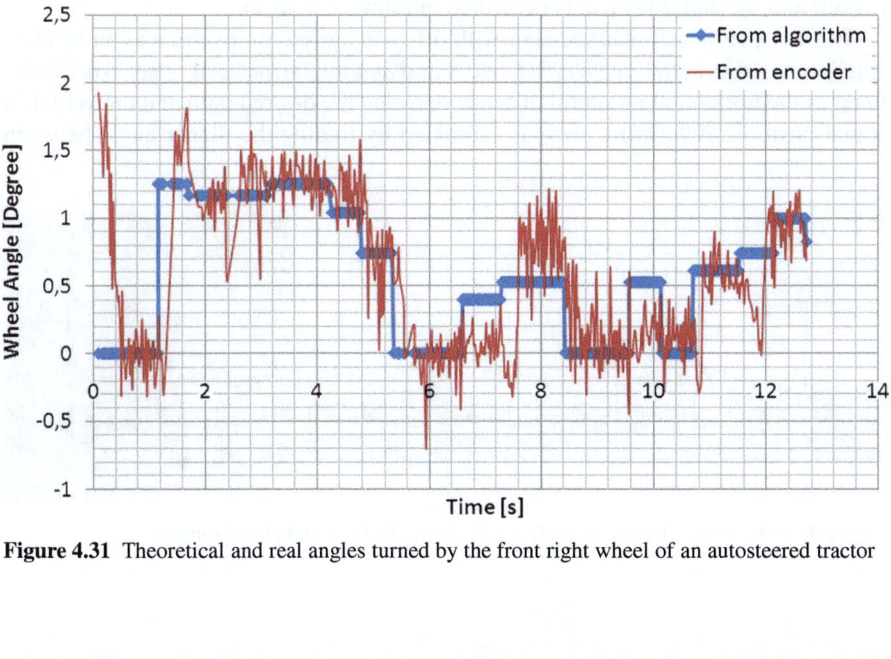

Figure 4.31 Theoretical and real angles turned by the front right wheel of an autosteered tractor

The graph of Figure 4.31 shows the wheel angle ϕ estimated by the vision-based navigation algorithm, as well as the actual angle ϕ_r turned by the front right wheel of the tractor, as measured with an optical encoder. Further details of this case study can be found in [3, 5].

4.6 *Case Study II*: Automatic Guidance of a Tractor with Sensor Fusion of Machine Vision and GPS

The set of images included in Figures 4.24 and 4.25 illustrate the difficulties that some of vision systems must cope with in the field. Chapter 3 and Figure 4.3, on the other hand, reveal the challenges of navigating in the absence of local awareness, when issues such as guess rows, bias, drift and signal drops can induce grave errors in the vehicle's course. A synergic approach where both kinds of perception – global and local – are merged can result in better and safer navigational commands. In fact, both technologies complement each other, and local perception works adequately where global localization is weak and *vice versa*. Nevertheless, the signals need to be combined carefully in such a way that the most reliable signals are used all the time. The way to join both sources of data is through the concept of *sensor fusion*, and different techniques can be employed to achieve this goal. This section describes the fundamentals of fusing global and local perception, as well as the geometry involved in the execution of guidance commands. The local perception sensor was a video camera mounted under the DGPS receiver (Figure 4.32b) and located at the front of the tractor cabin. The schematic of Figure 4.32a gives a basic idea of the system architecture. A detailed explanation of the core algorithms based on the Kalman filter and fuzzy logic is included in Chapter 8 (Section 8.2 on sensor fusion) and the entire project is described in references [6, 7].

The navigation errors associated with off-road agricultural vehicles can be classified into two general types: *offset errors* and *heading errors*. An *offset error* is defined as the horizontal or lateral distance between the desired path and the vehicle's actual position. A *heading error* is estimated by finding the planar angle between

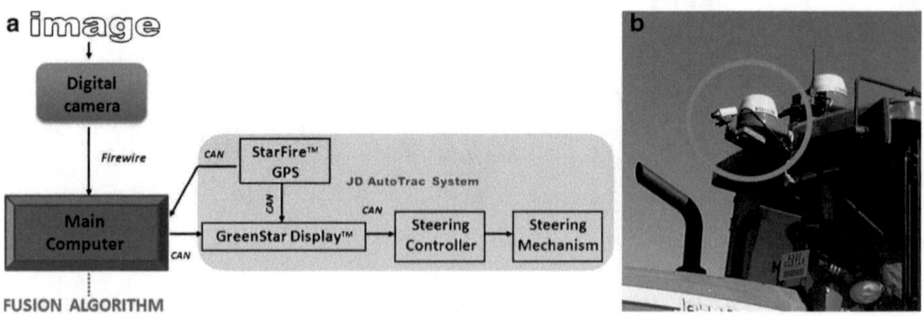

Figure 4.32 System architecture: (**a**) block diagram; (**b**) local and global sensors

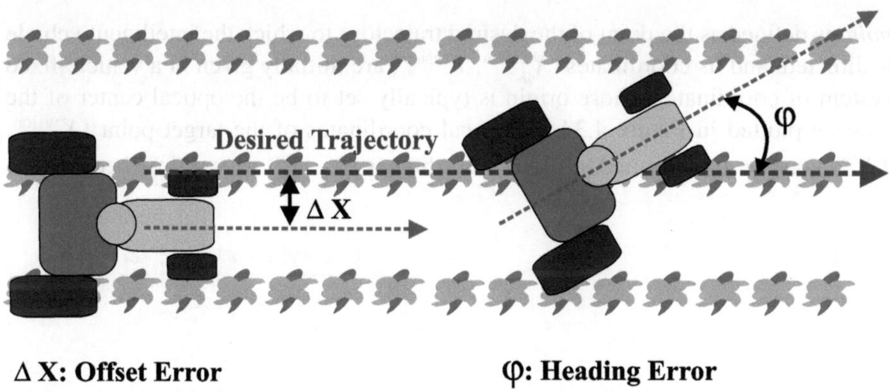

Figure 4.33 Principal navigation errors for intelligent vehicles: offset (ΔX) and heading (φ)

the orientation of the targeted position and the current orientation of the vehicle. Given that offset and heading errors are independent of each other, a vehicle can be misplaced, misaligned, or both. Figure 4.33 describes offset and heading errors for intelligent off-road vehicles graphically.

In order to compute offset and heading errors, the navigation engine needs to possess an estimate for the *optimal trajectory*. Globally, this is traced by GPS coordinates deduced from calculating new passes from previous ones, or from a complete map stored in the main computer. Locally, the trajectory is perceived in real time by the camera, so its field of view needs to be set up such that enough information is always available to determine the position of the target point. This *target*

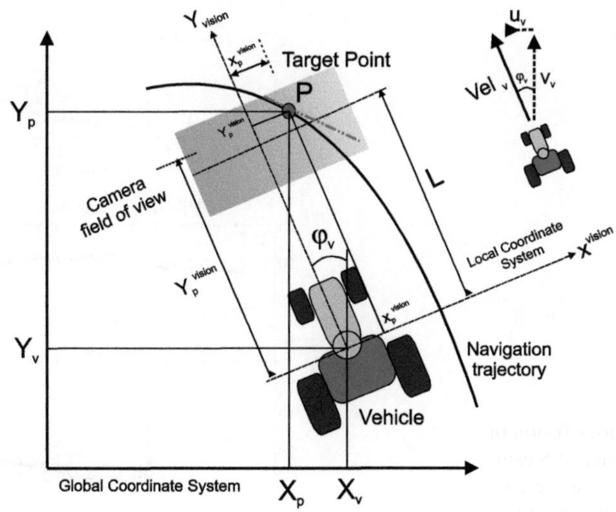

Figure 4.34 Geometry of navigation for global–local sensor fusion: basic parameters

point is defined as the point of the desired trajectory to which the intelligent vehicle is directed, and its coordinates (X_p^{vision}, Y_p^{vision}) are initially given in a vehicle-fixed system of coordinates whose origin is typically set to be the optical center of the lens, as plotted in Figure 4.34. The local coordinates of the target point (X_p^{vision}, Y_p^{vision}) can be transformed to global coordinates (X_p, Y_p) in real time because the

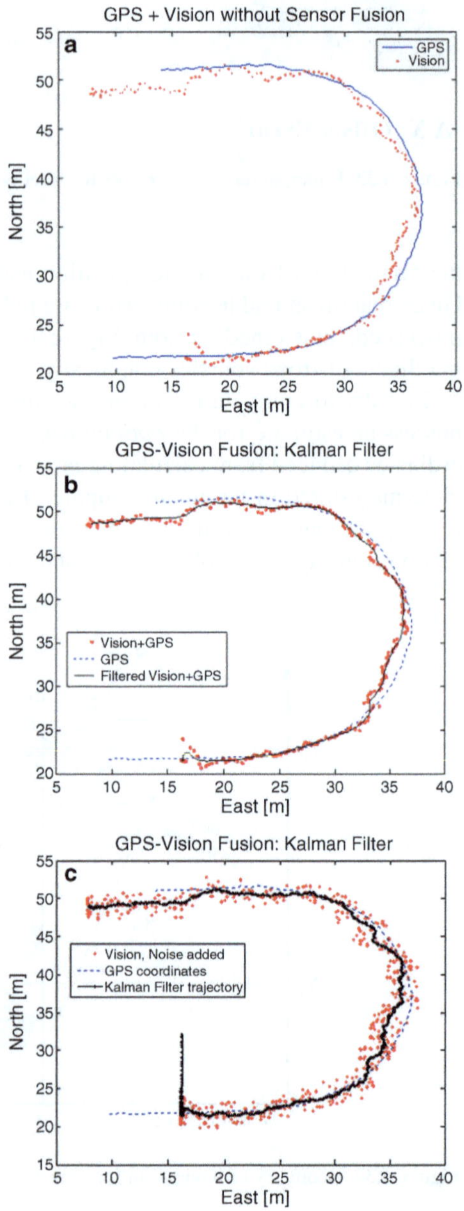

Figure 4.35 Sensor fusion of machine vision and GPS using the Kalman filter: (**a**) before filtering; (**b**) after filtering; (**c**) after filtering a very noisy estimate

GPS provides the global coordinates for the camera (X_v, Y_v), and consequently the transformation is immediate. The global coordinates of the desired trajectory obtained from both global and local sensors allow the calculation of heading and offset errors for each sensor in the global frame, and this permits the combined estimation of errors according to the sensor fusion philosophy implemented. In addition, offset and heading errors in the vehicle-fixed reference frame can also be found and then included in the fusion algorithm if necessary. The GPS receiver also provides the tractor heading φ_v and its forward speed Vel_v, as indicated in Figure 4.34. The look-ahead distance for the camera is represented by L in the diagram.

While the particular approaches that are followed to implement the fusion of machine vision and GNSS are completely different, the general philosophy employed in all of them is not so different: error estimates from two dissimilar sensors must be integrated according to the reliability of their estimation. The key element in this idea is the *quality of the estimate for each sensor*. *Kalman filter* fusion utilizes *noise covariance matrices* to model uncertainty in both the process and the sensor measurements [7]. The *fuzzy logic* approach, on the other hand, assigns a *quality index* to each sensor each time a measurement is issued [6]. The quality index establishes an instantaneous *mixing ratio* to decide, for example, that the reading coming from the camera has a weight of 70% for the offset error, with the remaining 30% being the contribution from the GPS. The definition of signal quality is crucial to the correct implementation of the fusion policy. In this case study, machine vision quality is based on a confidence index that depends on how accurately a template matches the detected crop rows, and the GPS quality is related to the number of satellites in the solution and other common quality indicators already provided by the GPS receiver.

One beneficial effect of merging sensor data is the smoothness of the resulting commands, which in turns leads to more stable navigation. When a tracking band that formed a curve with a diameter of 30 m was set up for a tractor to follow, the exact position of the band was determined only by the camera, and the detected profile was not very smooth for guidance (Figure 4.35a). After the Kalman filter was implemented to merge global and local information, the trajectory generated for the vehicle was smoother (Figure 4.35b). Even when the camera's estimate for the tracking band was heavily corrupted with computer-generated random noise, the final filtered trajectory was still acceptable, as demonstrated by Figure 4.35c.

References

1. Rovira-Más, F (2009) Recent innovations in off-road intelligent vehicles: in-field automatic navigation. Recent Pat Mech Eng 2(3)
2. Guo L, Zhang Q, Han S (2002) Agricultural machinery safety alert system using ultrasonic sensors. J Agr Saf Health 8(4):385-396
3. Rovira-Más F, Zhang Q, Reid JF, Will JD (2005) Hough-transformed-based vision algorithm for crop row detection of an automated agricultural vehicle. J Automob Eng 219D:999–1009
4. Reid JF, Searcy SW (1991) An algorithm for computer vision sensing of a row crop guidance directrix. Trans SAE 100(2):93–105

5. Rovira-Más F, Zhang Q, Reid JF, Will JD (2003) Machine vision based automated tractor guidance. Int J Smart Eng Syst Des 5:467–480
6. Rovira-Más F, Han S, Wei J, Reid JF (2005) Fuzzy logic model for sensor fusion of machine vision and GPS in autonomous navigation (ASABE Paper 051156). ASABE, St. Joseph
7. Rovira-Más F, Han S (2006) Kalman filter for sensor fusion of GPS and machine vision (ASABE Paper 063034). ASABE, St. Joseph

Chapter 5
Three-dimensional Perception and Localization

5.1 Introduction to Stereoscopic Vision: Stereo Geometry

The macroscopic world we live in is a *three-dimensional world* where three independent variables are sufficient to locate and delimit the objects we can perceive with our senses. *Length*, *width*, and *depth* provide enough information to bound the region of space occupied by an inanimate thing or a living being, and to unmistakably localize it. Human vision is probably the most precious sense that we have. Our two eyes, strategically positioned at the same height from the ground and separated by a particular horizontal distance from one another, allow the brain to compare the images recorded by both retinas in order to estimate how far surrounding objects are from the eyes. This is the fundamental principle of *stereoscopic vision* that is used by both humans and computerized perception. Instead of utilizing the brain to match the images from the left and right eyes and deduce the three-dimensional (3D) characteristics of the examined scene, a computer algorithm correlates a pair of digital images captured simultaneously with a *stereo camera*, identifying the same objects in both images and measuring their difference in position in pixels. This difference is called *disparity* and, together with other camera parameters, enables computers to recreate a real scene in a 3D virtual image. Mimicking the excellence of human vision has been a dream of those working in robotics and artificial intelligence. The principles of stereoscopy are not new: they were discovered in the nineteen century. Those who have visited the home of Abraham Lincoln in Springfield (Illinois, USA) will remember a stereoscope for the cross-eyed viewing of stereo photographs; this device created the illusion of depth – a novel entertainment of the time. However, more than one hundred years had to elapse from the construction of this stereoscope before robotic technology could take advantage of the stereo effect. Although monocular machine vision has been investigated intensely over the last three decades, only during the last ten years have stereo sensors reached the desired degree of popularity and dissemination. The reason for this delay is the complexity involved in finding the same object in both images simultaneously. At the end of the 1990s, processors had become fast enough to run algorithms that permitted the real-time correlation of stereo images with a high rate of success.

Therefore, soon after this, compact binocular cameras emerged in the fast-growing market of sensors and electronics, and applications of real-time stereo perception permeated into many fields, especially mobile robotics. However, this technological breakthrough did not impact on agricultural vehicles, except in some unusual cases, despite its extraordinary potential for use in intelligent off-road vehicles [1]. The capacity to map agricultural scenes in three dimensions is a significant step beyond the two-dimensional imagery currently available from remote sensing applications. The availability of real-time obstacle detection results in a great number of safe-guarding options for autonomous vehicles, whose navigation commands can also be determined after analyzing stereo images of trees or crop rows.

A three-dimensional scene can never be uniquely represented by a two-dimensional image because a point in the image plane can be fully characterized by its image coordinates (x_i^P, y_i^P), whereas the same point in space requires the knowledge of its 3D coordinates (x^P, y^P, z^P). Information is missing from a simple image of a 3D scene: the *depth*, or distance between the point and the image plane. A second image of the same scene taken to achieve the stereo effect allows the disparity to be calculated, which provides the missing dimension. This fact does not mean that monocular machine vision is not valuable; on the contrary, there are many applications that are particularly suited to single lens cameras, as demonstrated in Chapter 4. However, stereo vision brings new possibilities and an entire new way of perceiving things. As a matter of fact, both of these machine vision technologies complement each other, and very often one excels where the other encounters difficulties, and *vice versa*. The nature of each application is the deciding factor in whether to choose one or the other, although in general terms, stereo offers a more complete solution, as it adds a third coordinate to the other two that are already available with monocular vision. However, the introduction of this complexity obviously results in the need for heavier image processing computation, which must also be accounted for in the design of the perception engine.

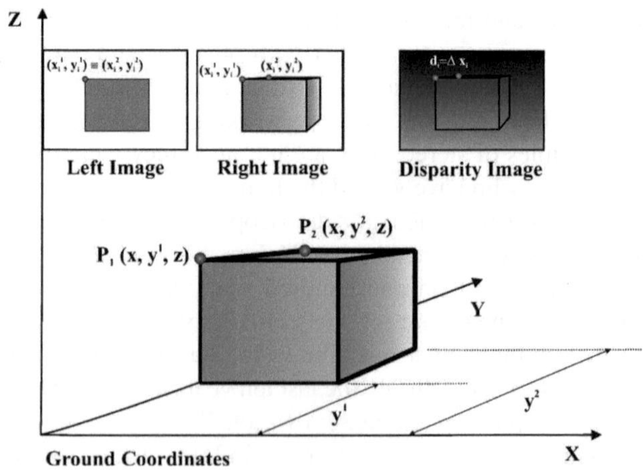

Figure 5.1 Conceptual representation of the generation of a 3D image

A *three-dimensional image* is composed by estimating the positions of points in planes parallel to the image plane (x and z 3D coordinates from x_i and y_i image coordinates) and then calculating the depth (y coordinate) from the disparity image. Let us illustrate the problem of rendering a 3D scene from planar 2D images through the intuitive drawing of Figure 5.1. In this 3D scene, P_1 and P_2 are two separate points on the sensed object, but they are mapped to the same position in the image domain of the reference 2D image (the left image of the pair). They differ in location solely by their depths (their y coordinates), which single 2D images cannot provide by themselves. If a second image (the right image) that complies with the constraints of stereo vision is available, given that points P_1 and P_2 will occupy different locations in the right image, their depths can be calculated by means of the disparity map, as both points will have different disparity values. Their 3D Cartesian coordinates will be given by P_1 (x, y^1, z) and P_2 (x, y^2, z). The following paragraphs of this section develop the mathematical relationships necessary to perform this transformation and generate 3D images with stereoscopic cameras.

The objective of stereo calculations is to create a 3D representation of a scene from a pair of stereo images that are captured simultaneously with a stereo camera. These stereo images are two conventional digital images (consisting of pixels) that are taken under special constraints that are required to achieve the stereo effect. This special configuration, in its most simplified form for achieving stereo geometry, forces both image sensors (electronic light-sensitive arrays known as *imagers*) to incorporate lenses of identical *focal length* f, and places the imagers a particular distance apart that allows the $X_i Y_i$ plane of the left image to be coplanar with that of the right image, meaning that the range or depth of any sensed point will be exactly the same for both images of the stereo pair. Furthermore, the horizontal axes (X_i) of both images must be collinear, which is known as the *epipolar constraint*. Figure 5.2 represents a stereo scene acquired by a schematized binocular stereo camera.

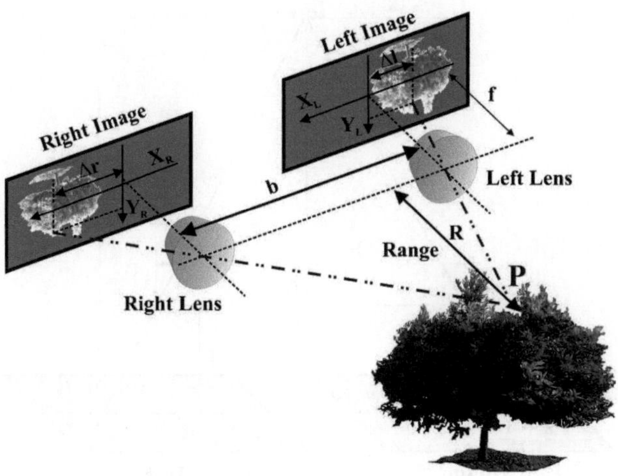

Figure 5.2 Principal parameters involved in stereoscopic vision

The pictorial of Figure 5.2 includes all of the key parameters involved in stereo calculations. The distance between the focal centers of the left and right lenses is defined as the *baseline* of the stereo camera, and represented in the figure by b. The focal length of both camera lenses (it must be the same for both) is indicated by f, and finally the *range* or distance to the sensed object is denoted by R. The essential operation in stereo viewing is the calculation of *disparity maps*, which involves *triangulation*. In order to generate a 3D representation of the tree of Figure 5.2, two simultaneous images (left and right) need to be taken. As a result, any point on the tree, say point P, will be mapped at different locations in the stereo images: (x_L, y_L) for the left image and (x_R, y_R) for the right image. The epipolar constraint of stereo vision allows the X coordinates to be compared. If the horizontal locations of the projections of point P onto the image planes are represented by Δr for the right image and Δl for the left image, their disparity d will be given by the difference between them, as expressed in Equation 5.1:

$$d = \Delta r - \Delta l \,. \tag{5.1}$$

Both distances (Δr and Δl) are measured in pixels, and their signs depend on the position of the object of interest and the system of coordinates adopted for the left and right images. The disparity d is the fundamental parameter used to calculate the range R by triangulation. Figure 5.3 depicts the main geometrical parameters involved in the calculation of the range for point P, where Δr and Δl are measured with respect to the focal centers of the left and right images. Recall that, according to the coordinate system chosen in Figures 5.2 and 5.3, Δl is negative and Δr is positive; had the X-axis been defined in the opposite direction, the signs of Δr and

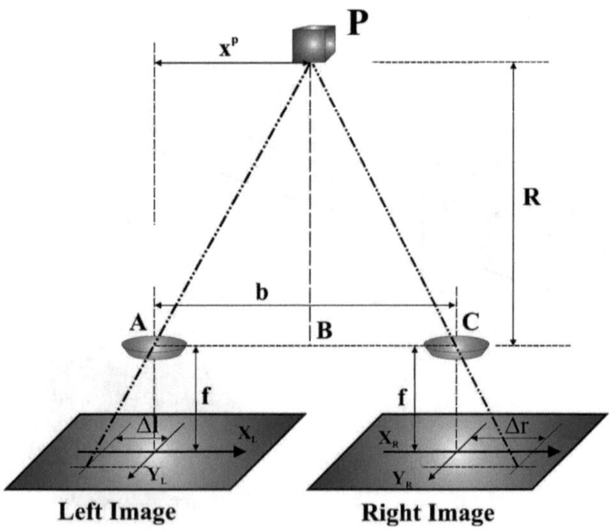

Figure 5.3 Calculation of ranges from disparities

Δl would have been inverted, but the magnitude of the disparity would have been the same, equivalent to the addition of both quantities.

The baseline (b), represented in Figure 5.3 by the distance AC parallel to the image planes, can be calculated with Equation 5.3 given the similarity of the triangles of Equation 5.2, as established in the scheme of Figure 5.3.

$$\left\{ \begin{array}{lcl} \dfrac{R}{\overline{AB}} = \dfrac{f}{|\Delta l|} & \rightarrow & \overline{AB} = \dfrac{|\Delta l| \cdot R}{f} \\ \dfrac{R}{\overline{BC}} = \dfrac{f}{|\Delta r|} & \rightarrow & \overline{BC} = \dfrac{|\Delta r| \cdot R}{f} \end{array} \right\} \tag{5.2}$$

$$\begin{aligned} b = \overline{AC} = \overline{AB} + \overline{BC} &= \frac{|\Delta l| \cdot R}{f} + \frac{|\Delta r| \cdot R}{f} \\ &= (|\Delta l| + |\Delta r|) \cdot \frac{R}{f} = (\Delta r - \Delta l) \cdot \frac{R}{f} = \frac{d \cdot R}{f} \end{aligned} \tag{5.3}$$

Equation 5.3 provides an expression for the range as a function of the baseline, the focal length, and the disparity. However, note that a conversion factor between pixels and distances must be included to ensure that this relationship is homogeneous with respect to units. This factor is the *size of the pixels* that form the sensor array – that is, the sensor resolution – which is represented by w (mm/pixel) in the final expression for the range, Equation 5.4. This equation is easy to apply; for example, a stereo camera with a baseline of 90 mm and square pixels with sides of 0.0075 mm will yield a range of 9 m for a disparity of 10 pixels.

$$R \text{ [mm]} = \frac{b \text{ [mm]} \cdot f \text{ [mm]}}{d \text{ [pixel]} \cdot w \left[\text{mm} \cdot \text{pixel}^{-1}\right]} \tag{5.4}$$

If the range corresponds to the Z-axis in a generalized camera coordinate system such as the one shown in Figure 5.4, Equation 5.4 will provide the z^{P} coordinate of the targeted point P directly, but the two remaining real-world coordinates x and y will need to be calculated too. The procedure used to determine them is identical to the triangulation carried out to deduce the range, but the value of the range R and the image coordinates $(x_{\mathrm{i}}^{\mathrm{P}}, y_{\mathrm{i}}^{\mathrm{P}})$ registered in the reference image (left or right according to the specific camera implemented in the vehicle) are used instead. The final expression that converts a pixel from image space to a point in 3D space is provided by Equation 5.5:

$$\begin{aligned} x^{\mathrm{P}} &= \frac{x_{\mathrm{i}}^{\mathrm{P}} \cdot w \cdot R}{f} \\ y^{\mathrm{P}} &= \frac{y_{\mathrm{i}}^{\mathrm{P}} \cdot w \cdot R}{f} \\ z^{\mathrm{P}} &= R\,. \end{aligned} \tag{5.5}$$

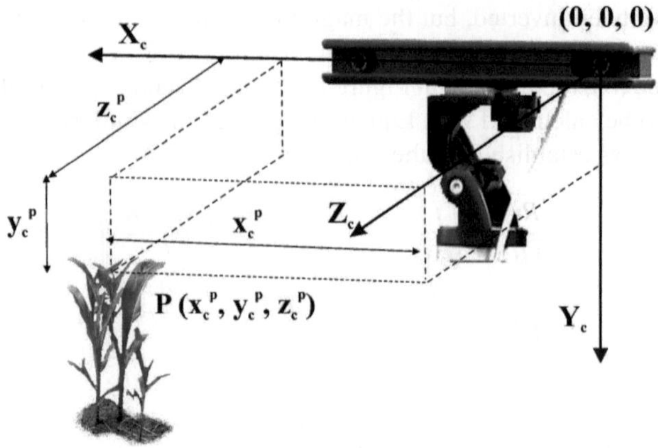

Figure 5.4 Definition of camera coordinates for binocular stereo sensors

After the perception system has been configured by choosing the optimum camera parameters, such as the baseline, lenses, and imaging sensor, these parameters tend to remain fixed, and so Equation 5.5 can be simplified by introducing the coordinate transformation constant T defined in Equation 5.6. The equation used to transform from image coordinates (x_i^P, y_i^P) in pixels to 3D coordinates in length units (x^P, y^P, z^P) is shown in matrix form in Equation 5.7.

$$T = \frac{w}{f} \ [\text{pixel}^{-1}] \tag{5.6}$$

$$\begin{bmatrix} x^P \\ y^P \\ z^P \end{bmatrix}_{\text{meters}} = R_{\text{meters}} \cdot \begin{bmatrix} T & 0 & 0 \\ 0 & T & 0 \\ 0 & 0 & 1 \end{bmatrix}_{\text{pixel}^{-1}} \cdot \begin{bmatrix} x_i^P \\ y_i^P \\ 1 \end{bmatrix}_{\text{pixel}} \tag{5.7}$$

Once the main camera parameters have been selected for a particular off-road application (Section 5.4), it is important to get an estimate for the smallest change in range that the system can detect. In other words, what is the change in range when the disparity varies by one unit? Disparities are commonly measured in pixels, but high-performance cameras commonly work at subpixel levels to enhance precision. The variation of the range with respect to the disparity can be found by differentiating Equation 5.4, as detailed in Equation 5.8:

$$\frac{\partial R}{\partial d} = -\frac{b \cdot f}{w} \cdot \frac{1}{d^2} = -\frac{b \cdot f}{w} \cdot \frac{1}{\frac{b^2 \cdot f^2}{w^2 \cdot R^2}} = -\frac{w}{b \cdot f} \cdot R^2 \tag{5.8}$$

Replacing the differentials of Equation 5.8 by finite increments and neglecting the negative sign, the algebraic equation that allows the range resolution to be estimated as a function of the disparity for a given stereo system is shown in Equation 5.9:

$$\Delta R = \frac{w}{b \cdot f} \cdot R^2 \cdot \Delta d \quad \Rightarrow \quad \Delta R = \frac{T}{b} \cdot R^2 \cdot \Delta d \tag{5.9}$$

The expression found for the range resolution, and given in Equation 5.9, implies that wide baselines, long focal lengths, and sensor arrays of tiny pixels are favorable for achieving small ΔR, which results in more accurate outcomes for a given uncertainty Δd. However, the range resolution deteriorates with the square of the range, which means that losses in range resolution grow rapidly as the tracked objects get further from the camera. Safeguarding applications, for example, will demand acceptable range estimations in the vicinity of the vehicle.

The objective of the transformation obtained by applying Equation 5.7 is to calculate all of the coordinates that comprise a real scene in a 3D Cartesian system of coordinates. Given that off-the-shelf stereo cameras output data in this format, engineers working on agricultural robotics must direct their efforts into deploying high-performance vehicles rather than improving the inner workings of particular sensors – something that commercial firms usually do excellently; the image coordinates and the image coordinate systems are secondary. On the contrary, the coordinate system of the 3D point clouds that are generated by the stereo camera and used by roboticists and vehicle designers is critical. The coordinates $(x^{\mathrm{P}}, y^{\mathrm{P}}, z^{\mathrm{P}})$ obtained in Equation 5.7 for point P are known as *camera coordinates*, and should be more appropriately labeled $(x_{\mathrm{c}}^{\mathrm{P}}, y_{\mathrm{c}}^{\mathrm{P}}, z_{\mathrm{c}}^{\mathrm{P}})$. Camera coordinates are shown in Figure 5.4 for a camera whose reference lens is the left one (seen from behind), as indicated by the origin O_{c} (0, 0, 0). In this coordinate system, the $X_{\mathrm{c}}Y_{\mathrm{c}}$ plane is coincident with the plane of the imagers, the range Z_{c} is perpendicular to the $X_{\mathrm{c}}Y_{\mathrm{c}}$ plane, the Y_{c} axis follows the vertical dimension of the reference image (increasing downwards), and the X_{c} axis coincides with the horizontal dimension of the reference image (increasing in positive values from left to right). The camera coordinates have their origin at the optical center of one of the lenses, which is arbitrarily decided by the camera designer.

The camera coordinates shown in Figure 5.4 are the standard ones in which the points of 3D clouds are expressed. However, they are totally dependent on the camera's position and orientation. Moving the sensor from the top of a harvester cabin pointing downwards to the picking header would lead to a total change in the system of reference: not simply a simple lateral shift but also a rotation, as the camera would probably look straight out rather than down. In order to avoid this dependency, a new system of coordinates needs to be defined. These coordinates must be independent of the position and orientation of the sensor. The specific applications pursued eventually define the particular characteristics of this system, but some general properties can be mentioned here. As one of the basic properties of agricultural robotics is that intelligent vehicles need to be "grounded," such *ground coordinates* can be defined as follows. The Z-axis indicates the height of the object, the Y-axis provides the distance between the camera and the object, and the X-axis registers horizontal positions with respect to the center of coordinates. The origin of the ground coordinates is placed on the ground ($Z = 0$), and its exact position in relation to the vehicle will be defined by each application. The six case studies at

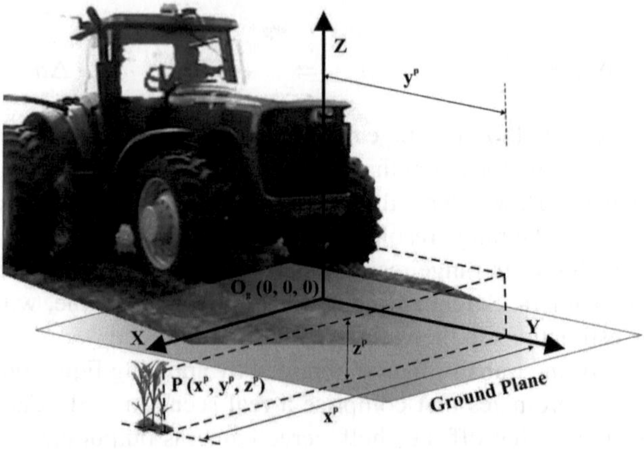

Figure 5.5 Ground coordinates for an intelligent off-road vehicle

the end of this chapter provide several examples of ground coordinate systems and their transformations from camera coordinates. Figure 5.5 shows a general ground coordinate system and an object referenced to it.

5.2 Compact Cameras and Correlation Algorithms

When assembling a stereo perception system, the first question that may come to mind is whether to purchase a compact off-the-shelf sensor or to build your own device by coupling together two identical digital cameras. Both solutions are feasible and equally affordable. Furthermore, commercial software for image correlation runs well in custom-made stereo heads, which has encouraged many researchers to mount their own stereo cameras. Nevertheless, compact solutions offer so many options that they likely cover most of the requirements of off-road environments. However, there are sometimes special needs that demand customized designs, such as the extra-wide baselines sometimes used in planetary exploration [2], images of a very high resolution, and specific spectral bands outside the visible light range. Figure 5.6 shows several compact stereo cameras (a and b) and a customized stereo rig consisting of two multispectral cameras (c).

One delicate aspect of stereo perception is the *calibration* of the cameras. Both customized and compact sensors need to go through camera calibration routines before they can be used effectively. However, some compact solutions offer *pre-calibrated* cameras that do not need to be calibrated by the user because they have been calibrated by the manufacturer before their deployment. The disadvantage of this alternative is that neither the baseline nor the lenses can be changed, and if the camera ever suffers an accident and the relative position between the lenses varies, even by a very small amount, it would have to be recalibrated. On the other hand,

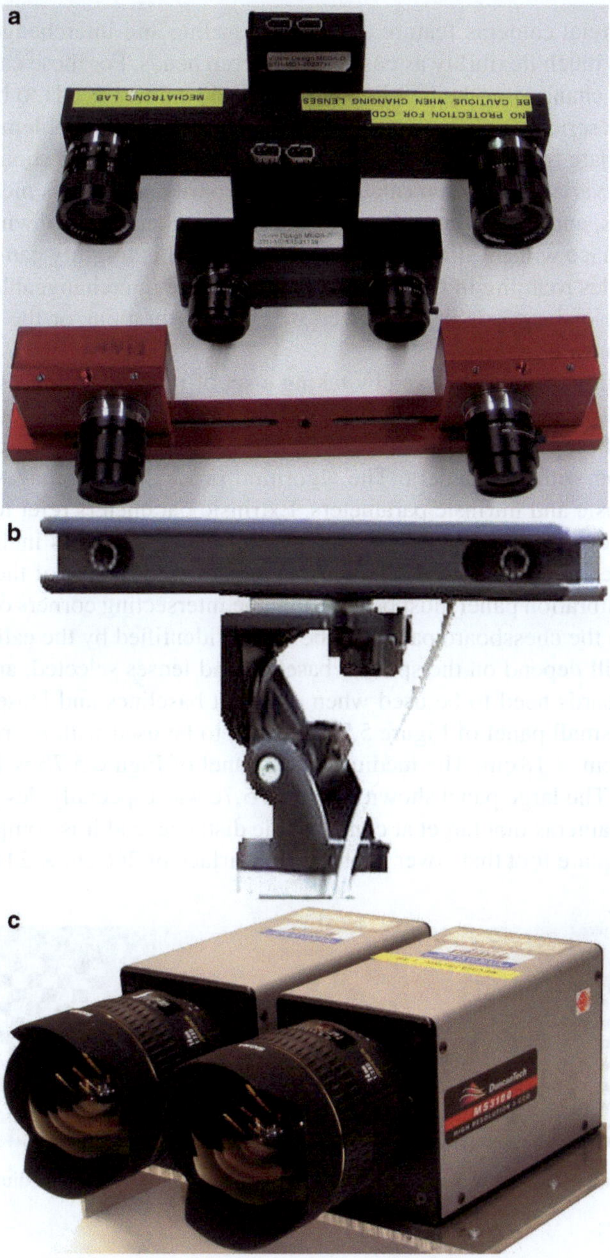

Figure 5.6 Stereo cameras: commercial sensors (**a, b**) and customized rig (**c**)

some commercial cameras feature a variable baseline and interchangeable lenses, which give as much flexibility as customized stereo heads. For these cameras, every time a lens is changed or the baseline is altered, the camera needs to be calibrated. This is a very serious matter (more than it may seem). Many problems with stereo sensing originate from deficient or obsolete calibration data. We should not forget that the basic variable in stereo calculations is disparity, and this is measured to the level of pixels, and very often subpixels. Removing a lens and screwing it in again in the same place will invalidate the calibration file. This fragility can pose a problem for vehicles roaming in harsh environments where interchangeable lenses may become loose or be partially unscrewed by engine vibrations or the suspension's response to uneven terrain.

Camera calibration is achieved by taking a set of images of a chessboard panel, where every image shows the panel at a different position or orientation. There are many calibration programs on the market, and commercial compact cameras tend to include one with the camera. The algorithm processes the images to estimate the key extrinsic and intrinsic parameters. Extrinsic parameters refer to the relative position and orientation of the imagers, and intrinsic parameters include the geometrical center of the light-sensitive sensors and the aberrations of the lenses. The size of the calibration panel must be such that the intersecting corners of the squares that comprise the chessboard panel can be clearly identified by the calibration software. This will depend on the specific baseline and lenses selected, and so several calibration boards need to be used when different baselines and lenses are implemented. The small panel of Figure 5.7a is meant to be used with short ranges, and measures 23 cm × 18 cm. The medium-sized panel of Figure 5.7b is 1 m high and 75 cm width. The large panel shown in Figure 5.7c was especially designed to calibrate stereo cameras that target at considerable distance, and it is composed of tiles of area one square foot that cover a total panel surface of 366 cm × 244 cm.

Figure 5.7 Calibration boards used for stereo cameras: (**a**) small board; (**b**) medium-sized board; (**c**) large board

When the stereo camera is not pre-calibrated, it can be calibrated before it is mounted on the vehicle or after it is fixed to it. Our experiences indicate that the camera should be fixed very tightly at its final location, the baseline and pair of lenses to be used in the field should be chosen, and the camera should be properly orientated before the calibration routine is performed. The less the camera is manipulated after calibration the better. Given the sensitivity of a stereo camera to mechanical stresses and the importance of not changing its position with respect to the vehicle, it is very important to employ a high-quality tripod to attach the camera body to the vehicle chassis. Investing in this apparently insignificant piece of equipment might initially seem a waste of resources, but in the long run it is not. The robust tripod holding the camera in Figure 5.8b costs over $500 (2008 price), but it maintained the required camera tilt angle after multiple tests and many kilometers driven on both paved roads and orchard terrains. Apart from achieving a solid linkage between camera and vehicle, the position of the camera in the vehicle should also be discussed. Just as for the monocular cameras described in Chapter 4, the sensor's position will largely be determined by the specific application it is used for. In fact, the preferred locations for stereo sensors are quite similar to those discussed in the case of monocular vision. Figure 5.8a shows a stereo camera mounted on a purpose-built frame at the front of a tractor used for stereo-based crop tracking capabilities. The variable baseline stereo rig of Figure 5.8b was placed at the center of the cabin of a robotized tractor, and the stereo sensor of Figure 5.8c was attached to the front of a utility vehicle conducting 3D mapping operations. The stereo system analyzed in *Case Study I* features a stereo camera located on the side of a harvester corn header.

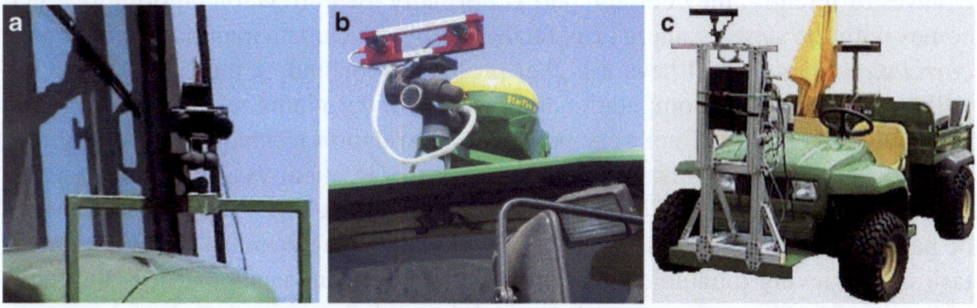

Figure 5.8 Alternative positions for stereo cameras on intelligent vehicles: (**a**) purpose-built frame; (**b**) center of the cabin; (**c**) front of vehicle

After the camera has been properly set up on the intelligent vehicle and carefully calibrated, we are ready to start grabbing stereo images and registering stereo information about the scene within the field of view of the camera, which is defined by the choice of baseline and lenses. The output of the stereo camera is the disparity image. What happens between the acquisition of the raw images (left and right) and the generation of the disparity image will largely depend on the *correlation algorithm* installed in the perception computer that controls the stereo system. The objective of the correlation algorithm is to match the left and right images to find the same objects in both images and, based upon their relative positions in the image domain, to measure horizontal differences in order to calculate the disparity according to the procedure of Equation 5.1 and Figure 5.2. The comparison between stereo images is not realized by scanning the whole image, but rather a small window, the size of which is typically set by the user. Figure 5.9 illustrates this process

Figure 5.9 Concept of window correlation for stereo matching between left and right images

where two stereo images are scanned sequentially by a running window. Commercial software for stereo processing usually incorporates libraries to carry out this correlation. In reality, each robotic application requires a different analytical routine for the stereo data acquired, and once the 3D data has been extracted, each system behaves differently; the common part is normally the correlation algorithm, which comes with the stereo camera or is facilitated by the camera manufacturer. A large *correlation window* will have a higher probability of finding the right objects but will also require more computational resources, undermining real-time capabilities if the processor used is not very powerful. Correlation techniques aside, filtering routines try to remove mismatches due to failures in attempts to match features in the left and right images. Correlation mismatches and electronic noise will always be part of the stereo process, and reducing or eliminating them is such an important task for achieving reliable navigation and mapping commands that it is the focus of Section 5.3. Many efficient and successful correlation algorithms have been published and disclosed, and so it does not make much sense for a stereo and robotics practitioner to invest in improving and refining them unless a satisfactory solution cannot be found in the machine vision community. The census algorithm [3], for example, uses a non-parametric summary of local spatial structure for pixel correlation, and was implemented with the stereo camera shown in Figure 5.6b. The sensors depicted in Figure 5.8a and b calculate disparity maps based on the procedure discussed in [4].

As discussed in the introductory section of this chapter (5.1), the major advantage of stereoscopic vision over monocular machine vision is the availability of ranges, or depths, in real time. However, this advantage could turn into a dangerous perceptual tool if the estimated ranges are not reliably determined; for example, if they are measured outside the expected field of view of the camera. As a matter of fact, ranges are reliably determined in a relatively narrow band, which often limits the applicability of stereo vision in intelligent vehicles. Take, for example, the two perception challenges faced by the intelligent tractor of Figure 5.10. In semiautonomous operation, the mowing tractor needs to find a guidance directrix at a certain look-ahead

Figure 5.10 Multi-range perception needs for vehicles traversing off-road environments

distance; on the other hand, it can never stop safely monitoring its surroundings for unexpected obstacles. The difficulty arises when one particular stereo system must behave reliably at both ranges. This problem motivated the development of *bifocal stereo*.

Bifocal lenses are special lenses – usually mounted on prescription spectacles – which consist of two areas with different eyesight corrections. This special design lets people see both near and far without needing to change their glasses. A computer vision camera cannot mimic this idea because each lens has a unique focal length f; therefore, the only way of achieving bifocal perception with machine vision is by using at least two lenses with different focal lengths to cover a greater field of view. The depth and coverage angle of each individual field of view will obviously depend on the choice of lenses, but the combined field sensed ahead of the camera should be constructed by overlapping the fields of view, as illustrated in Figure 5.11. Notice that shorter fields of view will cover wider angles, and far-reaching fields acquired with telephoto lenses will typically be narrow.

In addition to perceiving ahead of the vehicle, since the motivation for bifocal vision was to extend range capabilities, the perception needs to be in stereo as well. For this reason, the only way to assure long-range and short-range detection simultaneously in a unique sensor is to mount two stereo cameras in a common perception head denoted the *bifocal head*. Although both sensing units can work independently, their frame rates must be sufficiently high to capture all of the images almost concurrently. When images are initially acquired, their resulting 3D point clouds naturally have different systems of coordinates, as each camera will record data in its own camera coordinate system. However, the final 3D cloud should possess a unique origin and system of coordinates, which necessitates translating one frame and transforming all of the coordinates to the final system of reference. Figure 5.12a shows a proposed design of a *bifocal stereo camera*. This prototype defines the unique center of the camera coordinates as being at the optical center of the left lens of the short baseline camera. This arbitrary choice results in the need to transform all of the coordinates of the points sensed by the long baseline camera according to

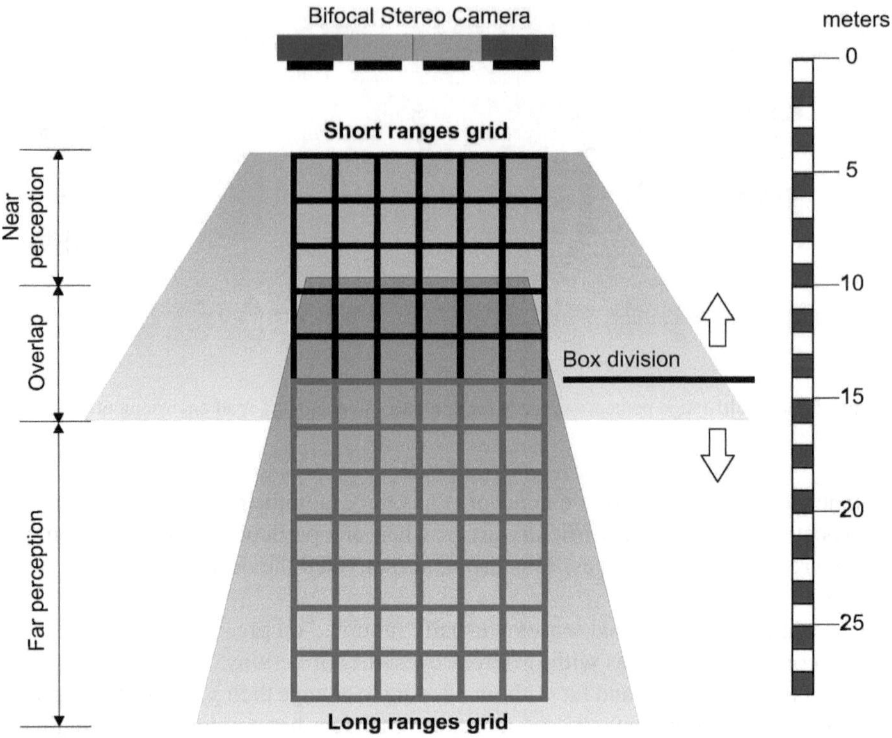

Figure 5.11 Composition of fields of view when expanding stereo perception capabilities

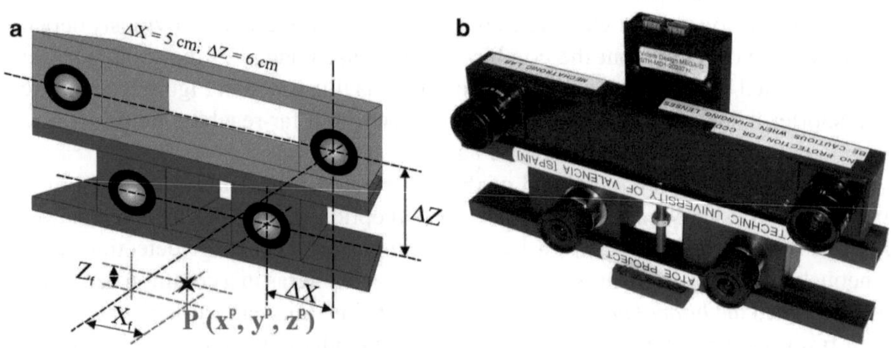

Figure 5.12 Bifocal stereo camera: conceptual design (**a**) and real head (**b**)

Equation 5.10, where (x_f, y_f, z_f) represent the ground coordinates acquired by the far-range camera (long baseline), and (x^p, y^p, z^p) are the ground coordinates of the merged 3D cloud. The design presented in Figure 5.12a was physically embodied in the bifocal head represented in Figure 5.12b, developed and tested in *Case Study V* in the research project reported in [5]. This stereo sensor coupled an 11 cm baseline camera equipped with 4 mm lenses with a 22 cm baseline camera perceiving through 16 mm lenses.

$$\begin{bmatrix} x^{\mathrm{p}} \\ y^{\mathrm{p}} \\ z^{\mathrm{p}} \end{bmatrix} = \begin{bmatrix} x_{\mathrm{f}} \\ y_{\mathrm{f}} \\ z_{\mathrm{f}} \end{bmatrix} + \begin{bmatrix} -\Delta X \\ 0 \\ \Delta Z \end{bmatrix} \tag{5.10}$$

5.3 Disparity Images and Noise Reduction

A *disparity image*, or *disparity map*, is the graphical representation of the stereovision image disparities calculated using Equation 5.1, correlation algorithms such as [3,4], and a scanning process similar to that shown in Figure 5.9. After assigning one of the images of the stereo pair as a reference – say the left image – every pixel is represented by an intensity value that is proportional to its disparity. The largest disparity value in the disparity map will correspond to the closest point that the system can detect, which depends on camera parameters such as the baseline and lenses used. Zero disparity will correspond, theoretically speaking, to objects located at infinity, given by the furthest match. However, objects that are very distant are not accurately perceived in the images, especially if short baselines and wide-angle lenses are used, and therefore the ranges calculated from disparities of close to zero are not very interesting for the kind of perception used in intelligent off-road vehicles. When a disparity image is examined, it should never be forgotten that the image consists of false colors, and as a result the intensity values portrayed are independent of the illumination pattern of the scene. Light tones imply either close objects or far objects, depending on the convention employed, but never indicate well-illuminated areas.

The disparity image is the primary source from which point clouds are fabricated; hence, the higher the quality of the disparity, the better the perception achieved with the stereo camera. A very useful first step towards generating good-quality disparity images is to conduct a meticulous *camera calibration* after ensuring that the camera is robustly fixed at its final location on the vehicle. However, regardless of how well the camera has been calibrated, two facts need to be considered: first, pixel correlation is conducted by a computer algorithm, which – just like any algorithm – will be prone to errors; second, the texture and quality of illumination of the scene will play a vital role in the efficiency of pixel matching. As a result, not every pixel will be correctly correlated. To begin with, certain boundary areas in the stereo images cannot overlap due to the separation between the lenses. Low-texture areas will impede the unique identification of scene features. Thus, the disparity image will have "blind" areas where no disparity information is stored. These non-informative areas are typically mapped in black and are the result of the filtering routines implemented by the correlation software. The areas that are correlated with low reliability and which are suppressed from the disparity map by the matching algorithm result in sparse 3D clouds that render a poor representation of the scene; however, the information conveyed is not misleading. A harder problem to cope with, though, is

Figure 5.13 Actual (**a, c, e**) and disparity (**b, d, f**) images carrying mismatching pixels

caused by wrong matches that pass through the correlation filter embedded in the algorithm; these carry wrong information which is then transferred to subsequent results. These mismatches can be regarded as *disparity image noise*, and this needs to be reduced or eliminated before processing the point clouds. The disparity images of Figure 5.13 are contaminated with noisy mismatches, some of which are not apparent at first sight. The pixels representing the clouds in the sky, for example, have disparity values that produce wrong estimations because the real positions of the clouds in the sky are well outside the detection range of the camera. Other times, mismatches are more evident, as they appear as isolated blobs of intensity that are very different from those of surrounding pixels. The remainder of this section deals with procedures used to lessen the effect of erroneous disparity values. Although disparity maps are intermediate stages in the stereo process, they hold key information and therefore determine the perceptive quality of the final outcome. For this reason, it is always helpful to display them and often to manipulate them to extract potential problematic mismatches. Processing a disparity image is easier and faster than dealing with the complete 3D cloud.

An effective way to avoid the negative consequences of meaningless outliers is to define a space of logical coordinate placement that agrees with the field of view corresponding to the camera settings selected. This space of acceptable point cloud localization is denominated the *validity box*, and all of the points positioned outside the box are discarded. Generally speaking, the concept of a validity box is an effective way of filtering field images, as the points eliminated tend to be small in number but large in error. Figure 5.14 illustrates the concept of a validity box (V_{BOX}) for an off-road vehicle traversing an orchard, and its mathematical formulation is given in Equation 5.11. According to the scene depicted in Figure 5.14, in order for a point to be considered valid, it must be located above the ground plane Ω_1 but below the height-limiting plane Ω_5, laterally situated between planes Ω_2 and Ω_3, and closer to the camera than plane Ω_4. Under these constraints, the wrongly determined clouds of Figure 5.13 are no longer a problem; nor are negative heights (which are unlikely) or points that possess unreliable ranges for the camera settings implemented. Plane Ω_1 does not need to be located at ground level, and it can be either positive or negative. Plane Ω_4, on the other hand, can only be positive and must be separated from the camera by a distance L_2 that is larger than the minimum detectable range.

$$
\begin{aligned}
&V_{\mathrm{BOX}} \stackrel{\mathrm{def}}{\longrightarrow} L_1 \times L_2 \times L_3 \\
&\text{Boundary Planes: } \begin{cases} \Omega_1 \equiv & z = 0 \\ \Omega_2 \equiv & x = \frac{-L_1}{2} \\ \Omega_3 \equiv & x = \frac{L_1}{2} \\ \Omega_4 \equiv & y = L_2 \\ \Omega_5 \equiv & z = L_3 \end{cases} \\
&P\ (x^{\mathrm{P}}, y^{\mathrm{P}}, z^{\mathrm{P}}) \subset V_{\mathrm{BOX}} \text{ if } \begin{cases} \frac{-L_1}{2} < x^{\mathrm{P}} < \frac{L_1}{2} \\ 0 < y^{\mathrm{P}} < L_2 \\ 0 < z^{\mathrm{P}} < L_3 \end{cases}
\end{aligned}
\tag{5.11}
$$

The removal of outliers using the validity box is an operation applied to the 3D point cloud after the disparity image has been processed to estimate the coordinates of all of the points with a non-filtered disparity value. Nevertheless, conventional image processing tools can be applied to the disparity image before the point cloud is generated, with the purpose of smoothing abrupt changes in intensity (represent-

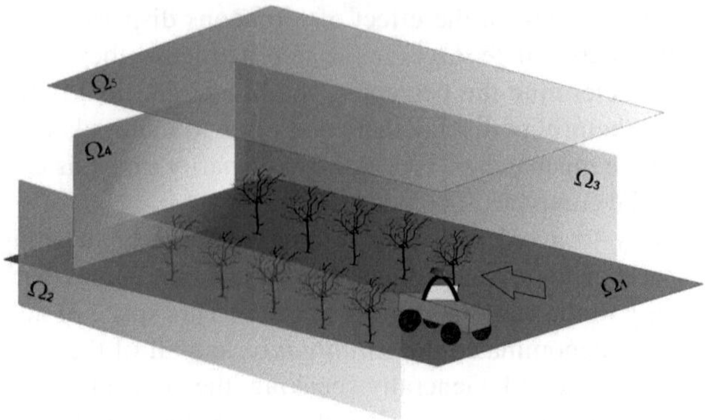

Figure 5.14 Concept of a validity box for filtering out outliers

ing disparity) and therefore potentially noisy patches. Given that intensity changes can be seen as frequency changes in an image, the application of *frequency* or *spectral analysis* could cast some light on the problem of mismatching. The *spatial frequency* of a digital image is an attribute that indicates how its *brightness* changes in the image space. This image space can be transformed into a different space by means of a mathematical function that enhances such brightness changes. Two popular transformations that emphasize the frequency distribution of image intensities are the *discrete Fourier transform* (DFT) and the *discrete cosine transform* (DCT). The DFT is the discrete version of the *Fourier transform*, a trigonometric function that enhances brightness changes when applied to an image. The mathematical expression for the DFT, applied to a digital image of resolution $H \times V$, is provided in Equation 5.12. Note that the final expression of the DFT separates out real and imaginary parts. One way to avoid imaginary numbers is to apply the DCT, a variant of the DFT that not only gives only real values but is also faster to execute. Equation 5.13 gives the basic formula for transforming an image of dimensions $H \times V$, where $I(b, a)$ is the brightness value of the pixel located at position (b, a).

$$\begin{aligned}
\Im(L, K)_{\text{Real}} &= \frac{1}{H} \sum_{a=1}^{H} \left[\left(\frac{1}{V} \sum_{b=1}^{V} I(b, a) \cos \frac{2\pi L b}{V} \right) \cos \frac{2\pi K a}{H} \right. \\
&\quad \left. - \left(\frac{1}{V} \sum_{b=1}^{V} I(b, a) \sin \frac{2\pi L b}{V} \right) \sin \frac{2\pi K a}{H} \right] \\
\Im(L, K)_{\text{Imag}} &= \frac{-1}{H} \sum_{a=1}^{H} \left[\left(\frac{1}{V} \sum_{b=1}^{V} I(b, a) \cos \frac{2\pi L b}{V} \right) \sin \frac{2\pi K a}{H} \right. \\
&\quad \left. + \left(\frac{1}{V} \sum_{b=1}^{V} I(b, a) \sin \frac{2\pi L b}{V} \right) \cos \frac{2\pi K a}{H} \right]
\end{aligned} \tag{5.12}$$

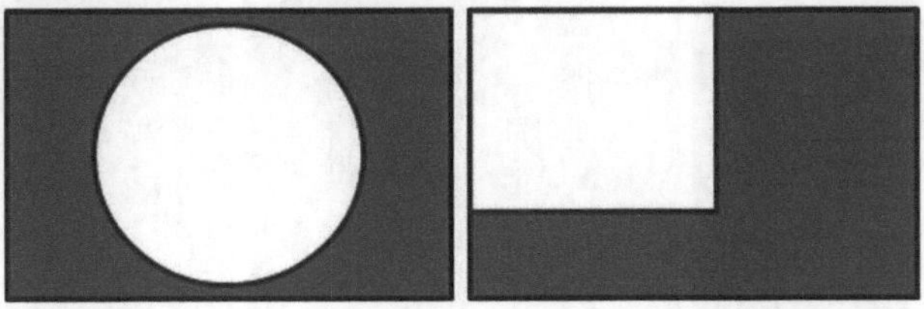

Figure 5.15 Low-pass filters for different frequency transforms: (**a**) DFT; (**b**) DCT

$$C\ (L,K)=\sum_{a=1}^{H}\sum_{b=1}^{V} I\ (b,a)\cdot\cos\left(\frac{(2\,a+1)\cdot L\cdot\pi}{2\cdot H}\right)\cdot\cos\left(\frac{(2\,b+1)\cdot K\cdot\pi}{2\cdot V}\right) \qquad (5.13)$$

As important as the generation of the frequency spectrum with the DFT or the DCT is its interpretation. The results of applying Equation 5.12 have a very large dynamic range, and visualization improves if a logarithmic transformation of the magnitude of the complex function F(L, K) is carried out. In addition, the DFT spectrum is easier to handle if the transform is shifted in such a way that the lowest frequencies are mapped to the center of the image and grow radially. The frequency value in the exact center of the shifted space represents the lowest frequency, which corresponds to the average intensity of the image or DC value. With such a distribution of frequencies, it is very easy to remove the high frequencies of the original image by constructing a *low-pass filter* and applying it to the DFT spectrum, as shown in Figure 5.15a. The frequency space for the DCT places low frequencies in the upper left corner, meaning that values increase down and to the right. Unlike the DFT, the DCT should never be shifted in order to preserve high frequency information. An appropriate low-pass filter for the DCT is shown in Figure 5.15b. A more detailed analysis of the use of the DFT and the DCT to smooth disparity images can be found in [6].

Spectral analysis is usually applied as a filtering technique as follows. First, the image space is transformed to the frequency space through the application of the DFT or the DCT, following Equations 5.12 and 5.13, and by shifting spectra when appropriate. Second, a low-pass filter of the type depicted in Figure 5.15 removes high frequencies to smooth the original image. Finally, an *inverse transform* of the filtered spectrum needs to be conducted to retrieve the final results in the image domain, from which further calculations can be realized. Disparity images are by their very nature depth images, and changes in depth in agricultural and orchard scenes tend to be gradual unless an object suddenly appears in front of the camera. This feature of disparity images helps to identify big jumps in pixel intensity – as potential outliers within the field of view of the camera, where the validity box filtration has no effect. The application of frequency analysis techniques in order to smooth disparity images shows great potential when intense noise is present in the original images, but may not be so useful for good-quality disparity images where noise and

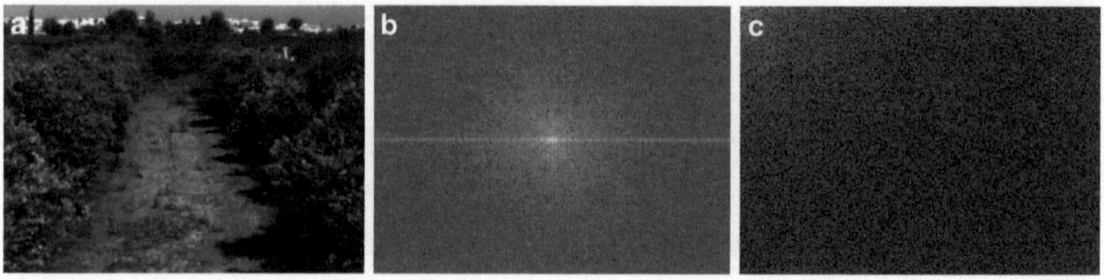

Figure 5.16 Frequency space of a typical orchard scene (**a**) after applying the DFT (**b**) and the DCT (**c**)

outliers are not problematic. Rather, it can create contour artifacts which result in filtered images that do not improve the initial unfiltered image. Figure 5.16a is an RGB image of a typical orchard scene taken with a stereo camera located on a tractor cabin (Figure 5.8b). Its corresponding disparity image is displayed in Figure 5.17a. The frequency spaces for the disparity image when the DFT (Figure 5.16b) and the DCT (Figure 5.16c) were applied are also shown. After the removal of high frequencies with low-pass filters of the type depicted in Figure 5.15, the DFT-filtered disparity map provided in Figure 5.17b or the DCT-filtered disparity image given in Figure 5.17c is obtained. Visual inspection of the three disparity images does not reveal any advantage of the filtered images (b and c) over the original (a). What is an advantage, however, is the generation of high-quality disparity images, which guarantee useful 3D point clouds.

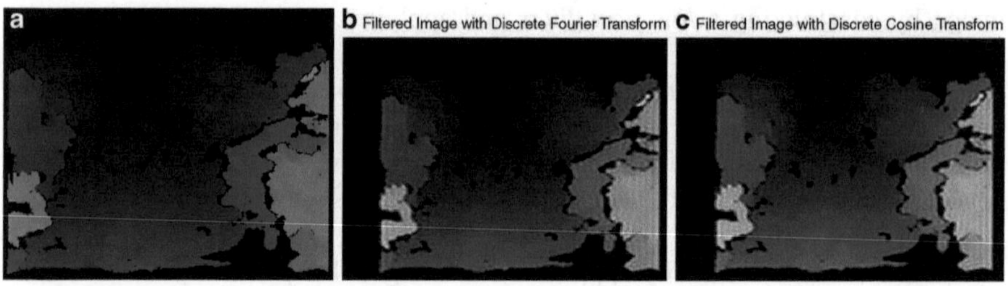

Figure 5.17 Disparity images of an orchard scene: (**a**) original unfiltered; (**b**) DFT filtered; (**c**) DCT filtered

5.4 Selection of Basic Parameters for Stereo Perception: Baseline and Lenses

Changing lenses or baselines in the field is not convenient. This is because, apart from the inconvenience of having to reach out to the stereo sensor and perform the adjustments when the weather is rough, any such change immediately demands a calibration test. Other parameters, such as the imager size and resolution, need to be determined before the stereo sensor is purchased. Each particular application will require the examination of a portion of space ahead of the camera, so, for instance, the vehicle represented in Figure 5.10 needs to reach autoguidance look-ahead distances of, say, 15 m. Once the perception needs of an intelligent vehicle have been defined, the following stage in the *stereo system design* is the selection of camera parameters that best cover the required field of view. After the electronic image sensors (imagers) have been chosen, the other two crucial parameters to select are the baseline (the separation between the lenses) and the focal length of the lenses. Unlike stationary nature of imagers, many stereo cameras allow the baseline to be modified and the lenses to be changed, but the imagers are internal elements of the camera and are not user-accessible after the camera has been manufactured. Nevertheless, an educated choice of baseline and lenses usually leads to an appropriate coverage of the desired field of view. Often, the application of theoretical equations to estimate potential range intervals (such as Equation 5.4) needs to be backed up by field tests to determine the real boundaries of the field of view, as multiple factors such as lens quality or correlation inaccuracies may intervene in the final results. A software application such as the example reproduced in Figure 5.18 can be helpful for assessing the perception qualities of the selected sensor. Once the parameters have been verified, the camera can be installed in the vehicle for continual use without the need to alter the optimal design parameters.

The graphical user interface (GUI) of the customized application shown in Figure 5.18 allows the registration of the original and the disparity images as well as 3D

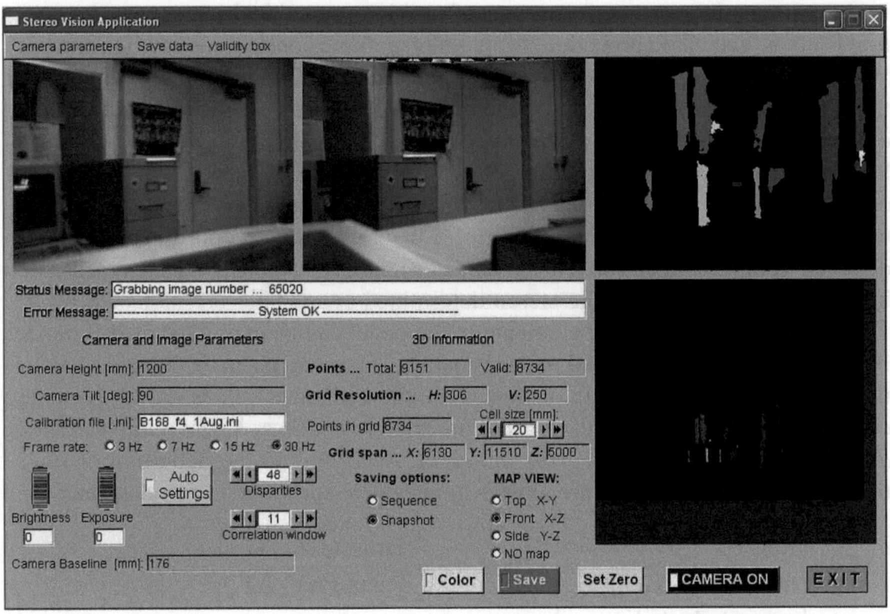

Figure 5.18 Stereo application software for evaluating fundamental camera parameters

point clouds and the density grids developed in Section 5.5. However, when comparing different baseline–lens combinations, the same scene should be studied by each combination, so that the results can be contrasted. The test panel of Figure 5.19 has been successfully used to analyze the performances of various camera assemblies. It consists of nine cubes in a 3×3 matrix arrangement with a regular pattern on the side of each cube that adds sufficient texture for the correlation algorithm to locate them indistinctly. Notice that two camera positions are considered: 6 m for short ranges and 12 m for medium-length ranges.

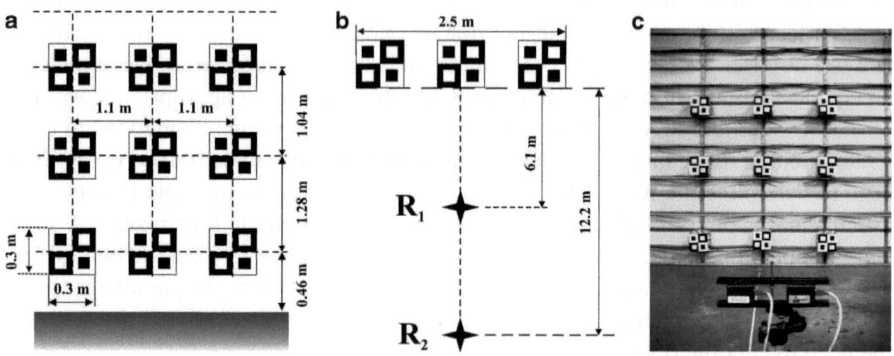

Figure 5.19 Test bench for evaluating the perceptive qualities of stereo cameras

A numerical evaluation of the perceptive qualities of a stereo camera is essential in order to compare the performances of multiple stereo devices, as visual inspection is not a systematic procedure from which definite conclusions can be extracted. When a stereo image is processed, the point cloud will facilitate the evaluation of the stereo-estimated coordinates and dimensions of the scene. However, these coordinates and dimensions are already known exactly, as the test panel (Figure 5.19) has been defined and measured. Comparing the actual distances with the image-estimated lengths provides a means to establish a fair comparison. The experimental setup, together with the definition of a favorable ground coordinate system, is depicted in Figure 5.20. The real coordinates measured on the stage are represented by X, Y, and Z, whereas the stereo-determined coordinates are indicated by X', Y', and Z'. The sizes of the elements in the panel can be expressed as differences in coordinates, both measured and estimated. The *relative errors* defined in Equation 5.14 provide an estimate of how close the estimated values are to the actual ones. Therefore, ε_x represents the *relative error in width*, ε_y is the *relative error in depth*, and ε_z is the *relative error in height*. According to the definition of Equation 5.14, the relative errors are always nonnegative and given in percentages.

$$\varepsilon_X = \left| \frac{\min(\Delta X, \Delta X')}{\max(\Delta X, \Delta X')} - 1 \right| \cdot 100\,; \quad \varepsilon_y = \left| \frac{\min(\Delta Y, \Delta Y')}{\max(\Delta Y, \Delta Y')} - 1 \right| \cdot 100\,;$$
$$\varepsilon_Z = \left| \frac{\min(\Delta Z, \Delta Z')}{\max(\Delta Z, \Delta Z')} - 1 \right| \cdot 100 \tag{5.14}$$

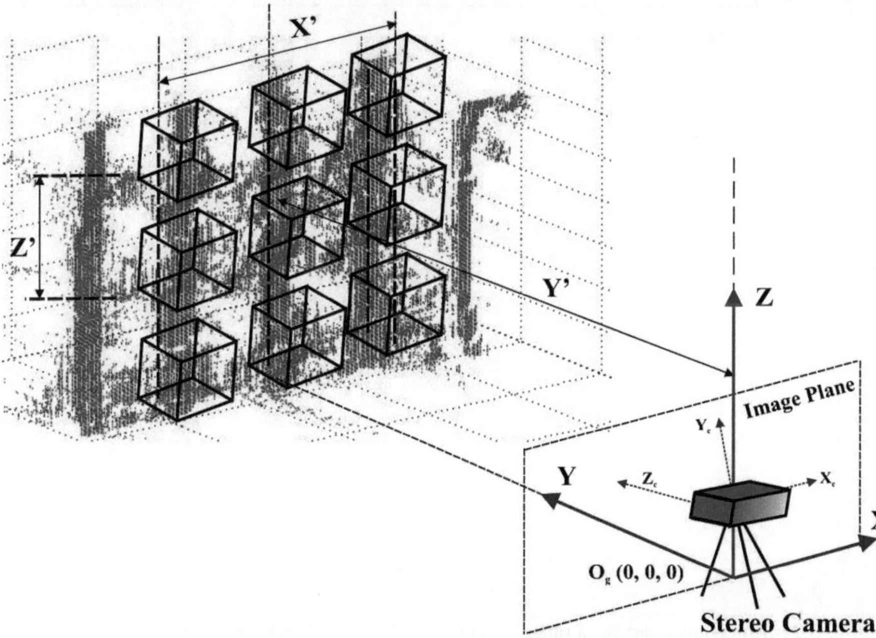

Figure 5.20 Experimental setup for evaluating the perceptive performances of stereo cameras

The three equations included in Equation 5.14 quantify the deviations along the three Cartesian axes outlined in Figure 5.20. However, each relative error assesses the accuracy of the measurements in one direction (X, Y, or Z). As coordinates X and Z define a plane perpendicular to the depth of the scene, according to the coordinate system defined in Figure 5.20, an efficiency index for the estimation of flat objects (where depth is not very relevant) can be defined: the *planar efficiency*, mathematically expressed by Equation 5.15. When the evaluation of depth measurements needs to be included, the three dimensions of the scene are considered and we talk of the *stereo efficiency*, which can be defined numerically via Equation 5.16. Both indices are positive and expressed in percentages.

$$\eta_{2D} = (1 - 0.01 \cdot \varepsilon_x) \cdot (1 - 0.01 \cdot \varepsilon_z) \cdot 100 \tag{5.15}$$

$$\begin{aligned} \eta_{3D} &= (1 - 0.01 \cdot \varepsilon_x) \cdot (1 - 0.01 \cdot \varepsilon_z) \cdot \left(1 - 0.01 \cdot \varepsilon_y\right) \cdot 100 \\ &= \eta_{2D} \cdot \left(1 - 0.01 \cdot \varepsilon_y\right) \end{aligned} \tag{5.16}$$

The efficiency indices of Equations 5.15 and 5.16 were used to evaluate the behaviors of multiple camera assemblies. Three baselines (103, 150, 194 mm) and five lenses (2.8, 4, 8, 12, 16 mm) were tested in various combinations according to the protocol illustrated by Figures 5.19 and 5.20. The results of this evaluation are graphically depicted in the efficiency charts of Figure 5.21 (for a range of 6 m) and Figure 5.22 (for a range of 12 m). Such *efficiency charts* represent the planar

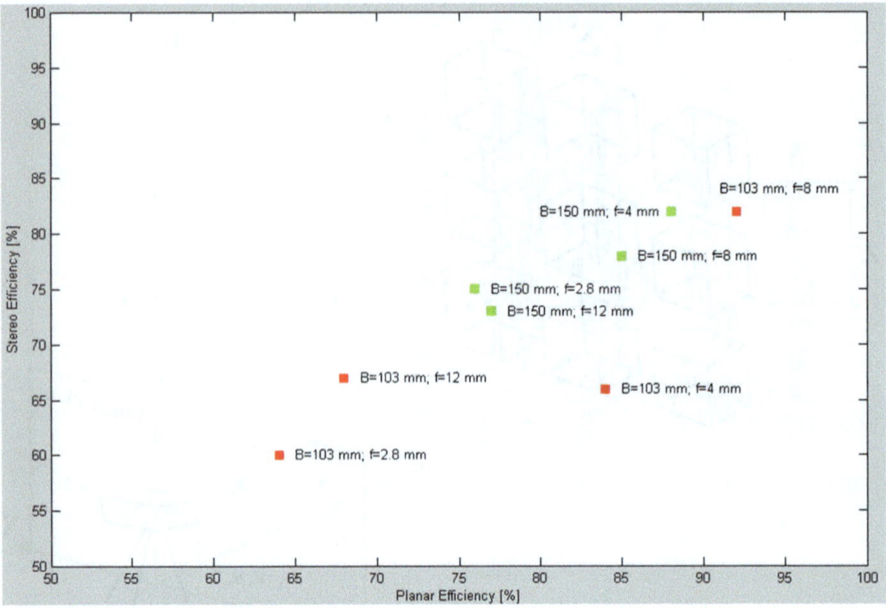

Figure 5.21 Efficiency chart for a range of 6 m

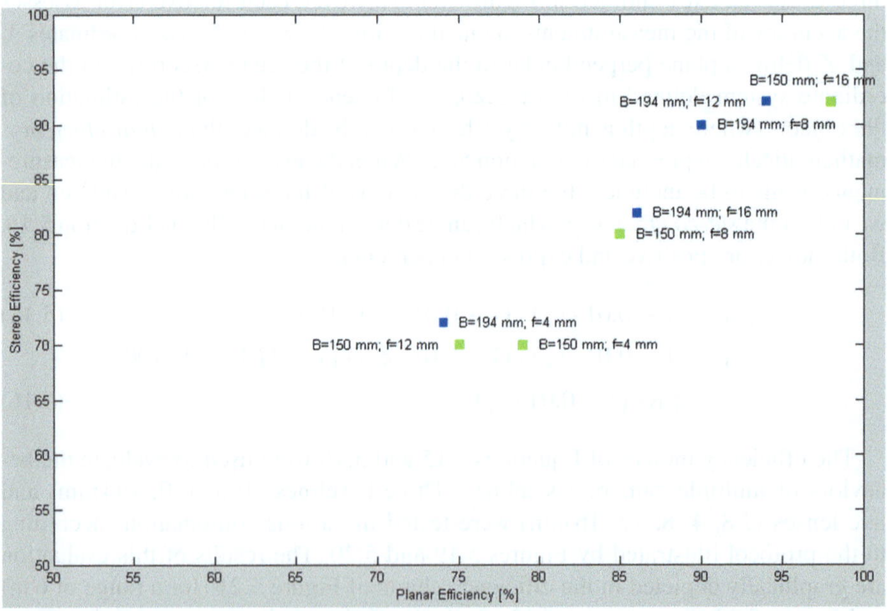

Figure 5.22 Efficiency chart for a range of 12 m

efficiency on the abscissa and the stereo efficiency on the ordinate. According to Figure 5.21, the preferable lenses for a range of 6 m have 8 mm focal lengths, and can be paired with either 10 or 15 cm baselines, although the latter performed better for a range of 6 m. Doubling the range (Figure 5.22) resulted in positive combinations of a 15 cm baseline with 16 mm lenses, and 19 cm baselines with either 8 or 12 mm lenses. A detailed description of these experiments is included in [7].

5.5 Point Clouds and 3D Space Analysis: 3D Density, Occupancy Grids, and Density Grids

The natural output from a stereo camera is a three-dimensional set of points denominated the *3D point cloud*. This set of points represents unstructured data for which there is no generic theory. The cloud is built by calculating the three coordinates of every pixel in the disparity image with a valid disparity magnitude. Other properties associated with each pixel can also be registered, such as the color code (RGB) or the intensity value for a given electromagnetic band. Given that the number of pixels is finite and depends on the image resolution, the set of points will be discrete. This approach to representing reality – through a discrete group of points – raises some difficulties that need to be addressed before processing clouds and making decisions about perception and navigation. A first concern arises from the consistency of the disparity image, as wrongly correlated pixels will lead to meaningless locations. These *mismatches* are unavoidable, and the technique developed to handle and process 3D data must consider their presence to be quite likely. A second problem can derive from the unfortunate coincidence of *weak texture* and *poor illumination*. Agricultural scenes and off-road environments tend to be fairly rich in texture, which is a favorable condition for reliable image correlation (in contrast to indoor robotics, where man-made environments usually consist of flat and smooth surfaces such as walls and floors). Nevertheless, texture patterns are not helpful if the light intensity is too low for the camera to clearly make out the distinct spatial differences within the images. High-contrast conditions, as well as a frontal sun at dusk or dawn, can severely decrease the textural qualities of the input images. Weak textures lead to poor disparity images, which in turn result in sparse and noisy 3D clouds. One final difficulty is the need to handle *massive sets of data* in real time. Even if images of moderate resolution are used, such as 320 × 240, a large number of points (76,800) will have to be transferred among the processing routines. This problem worsens if several images need to be merged, as in the case of 3D mapping. All of these considerations have motivated and shaped the concept of 3D density as a technique for efficiently processing 3D clouds generated with stereo sensors.

The principal aim of analyzing the point clouds is *3D awareness*. Since scene features need to refer to physical locations, the selection of an advantageous system of coordinates is very important. By default, point cloud coordinates are referred to the camera coordinate system (Figure 5.4), but their dependency on camera position and pose make this frame inconvenient. The ground coordinate system (Figure 5.5), on the other hand, is more intuitive and efficient at representing and positioning objects in the scene, and so it is the preferred coordinate system for analyzing 3D information. Consequently, an initial task will always be to transform the coordinates to the most adequate coordinate frame. The reliable detection of empty space is crucial to the safe navigation of intelligent vehicles. The recognition of solid space can only be achieved through the accurate examination of the point cloud. Intuitively, more points will gather around locations where the presence of an object is likely. However, a simple count of 3D points is misleading because the number of points in the cloud increases as the field of view of the camera widens. In fact, objects close to the camera are represented by more points than those further away. In some fashion, the number of 3D points associated with a determined object needs to be related to the actual space that they occupy, as far objects are necessarily represented in the disparity image by less pixels. Observations of nature suggest a concept based on the idea of *density*, a physical property that relates the mass of a matter to the volume that it occupies. Inspired by this idea, a new property of stereo vision 3D point clouds can be defined as the relationship between the number of detected points and the volume occupied by them. This specific property of stereo point clouds is denoted the *3D density* (d_{3D}), and is defined as the *number of stereo-matched points per volume unit*. The definition of the 3D density can be directly expressed in mathematical form according to Equation 5.17, where V quantifies the volume of space considered in the calculations, and N is the total number of stereo-matched points inside volume V. Notice that there is no relationship between the 3D density of an object and its physical density; they are only linked conceptually.

$$d_{3D} = \frac{N}{V} \tag{5.17}$$

The application of the concept of 3D density requires the estimation of volumes, which is not practical unless the space has been divided up regularly in previous d_{3D} calculations. An effective way of dividing the space ahead of the camera into regular volumetric portions is to quantize 3D environments into regular grids; knowing the volume of each cell, the number of points inside will provide its associated d_{3D}. The 3D density is, by definition, independent of the cell dimensions, in the same way that the density of a liquid is independent of the size of its container. Grids defined in order to utilize 3D densities are termed *density grids*. Figure 5.23a illustrates the concept of 3D density, and Figure 5.23b depicts the creation of a density grid that condenses 3D information into a three-dimensional density grid. The management of perceptual data (stereo vision, sonar, or laser range maps) by means of a discretized space is carried out using a similar concept: occupancy grids, also called *evidence grids* [8]. The main difference between occupancy and density grids is that the former follow a probabilistic approach whereas the latter are based on the concept of 3D density. Occupancy grids are applied to simultaneous mapping, and usually update information according to new evidence retrieved from the mobile robot; hence the name evidence grid. Each cell holds the probability of representing an obstacle. A density grid, on the other hand, provides a virtual representation of the captured scene with an application-determined level of quantization, which

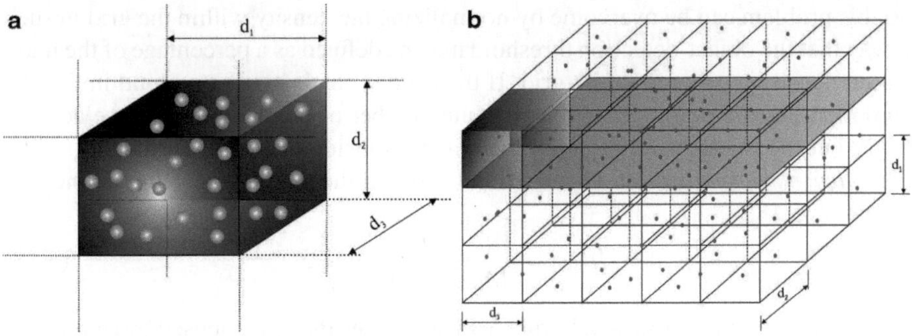

Figure 5.23 Concept of 3D density (**a**) and its embodiment through a density grid (**b**)

depends on the design of the grid and the object-empty space threshold selected. Density grids do not necessarily have to be two-dimensional or to render top-view representations of the sensed scenes.

Density grids can be two-dimensional (2D) or three-dimensional (3D), and when they are 2D they can be *front*, *top*, or *side grids*, depending on the view projected. In addition to configuring the grid, it is crucial to choose an adequate cell size: a large cell dimension might overlook important objects and scene details; on the other hand, a highly quantized grid will result in heavy computational loads that may collapse the perception computer. The optimum grid resolution will be determined by each particular application and the perceptual needs. An advantage of Equation 5.17 is the ability to compare densities among grids of different dimensions; the cells will be different but the d_{3D} will be equivalent.

Once an estimation of the volume becomes feasible, due to the design of suitable density grids for example, we need to focus on the other basic variable of Equation 5.17: the number of stereo-matched points. It is undoubtedly true that there is a direct link between N and the results of the stereo correlation process embodied in the disparity image. Furthermore, regardless of the correlation results, the maximum number of points enclosed in the disparity map, and therefore limiting the size of the point cloud, depends on the resolution of the images acquired by the stereo camera. So, for example, an image of resolution 1280×960 will naturally yield more populated clouds than an image of resolution 320×240. This obvious fact poses a dilemma when operating with density grids, because the establishment of a threshold to discriminate objects from vacant space will necessarily be resolution dependent. In fact, not only will the threshold be affected by the image resolution, but it will also be influenced by factors such as uneven illumination; variations in light intensity over the same object will produce 3D clouds of varying 3D density. Low-light conditions impede the full perception of texture, with negative consequences for the generation of point clouds. This problem occurs among different images, so a vehicle can change its heading by 180° and the sun's direction can induce such a radical change in the illumination of the scene that the initial threshold set to discriminate objects in the density grid becomes completely useless. Fortunately, a change in illumination or resolution will affect to the entire image, and so this problem can be overcome by normalizing the density within the grid in such a way that the object-detection threshold may be defined as a percentage of the maximum density registered in the grid. If the sudden shade cast by a cloud in the sky impoverishes the point cloud, the absolute number of points will decrease accordingly but the relationship between the densities inside the grid will remain stable. This *density normalization* is carried out through the simple operation defined by Equation 5.18:

$$|d_{3\text{D}}| = \frac{d_{3\text{D}}}{\text{Max }(d_{3\text{D}})_{\text{image } n}}\,. \tag{5.18}$$

The fields of view drawn in Figure 5.11 show how the space sensed by stereo sensors widens as the range grows; that is, the further from the camera, the larger the region monitored. However, all of the rows in the image have the same amount of pixels. This fact means that the space assigned per pixel in the background is larger than that covered by the pixels in the foreground; however, since all of the rows in the image have the same number of pixels, the 3D density in the background will be necessarily less than that close to the camera. Therefore, there is a *lack of uniformity in the distribution of* $d_{3\text{D}}$ *inside every image*. The fact that the points in the cloud spread out as the range increases causes a loss of resolution as objects move away from the camera. Equation 5.9 provides an expression for the range resolution for a given disparity as a function of camera parameters and range. Interestingly enough, Equation 5.9 proves that the range resolution is not linearly related to the range; instead, it follows a *quadratic relationship*. Once the fundamental parameters that configure any given stereo camera have been selected, namely the baseline (b), the focal length (f), and the pixel size of the imaging sensor (w), they usually remain fixed. Therefore, every camera setting can be associated with a camera constant K_s, defined according to Equation 5.19. Although Equation 5.9 provides an expression for the range resolution, the distribution of 3D density with range can be related to the distribution of disparity with range, which is easily obtained by inverting Equation 5.9. When Equation 5.9 is inverted and the camera constant K_s is introduced, the resulting distribution of density with range is as given in Equation 5.20.

$$K_\text{s} = \frac{b \cdot f}{w} \tag{5.19}$$

$$\frac{\Delta\, d}{\Delta\, R} = K_\text{s} \cdot R^{-2} \tag{5.20}$$

The expression deduced in Equation 5.20 provides an estimate for how the 3D density may vary with the range if we assume the hypothesis that the *decrease in disparity with range properly models the decrease in* $d_{3\text{D}}$ *with range*. At this point, it is necessary to contrast these theoretical assumptions with actual data. The stereo camera shown in Figure 5.6b has a camera constant K_s of 110,000 pixel mm. Figure 5.24 shows a field image acquired with this camera, and the distribution of 3D density with range found for the point cloud of that scene. The equation for the regression curve that best fits the data acquired in the field is provided in Equa-

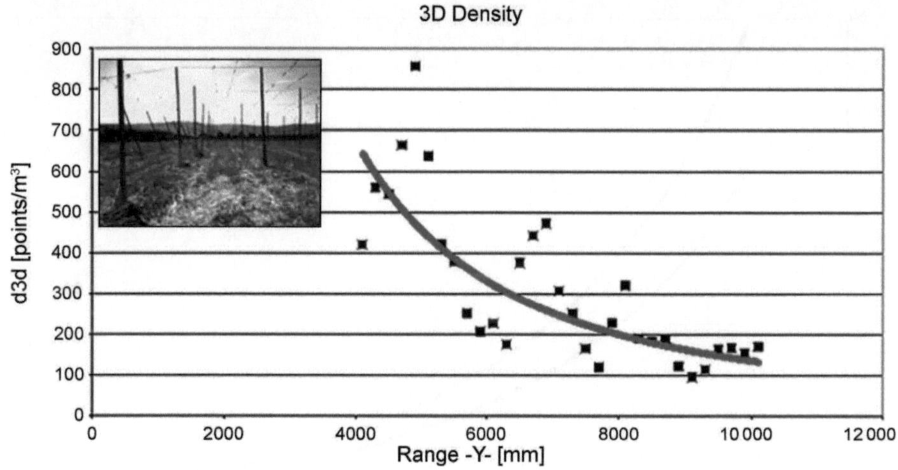

Figure 5.24 Distribution of 3D density with range for a field stereo image

tion 5.21.

$$d_{3D} = 10^9 \cdot Y^{-1.7481} \tag{5.21}$$

A comparison of Equations 5.20 and 5.21 suggests that both magnitudes behave in a similar fashion and that they are *proportional*. In fact, if a proportionality constant of 10^5 is introduced on the right-hand side of Equation 5.20, and the value of K_s is specified for this camera, Equation 5.22 provides a theoretical estimate of the distribution of d_{3D} with range for this camera. When both the theoretical expression of Equation 5.22 and the trend curve for the field data given in Equation 5.21 are plotted together in Figure 5.25, we can see that both curves are fairly close, and converge with increasing ranges.

$$d_{3D}^{Th} = 11 \cdot 10^9 \cdot Y^{-2} \tag{5.22}$$

The availability of a theoretical expression for the 3D density decay with respect to the range, such as Equation 5.22, allows us to compensate for the scarcity of cloud points representing far objects. In order to apply a unique threshold to the whole image, densities need to be normalized to an arbitrary *reference range*. Equation 5.23 is an example of an image normalized to a reference range set at 4.5 m and after applying the theoretical expression of Equation 5.22. This d_{3D} compensation was applied to avoid obstacles in *Case Study IV*, and in the project developed in [9]. Another d_{3D} compensation formula can be seen in Equation 5.47 of *Case Study VI*:

$$\begin{aligned}[d_{3D}(Y)]_{\text{COMPENSATED}} &= d_{3D} \cdot \frac{d_{3D}^{Th}(4500)}{d_{3D}^{Th}(Y)} = d_{3D} \cdot \frac{11 \cdot 10^9 \cdot 4500^{-2}}{11 \cdot 10^9 \cdot Y^{-2}} \\ &= 5 \cdot 10^{-8} \cdot Y^2 \cdot d_{3D}\end{aligned} \tag{5.23}$$

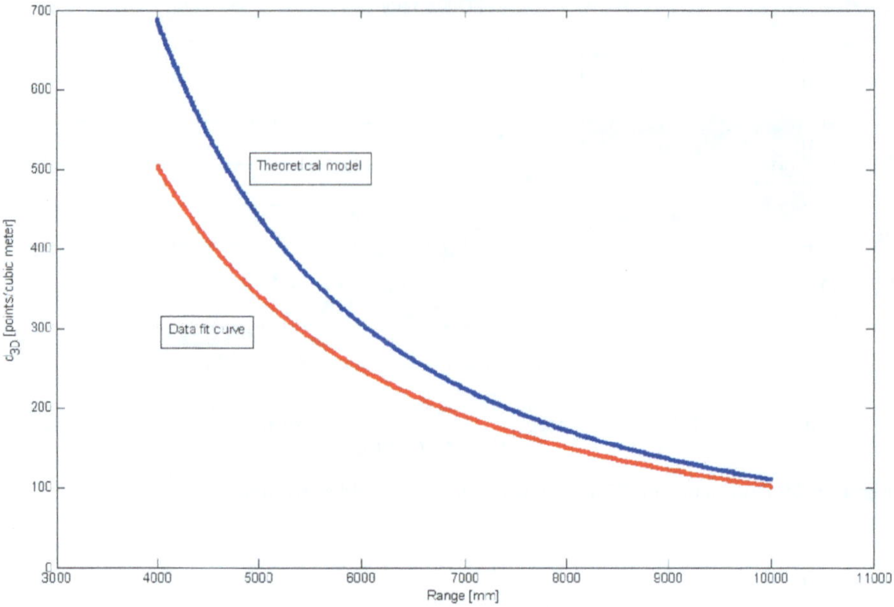

Figure 5.25 Theoretical and practical relationship between d_{3D} and stereo ranges

The calculation of 3D densities is intimately related to *density grids*, which are grids of cells where the d_{3D} is represented. Three-dimensional point clouds should be better managed with 3D density grids (Figure 5.23b), but the processing time allowed for real-time conditions usually results in a downgrade to 2D density grids. These two-dimensional grids accumulate point clouds in one dimension and offer greater resolution in the other two dimensions, as indicated in Figure 5.26. The design of the grid must serve the purpose of each application. So, for example, the *top-view grid* of Figure 5.26a is appropriate for detecting guidance features at a given

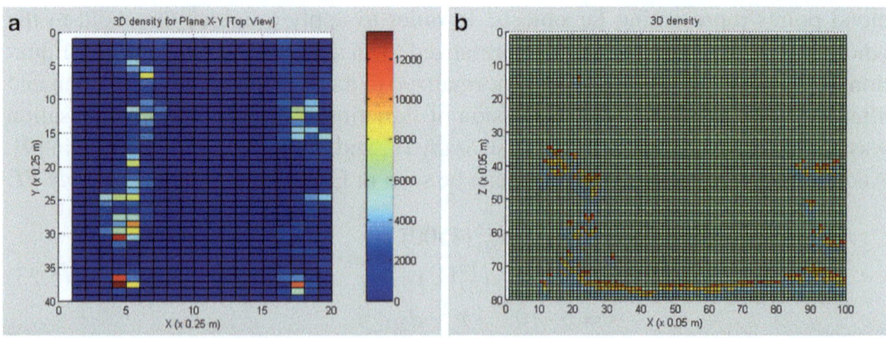

Figure 5.26 Two-dimensional density grids: (**a**) top-view grid; (**b**) front-view grid

look-ahead distance. The *front-view grid* of Figure 5.26b, on the other hand, is helpful for identifying potential lateral hazards when the vehicle moves in the field, such as protruding branches or tunnel-like agricultural constructions. The number of cells in the grid gives the resolution of the grid.

5.6 Global 3D Mapping

Information technology (IT) and geographical information systems (GIS) are new technologies that are helping to improve the management of production systems such as agricultural ones. These systems rapidly supply rich, up-to-date data to assist in regular decision-making strategies. Remote sensing techniques are employed continuously to support precision farming (PF) activities. However, none of these systems can provide the real-time actualization of three-dimensional centimeter-resolution terrain maps. GIS and remote sensing images typically provide two-dimensional spatial information, and rarely achieve sub-meter accuracies. An intelligent vehicle traversing a field autonomously requires an accurate description of the surrounding environment that neither GNSS receivers nor satellite imagery can provide. Terrain mapping operations that demand precise 3D awareness and high update rates cannot be successful without local perception. Stereo vision, in conjunction with global positioning systems, can provide this level of perception at the global level. This is the objective of *global 3D terrain mapping*.

Global 3D maps are generated in three key stages: the *acquisition* of 3D perceptual maps, the *orientation* of these 3D maps, and the *localization* of the 3D maps in a global reference. The architecture of the system developed must respond to these three needs. Figure 5.27 represents an example of architecture for a generic 3D mapping system where the perceptive unit is a binocular stereo camera, the orientation of local (vehicle-fixed) maps is estimated with a fiber-optic gyroscope (FOG), and the global position of the camera is obtained from a differential GPS.

The objects locally sensed with the stereo camera will be referred to a system of coordinates fixed to the vehicle; that is, they will be situated in a *local map* that travels with the camera. Local maps are useful for instantaneous perception and data processing, but they are difficult to relate to other local maps. A *global map* has a unique origin and fixed axes that provide a clear reference to all of the elements within its limits. The way to convert local coordinates to global coordinates is through a coordinate transformation that depends on the type of image acquired, but regardless of the transformation conducted, the pose (yaw, pitch, roll) and the global position of the camera (east, north, height) are required. In general, two types of images can be acquired for 3D mapping: aerial and ground images. *Aerial images* are those taken when the image sensor (imager) is approximately parallel to the ground, providing a top view representation of the scene. *Ground images* are taken when the camera is pointing ahead with the imagers oriented either perpendicular to the ground or inclined to it. An important feature of ground images is perspective (and vanishing points when parallel rows are captured). Directly related

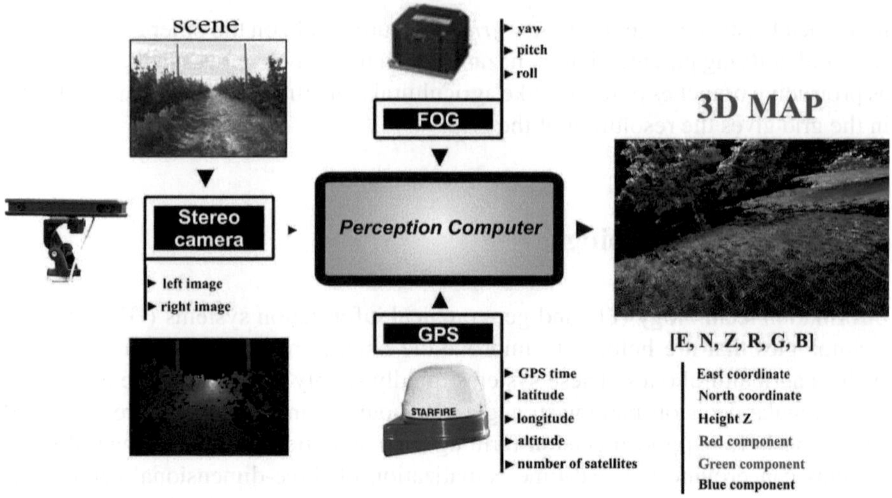

Figure 5.27 System architecture for a global 3D mapping platform

to the image type is the selection of mapping platforms. Aerial images can be captured with either ground vehicles or airborne equipment. The rows of potatoes in Figure 4.1a were photographed from a small horticulture tractor in Japan, but the image of corn of Figure 5.28a was captured from the remote-controlled helicopter of Figure 5.28b. Both figures are examples of aerial images. Ground images, on the other hand, are typically grabbed by cameras mounted on the front ends of ground vehicles. Tractors, harvesters, scouting robots, and utility vehicles are perfect candidates for mapping from the ground.

As well as influencing the choice of mapping vehicle, the type of images required determines the coordinate transformation needed to build the 3D map. Figure 5.29 includes all of the elements involved in the transformation of the coordinates of point P from image coordinates to ground coordinates. Note that the image plane and the ground plane are considered parallel, and the distance between them is labeled D in the pictorial of Figure 5.29. The transformation between both coordinate systems is very similar to that given in Equation 5.7. In fact, a small adjustment to relate the true height of P (z^{p}) with the separation between planes (D) is enough to get the ground coordinates of any point in the scene, as stated in Equation 5.24. This transformation yields the ground coordinates in a local (helicopter-fixed) system of reference, the origin of which is on the ground at the virtual projection of the optical center of the reference lens, as indicated in Figure 5.29. The following step is the transformation of the ground coordinates to a *global frame* that is not dependent on the vehicle's orientation and position.

$$\begin{bmatrix} x^{\mathrm{p}} \\ y^{\mathrm{p}} \\ z^{\mathrm{p}} \end{bmatrix}_{\text{meters}} = R_{\text{meters}} \cdot \begin{bmatrix} T & 0 & 0 \\ 0 & T & 0 \\ 0 & 0 & -1 \end{bmatrix}_{\text{pixel}^{-1}} \cdot \begin{bmatrix} x_{\mathrm{i}}^{\mathrm{p}} \\ y_{\mathrm{i}}^{\mathrm{p}} \\ 1 \end{bmatrix}_{\text{pixel}} + D \cdot \begin{bmatrix} 0 \\ 0 \\ 1 \end{bmatrix}_{\text{meters}} \tag{5.24}$$

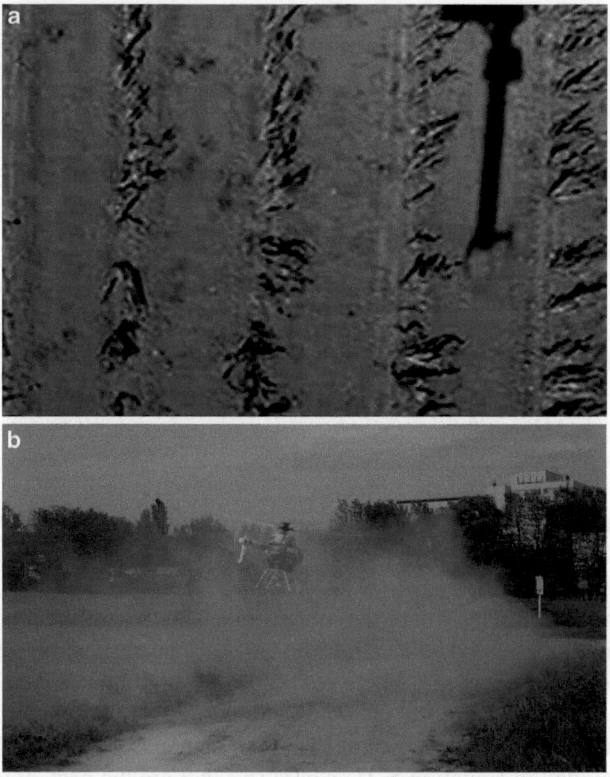

Figure 5.28 Aerial image of corn (**a**) taken from a remote-controlled helicopter (**b**)

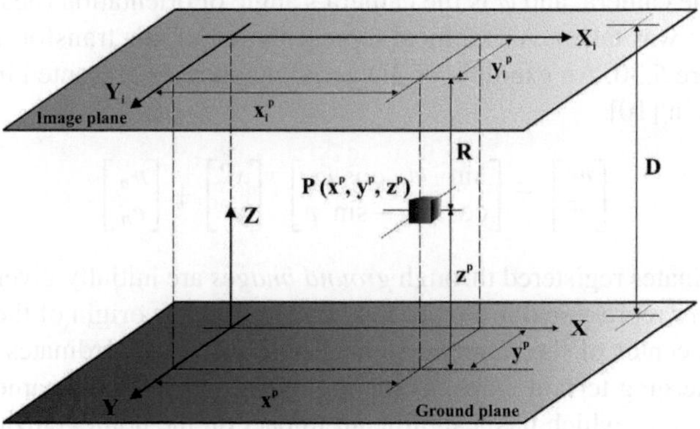

Figure 5.29 Transformation to ground coordinates for aerial images

Every stereo image results in a particular *local map* that is a building block of the *global map*, but before it can be assembled, each local map needs to be characterized by the global coordinates of its origin and the camera heading (yaw angle) when the image was taken. The pitch and roll angles of the camera are considered

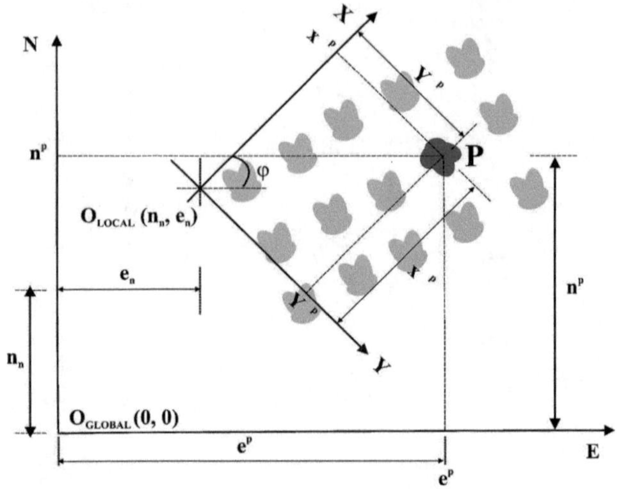

Figure 5.30 Transformation from local coordinates to global coordinates for aerial images

negligible for aerial images where the image plane and the ground are parallel. The usual dimensions of mapped fields allow the transformation of geodesic global coordinates to a *local tangent plane*, where *east* and *north* coordinates are more intuitive and convenient to use and represent. The definite transformation that converts any point P captured with the stereo camera into an element of the global terrain map $(n^{\mathrm{P}}, e^{\mathrm{P}})$ is provided in Equation 5.25, where (n_n, e_n) are the east–north global coordinates of the camera for image n, $(x^{\mathrm{P}}, y^{\mathrm{P}})$ are the local coordinates of a point P detected by the camera, and φ is the camera's angle of orientation (heading or yaw) when image n was taken. A graphical representation of this transformation is provided in Figure 5.30. An example of 3D aerial mapping is presented in *Case Study III*, as well as in [10].

$$\begin{bmatrix} n^{\mathrm{P}} \\ e^{\mathrm{P}} \end{bmatrix} = \begin{bmatrix} \sin\ \varphi & \cos\varphi \\ \cos\ \varphi & -\sin\varphi \end{bmatrix} \cdot \begin{bmatrix} x^{\mathrm{P}} \\ y^{\mathrm{P}} \end{bmatrix} + \begin{bmatrix} n_n \\ e_n \end{bmatrix} \tag{5.25}$$

The coordinates registered through *ground images* are initially given in the *camera coordinates* represented in Figure 5.4, which place the origin of the coordinates at the optical center of the reference lens. These camera coordinates are not convenient for merging terrain maps, as the inclination angle of the camera may vary among local maps, which has a significant impact on the homogenization of coordinates. A first step is therefore to transform the camera coordinates to a ground system of reference where the camera's inclination angle (ϕ) and height (h_{c}) do not impede the free combination of coordinates, such as the Cartesian frame presented in Figure 5.5. The transformation between both systems of coordinates can be executed with Equation 5.26, where $(x_{\mathrm{c}}, y_{\mathrm{c}}, z_{\mathrm{c}})$ are the camera coordinates, $(x^{\mathrm{P}}, y^{\mathrm{P}}, z^{\mathrm{P}})$ are the ground coordinates for point P, h_{c} is the camera height measured at the optical center of the lens from the ground, and ϕ is the camera's inclination angle.

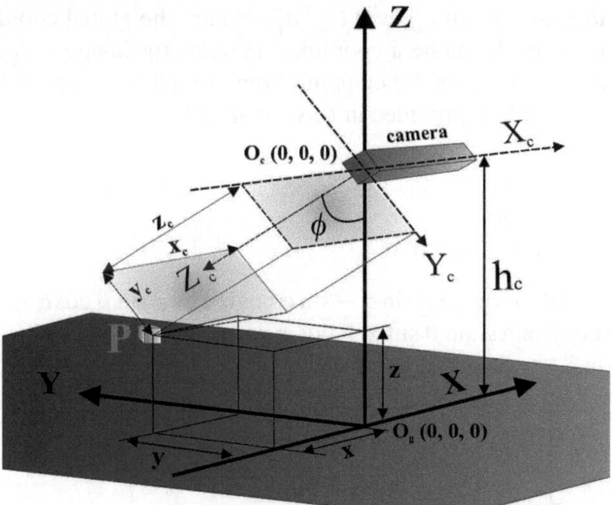

Figure 5.31 Transformation from camera coordinates to ground coordinates for ground images

Figure 5.31 explains how the camera coordinates relate to ground coordinates.

$$\begin{bmatrix} x^{\mathrm{p}} \\ y^{\mathrm{p}} \\ z^{\mathrm{p}} \end{bmatrix} = \begin{bmatrix} 1 & 0 & 0 \\ 0 & -\cos\phi & \sin\phi \\ 0 & -\sin\phi & -\cos\phi \end{bmatrix} \cdot \begin{bmatrix} x_{\mathrm{c}} \\ y_{\mathrm{c}} \\ z_{\mathrm{c}} \end{bmatrix} + h_{\mathrm{c}} \cdot \begin{bmatrix} 0 \\ 0 \\ 1 \end{bmatrix} \tag{5.26}$$

Just as in the construction of global aerial maps, several local maps must be fused in an orderly arrangement to build a global ground map. This can only be done if the global coordinates of the origin of every single local map are known. In contrast to aerial imagery, the image plane of a camera taking ground images will not generally be parallel to the ground, so the attitude of the vehicle (determined by its yaw, pitch, and roll) needs to be estimated for each image acquired. The aim is to create a 3D virtual map whose coordinates are given in a *north-east-height* frame with unique axes and origin. As the coordinate system follows a *local tangent plane* approach, the origin will be set arbitrarily based upon the users' convenience. The system architecture of Figure 5.27 includes an inertial measurement unit to provide the vehicle's pose in real time and a GNSS receiver that sends out positioning information at 5 Hz or more. This situation allows the transformation of every point detected with the stereo camera into global coordinates (E, N, Z) in real time. In addition to the coordinates, the system architecture proposed in Figure 5.27 records the RGB color code (*red*, *green*, *blue*) for each point when the stereo camera contains a color sensor to perceive the scene. The transformation in Equation 5.27 converts point n given in stereo-based local maps in ground coordinates $(x_n^{\mathrm{p}}, y_n^{\mathrm{p}}, z_n^{\mathrm{p}})$ to a globally referenced point n in general coordinates $(e_n^{\mathrm{p}}, n_n^{\mathrm{p}}, z_n^{\mathrm{p}})$, where d_{GPS} is the horizontal distance along the y-axis between the GPS receiver and the optical center of the reference lens, h_{GPS} is the height at which the GPS is mounted, α is the pitch, β is the roll, φ is the yaw (heading), and (e_i^c, n_i^c, z_i^c) are the global coordinates of the camera (the origin of the camera coordinate system) for image i. A complete description of this technique for 3D mapping from ground images is included in [11], and a practical example is provided in *Case Study III*.

$$\begin{bmatrix} e_n^{\mathrm{p}} \\ n_n^{\mathrm{p}} \\ z_n^{\mathrm{p}} \end{bmatrix} = \begin{bmatrix} e_i^c \\ n_i^c \\ z_i^c - h_{\mathrm{GPS}} \cos\beta \cos\alpha \end{bmatrix} + \begin{bmatrix} \cos\varphi\cos\beta & \cos\varphi\sin\beta\sin\alpha - \sin\varphi\cos\alpha & \cos\varphi\sin\beta\cos\alpha + \sin\varphi\sin\alpha \\ \sin\varphi\cos\beta & \sin\varphi\sin\beta\sin\alpha + \cos\varphi\cos\alpha & \sin\varphi\sin\beta\cos\alpha - \cos\varphi\sin\alpha \\ -\sin\beta & \cos\beta\,\sin\alpha & \cos\beta\,\cos\alpha \end{bmatrix} \cdot \begin{bmatrix} x_n^{\mathrm{p}} \\ y_n^{\mathrm{p}} + d_{\mathrm{GPS}} \\ z_n^{\mathrm{p}} \end{bmatrix} \tag{5.27}$$

Any of the fundamental sensors needed to generate the 3D global maps – the stereo camera, the GNSS unit, and the inertial measurement system – could fail during map generation, with negative consequences for the final results. Problems caused by the stereo camera can relate to the correlation algorithm, as discussed in Section 5.3, or simply to difficult images, as in the case portrayed in Figure 5.28b, where the dust provoked by the helicopter blades dramatically reduces visibility. Image noise can be compensated for with specific techniques such as the validity box or other previously described image analysis tools. A lack of visibility, however, forces the map builder to discard poor images.

GNSS information can be imprecise due to double-path issues, atmospheric errors, or signal blockages caused by copious vegetation. When this happens, an object detected by several images will not be represented by a unique entity; rather, several overlapping virtual embodiments of the same object will coexist, which practically

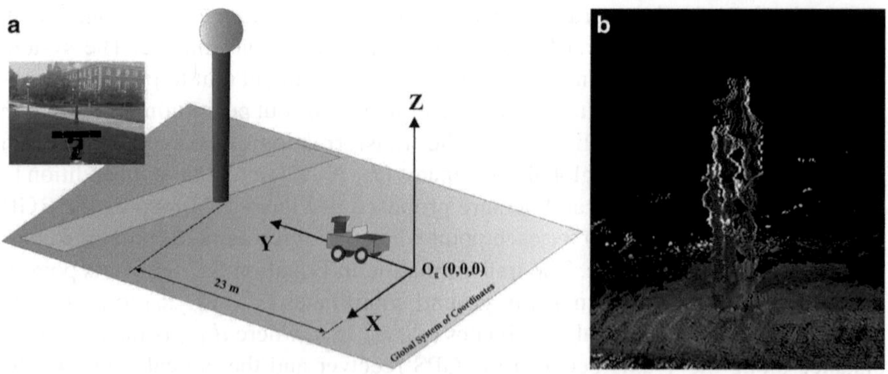

Figure 5.32 Erroneous global maps due to GNSS failures: (**a**) experiment; (**b**) 3D field map

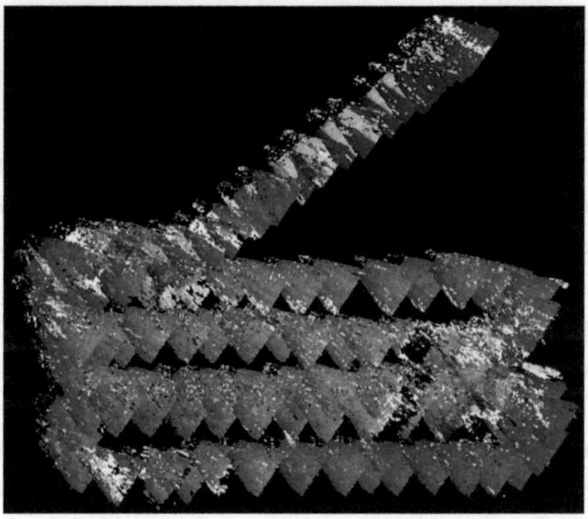

Figure 5.33 Erroneous global map due to IMU errors

ruins that portion of the global map. Figure 5.32a illustrates a situation that was specifically created to challenge the positioning sensor of a mapping vehicle. The vehicle was driven towards a lamp post and several local maps were obtained during the vehicle's course. When the local maps were fused into a unique global map, the post had multiple representations, as shown in Figure 5.32b.

The role played by the inertial measurement unit (IMU) is as important as those played by the other two sensors, so only accurate estimations are acceptable when assembling a global 3D map. When the IMU fails, the local maps are wrongly oriented and the global maps are totally meaningless. It is thus preferable to discard the local maps whenever any of the sensors output unreliable data. This procedure will generate empty patches, but the portions that are properly constructed will convey useful information. Furthermore, as suitable map portions have been safely determined, voids can be filled in later by other mapping missions. Obviously, such a procedure demands a quality filter for each of the fundamental sensors. Figure 5.33 gives an example of a global map spoilt by the failure of the IMU during map construction.

5.7 An Alternative to Stereo: Nodding Lasers for 3D Perception

The coherence of laser beams facilitates the generation of high-resolution range maps that have been successfully used for obstacle detection and mapping. Although most of the maps generated from laser sensors are two-dimensional, it is possible to configure the laser head, or lidar, in such a way that three-dimensional perception is feasible too. This would in theory position lidars as direct competitors to stereovision cameras, but in reality this is not the case due to the complexity of creating 3D maps from nodding lasers, even though the resolution achieved with the lasers tends to be higher. The laser head rotates in the horizontal plane where the range map is created. The angles covered typically range from 180 to 270°, and normal resolutions are about 1°. This procedure results in two-dimensional maps of the scene ahead of the beam emitter. In order to generate 3D maps, two approaches can be followed. The first solution utilizes a regular nodding laser attached to a vehicle where global coordinates and attitude estimates are available. This combination of sensors allows the transformation of vehicle-fixed local coordinates to global coordinates for all the points scanned by the laser as the vehicle travels. Merging the information from all of the scanned planes after converting it to global coordinates produces a 3D representation of the surroundings traversed by the vehicle, as reported in [12]. The second alternative is to make the laser head rotate in two perpendicular planes simultaneously. This procedure, while conceptually interesting, poses non-trivial challenges during the physical construction of a working unit. To begin with, the laser head must stand on a rotating platform whose movement is perfectly coordinated with the embedded rotating emitter inside the sensor. The slightest loss of synchronization between both angular movements would make the results useless, thus wasting the accuracy achieved by the laser. This double-nodding motion is usually achieved by using a mechanical device, and the precision of such a device may suffer when traversing rough agricultural environments. The angular position of the beam in both planes must be accurately tracked by feedback angular sensors, such as optical encoders, which in combination with the range allow the conversion of spherical coordinates into conventional Cartesian coordinates, in which 3D global maps are normally represented. Figure 5.34 illustrates the concept of a two-plane nodding laser for three-dimensional mapping.

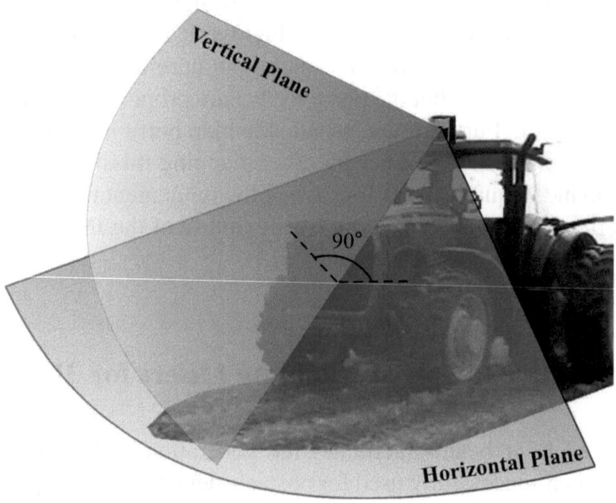

Figure 5.34 Nodding laser for rendering 3D agricultural maps

5.8 *Case Study I*: Harvester Guidance with Stereo 3D Vision

Harvesting is usually the last task of the season, but that does not mean it is the least important. As a matter of fact, harvesting is quite a delicate endeavor because it requires the occurrence of several circumstances: optimum vegetative estate (such as moisture content or degree of maturity), favorable weather conditions (such as a moderate wind speed or no rain), dry soil to allow machinery navigation, availability of loading trucks and drivers, *etc*. Meeting all of these constraints often results in compressed timeframes where operators spend long working days trying to reap as much crop as possible. This is a special situation that greatly benefits from vehicle automation. Driving assistance from a semiautonomous mode certainly relieves the combine operator of the fatigue provoked by holding the steering wheel and staring at the referencing rows continuously for several hours. Alleviating the driver leads to better working conditions, improvements in the task performed, and a significant reduction in the risk of a dramatic accident. This case study proposes the development of a stereo-based navigation system that provides combine drivers with steering assistance during regular corn-harvesting operations.

Harvesting corn is quite a rough activity for the delicacy of a digital camera. Dust, vibrations, and all kind of residues fly around the cutting header of the vehicle. Figure 5.8 shows different options for the best location of a stereo camera. Given that the objective of the algorithm is to detect the *cut–uncut edge*, which is always situated eccentrically with respect to the header centerline, camera placement close to the edge seems reasonable. However, affixing the sensor to the header puts it too close to the edge, and in addition to the potential adverse effects of dust and residues, the field of view is not favorable for stereo perception. Figure 5.35 illustrates this problem. Image (a) was taken with the camera mounted on the left side of the header, and image (b) shows the corresponding disparity image.

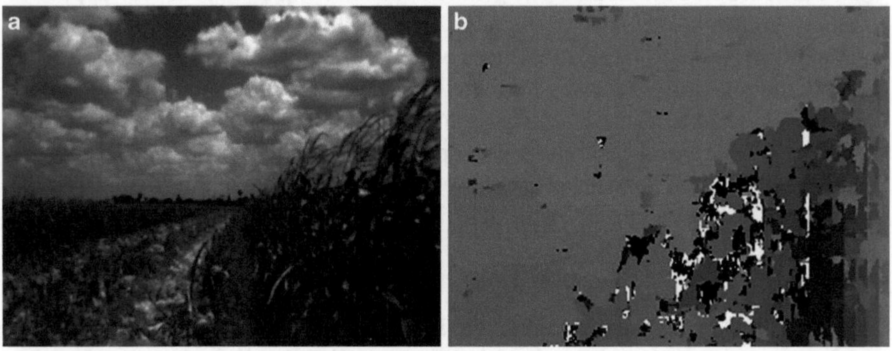

Figure 5.35 Difficult image (**a**) obtained when the camera is placed directly on the harvester header, and corresponding disparity image (**b**)

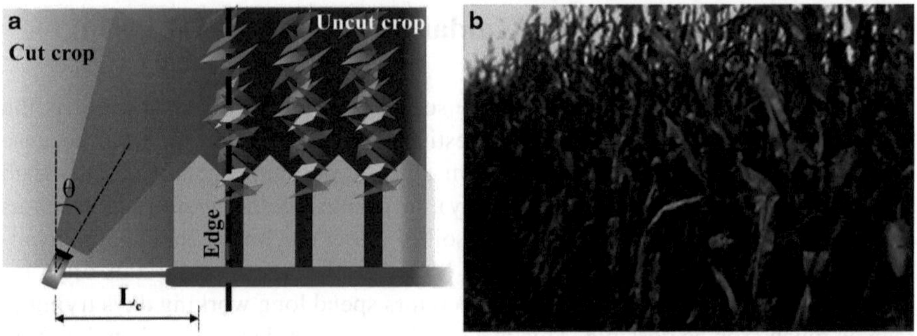

Figure 5.36 The camera's position and orientation on the corn harvester (**a**), and a sample image of the cut–uncut edge (**b**)

The solution proposed was to add an extension arm to the header and place the camera at its end, as indicated in Figure 5.36a. A sample image taken with a stereo camera featuring a 12 cm baseline and 6 mm lenses (BumbleBee, Point Grey Research Inc., Vancouver, Canada) is provided in Figure 5.36b. After testing several arrangements, the camera was fixed at a height of 1.66 m, an extension arm length L_c of 1.42 m, and a horizontal angle θ of 30°, as depicted in Figure 5.36a. As shown in the sample image, the horizontal field of view of the camera (42°) was sufficient to capture a significant portion of the edge row.

A reliable algorithm needs to be able to cope with complex situations, and a sparse edge might confuse the system if adjacent rows were captured by the camera. This situation actually happened in the field, but it was solved by implementing a *gradient operation.* The gradient enhances big changes, and the change from cut crop to standing crop is always greater than the gradient from the edge (first row) to the second row after the edge. Stereo data was processed with the density grids described in Section 5.5. The *gradient mask* of Equation 5.28 was applied to a density grid in the vicinity of the edge. The gradient mask enhances transitions, and therefore produces large differences between the maximum negative values and the maximum positive values, where the edge lies. This mathematical operation is particularly valuable when several rows are detected in the same image. Figure 5.37 shows just such a case, where the highest peak-to-peak difference in the 3D view (a) marks the position of the edge. The adjacent rows, noticeable in the 2D map (b), can easily be filtered or penalized by the algorithm before calculating the guidance commands.

$$\nabla d_{3D} = [\,-1 \quad -2 \quad 0 \quad 2 \quad 1\,] \tag{5.28}$$

The gradient increased the quality of edge detection, but the camera captures several frames per second, and not all of them are equally reliable for locating the edge. How can we establish a hierarchy based on the quality of the estimates? A *confidence index* can be defined to assess the reliability of the edge position in every processed image, so that every estimate is associated with a quality indicator that provides a confidence value for each solution. A confidence index (CI) was defined

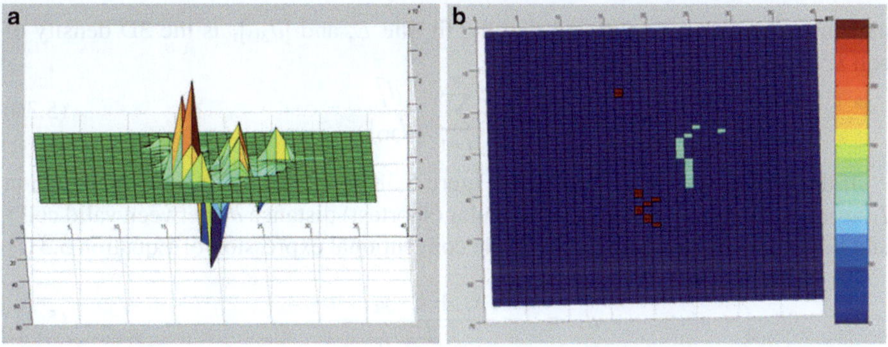

Figure 5.37 Edge enhancement by the application of a gradient mask: (**a**) 3D view; (**b**) 2D map

according to the expression given in Equation 5.29, where M_x is the moment index and C_x is the clustering index, both of which are defined in the following paragraphs.

$$\mathrm{CI} = M_x \cdot C_x \qquad \{M_x \in [0, 1]; \quad C_x \in [0, 1]\} \tag{5.29}$$

The confidence index is calculated on the density grid after applying the gradient operation. Only the *valid* cells that carry information about the edge will participate in the index calculation. These valid cells are above a certain threshold following the application of the gradient mask. The situation will be similar to the density grid of Figure 5.38, where three variables are known for each valid cell: x coordinate, y coordinate, and 3D density. The fit line L indicates the approximate position of the cut–uncut edge. Equation 5.30 gives the formula of the *moment* M_k calculated for image k, where n is the number of valid cells after thresholding, d_i is the perpen-

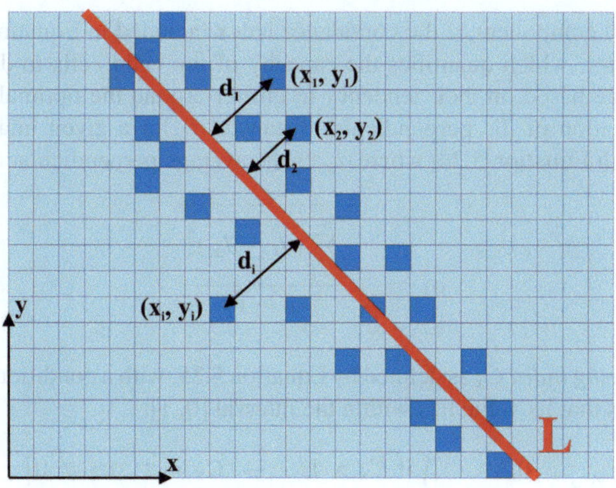

Figure 5.38 Calculation of the moment index M_x

dicular distance from valid cell i to the fit line L, and $[d_{3D}]_i$ is the 3D density of cell i:

$$M_k = \sum_{i=1}^{n} \frac{d_i^2}{[d_{3D}]_i} . \tag{5.30}$$

If the fit line represented by L in Figure 5.38 is defined by the general equation $L \equiv ax + by + c = 0$, the perpendicular (shortest) distance d_i between valid cell i and line L can be determined using the conventional expression of Equation 5.31:

$$d_i = \frac{|a \cdot x_i + b \cdot y_i + c|}{\sqrt{a^2 + b^2}} . \tag{5.31}$$

The moment index M_x is, by definition, a real number between 0 and 1. In order to normalize the moments calculated with Equation 5.30, the maximum value attainable for the moment M_k needs to be estimated for each image. The potential maximum value for the moment, $[M_k]_{\max}$, was estimated using Equation 5.32, where $d_{\max}$ is the maximum distance from cell i to line L considered in image k after thresholding, $[d_{3D}]_{\min}$ is the non-zero minimum density recorded among the n valid cells, and 2 is an arbitrary scaling factor:

$$[M_k]_{\max} = \frac{n \cdot d_{\max}^2}{2 \cdot [d_{3D}]_{\min}} \quad \{[d_{3D}]_{\min} \neq 0\} \tag{5.32}$$

The final expression for the *moment index* M_x is shown in Equation 5.33. It ranges between 0 and 1, and decreases when valid cells that hold low densities occur far from the fit line L, as the moment M_k will grow.

$$M_x = 1 - \frac{M_k}{[M_k]_{\max}} . \tag{5.33}$$

The second component of the confidence index included in Equation 5.29 is the *clustering index*, which quantifies the capacity of the valid cells to determine the cut–uncut edge based on their tendency to cluster around the optimal line L. The standard deviation of the perpendicular distances d_i for a given image k can be calculated with Equation 5.34, where d_{av} is the average perpendicular distance.

$$\sigma_k = \sqrt{\frac{\sum_{i=1}^{n} (d_i - d_{av})^2}{n-1}} \tag{5.34}$$

The *clustering index* C_x is defined by Equation 5.35, with a conditional constraint which ensures that its value lies within the interval [0, 1]:

$$C_x = 0.8 + \frac{0.2}{\sigma_k} \quad \begin{Bmatrix} \text{If } C_x > 1 & \rightarrow C_x = 1; \ \sigma_k \neq 0 \\ \text{If } \sigma_k = 0 & \rightarrow C_x = 1 \end{Bmatrix} \tag{5.35}$$

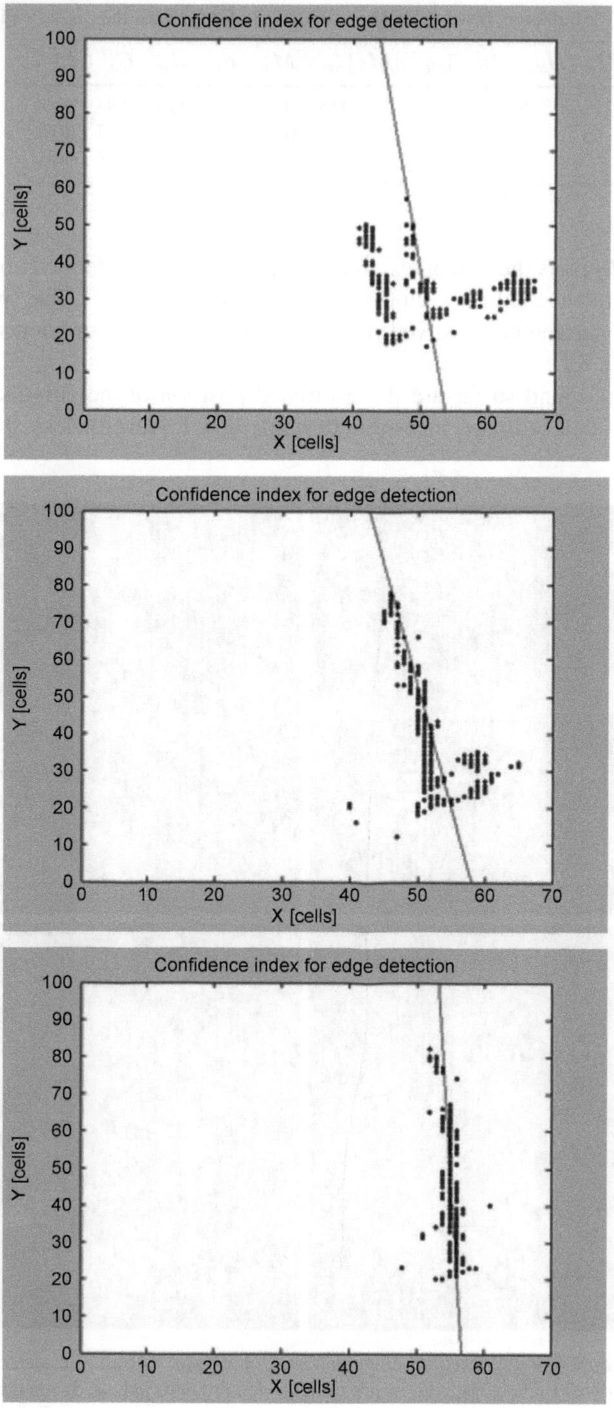

Figure 5.39 Confidence metrics for edge detection: (**a**) CI = 60%; (**b**) CI = 80%; (**c**) CI = 94%

Table 5.1 Basic parameters involved in the calculation of the CI for the cases of Figure 5.39

Case	n	M_k	d_{max}	$[d_{3D}]_{min}$	$[M_k]_{max}$	M_x	σ_k	d_{av}	C_x	CI
(a)	154	112	15	45	385	0.71	4.7	5.3	0.84	60%
(b)	168	33	13	34	417	0.92	2.9	1.5	0.87	80%
(c)	121	2	6	44	49	0.96	1.1	0.2	0.98	94%

Figure 5.39 plots the outcomes of the confidence metrics defined above for three different confidence indices CI: 60% (a), 80% (b), and 94% (c). The specific values of all of the parameters involved in the calculation of the confidence indices are listed in Table 5.1.

The results found so far are the estimated position of the cut–uncut edge and the quality of that estimate, measured through the CI. Nevertheless, the datum that

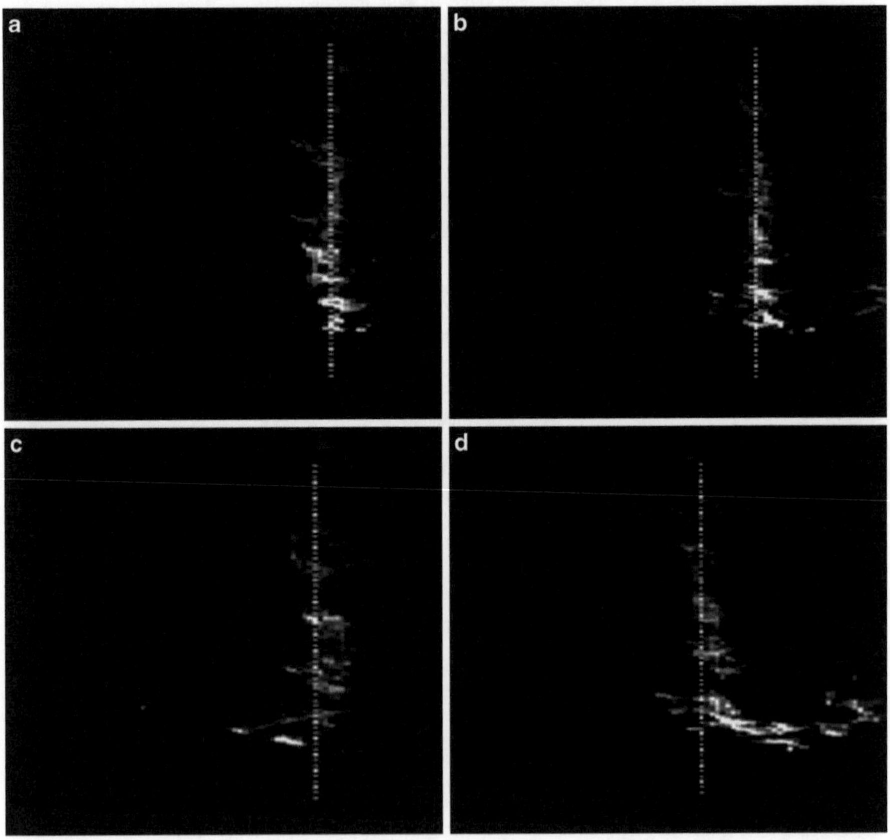

Figure 5.40 Results for the stereo-based guidance of a harvester with an edge-detection algorithm: **(a, b)** difficult images where the edge is not the only feature detected; **(c, d)** regular images where the edge is the only component containing cells with high 3D density

needs to be sent to the harvester controller is the offset of the vehicle. The *offset* can be defined as the lateral distance between the current position of the vehicle and its optimal position determined by the approximate location of the cut–uncut edge. Offsets were computed by analyzing the cumulative densities of rows of cells in the density grids and applying a gradient to the profiles found. The algorithm was tested on a John Deere 9660 STS corn harvester (Deere and Company, Moline, IL, USA) in October 2005. The combine was autonomously guided through a cornfield in central Iowa (USA) at speeds in the range 2.5–4 km/h. Figure 5.40 shows several graphical results of the edge detection algorithm. The first two images (a and b) are difficult images where the edge is not the only feature detected, but the algorithm still found the correct edge location. The other two examples (c and d) show the algorithm's response for regular images where the edge was basically the only component holding cells with high 3D density. Further information on this application can be found in [13]. An alternative solution would be to place the camera in the center of the vehicle to track crop rows as guidance features. This approach is developed in *Case Study VI*.

5.9 *Case Study II*: Tractor Guidance with Disparity Images

The objective set for *Case Study II* is the autonomous guidance of an agricultural vehicle, but unlike *Case Study I*, the stereo system was implemented on a medium-sized tractor, and the perception algorithm extracted feature locations directly from the disparity images rather than the 3D point clouds. As previously explained in Chapter 4, any operation on the input image that is carried out using the hardware saves computer resources for other processing tasks. This philosophy becomes more critical for real-time applications, such as the automatic steering system of this study case. It was concluded in Section 5.3 that the richer the disparity image, the better the 3D perceptive qualities of the system. However, the approach proposed here contradicts that statement to some degree, as it discerns the background soil by manipulating disparity images. This manipulation is done by turning the lens aperture ring and thus opening the lens diaphragm such that light saturates the soil in the stereo images. This saturation completely eliminates soil texture, and consequently the correlating algorithm filters out pixels relating to the ground. Conversely, crop features maintain their texture patterns and appear with valid disparity values in the disparity image, which provide their 3D position in the real scene. The key step is, therefore, to define the appropriate settings of the lenses to capture the crop rows and avoid the background. Figure 5.41 illustrates this idea. Image (a) is the aperture-biased initial image, and (b) is its corresponding disparity map, where the main rows are visible but the background soil has been eliminated. Compare this image with (c), which is the conventional disparity image obtained when the lens aperture is set for normal use. The 3D cloud is obviously much richer in the latter case, but the positions of the rows cannot be extracted directly from this disparity image.

Figure 5.41 Exposure-based segmentation for identifying crop rows in disparity images: (**a**) aperture-biased initial image; (**b**) corresponding disparity map; (**c**) conventional disparity image obtained when the lens aperture is set for normal use

Section 4.3.3 discussed the importance of limiting the area that needs to be processed by selecting a region of interest (ROI), thus reducing computing time and enhancing reliability. The project developed here implements regression techniques, and is therefore quite sensitive to the presence of outliers. For this reason, limiting the area processed by the machine vision algorithm is an essential task. Unlike the unique ROI shown in Figure 4.10, the analysis of the disparity image is applied to several areas of critical information, which are denoted *multiple regions* of *interest* (MROI). The criteria used to define the MROI can be enunciated as follows: every ROI should cover one crop row in such a way that its entire width is confined to the window. In reality, the number of regions will depend on the camera settings and field structure, but it will usually lie in the range 2–5. Evidently, in-field calibration is strictly required to determine the number and positions of the windows that best prepare the image for further analysis. Figure 5.42a is an example of a three-window MROI. Figure 5.45, on the other hand, shows an alternative design with two regions that actually guided a tractor in a soybean field. It is obvious that a fit line cannot be found directly by applying regression techniques to the whole row filling each ROI. Thus far, processing has been limited to a two-dimensional image, the disparity image, and so the techniques presented in Chapter 4 for monocular vision images can be applied here. The midpoint encoder (Figure 4.14), in particular, is particularly efficient at concentrating the information held by the blob represent-

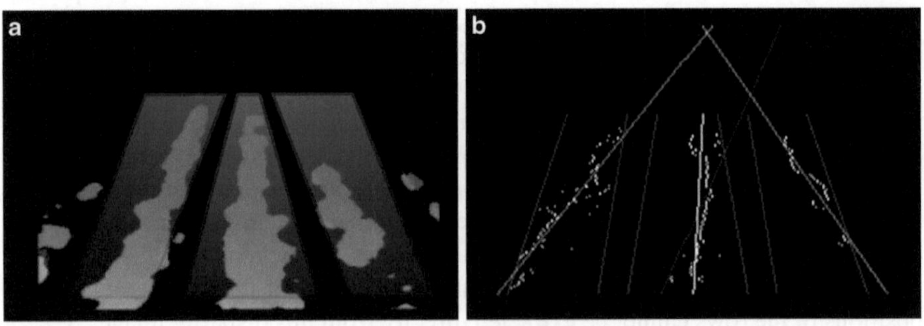

Figure 5.42 Selection of MROI (**a**) and window processing with regression analysis (**b**)

ing the row into a simplified set of points that indicate the centerline of the row on which the regression equations can be applied. Each region of interest j will be associated with a regression line whose general equation is given by Equation 5.36, where x_j and y_j are Cartesian coordinates that follow the conventional image space coordinate system but refer to window j, m_j is the slope of the fit line, and b_j is its y-intercept, which is only applicable to window j. The slope and y-intercept for the regression line found for window j are provided in Equations 5.37 and 5.38, respectively, where n_j is the number of midpoint encoded pixels of general coordinates (x_{ij}, y_{ij}). Figure 5.42b shows the result of applying the midpoint encoder followed by regression analysis to the disparity image of Figure 5.42a. Note that the central ROI yields an erroneous regression line, which is then discarded in order to calculate the central path by simply averaging the fit lines found for the two lateral windows.

$$y_j = m_j \cdot x_j + b_j \tag{5.36}$$

$$m_j = \frac{n_j \cdot \sum_{i=1}^{n_j} x_{ij} \cdot y_{ij} - \sum_{i=1}^{n_j} x_{ij} \cdot \sum_{i=1}^{n_j} y_{ij}}{n_j \cdot \sum_{i=1}^{n_j} x_{ij}^2 - \left[\sum_{i=1}^{n_j} x_{ij}\right]^2} \tag{5.37}$$

$$b_j = \frac{\sum_{i=1}^{n_j} y_{ij}}{n_j} - m_j \cdot \frac{\sum_{i=1}^{n_j} x_{ij}}{n_j} \tag{5.38}$$

The side trend lines plotted in Figure 5.42b efficiently locate the positions of the crop rows detected in the disparity image. However, there is no guarantee that this will happen for all of the windows of images grabbed by the stereo camera, as seen for the central window above. Indeed, it is relatively common to get an erroneous line in one of the regions of interest. In order to prevent mixing correct data with poorly determined trend lines, a minimum user-selected correlation factor r^2 must be fulfilled by any fit line before it is allowed to participate in the estimation of the guiding path. Lines that do not satisfy this evaluation criterion are discarded; the remaining lines are used to compose the perspective path that guides the tractor. The minimum correlation factor set in this case study was 0.7. Given that regression lines may be disqualified before the *central line* is estimated, a set of rules must be defined for the optimal combination of the eligible lines [14].

The central line sets the desired trajectory for the vehicle, but the tractor needs to calculate the autosteering commands. To do so, the position of the target point must be determined. The *target point* is the point in space ahead of the vehicle towards which the tractor is guided to complete a task. In this project, the target point is chosen at the intersection of the center path with the upper boundary of the regions of interest. Theoretically, the target point will have a position in image space, T $(x_i^{\mathrm{T}}, y_i^{\mathrm{T}})$, and a corresponding 3D location in world space $(x^{\mathrm{T}}, y^{\mathrm{T}}, z^{\mathrm{T}})$. In reality, overexposed images such as the guidance scene of Figure 5.41 possess significant filtered areas, typically mapped in black, where no disparity information is avail-

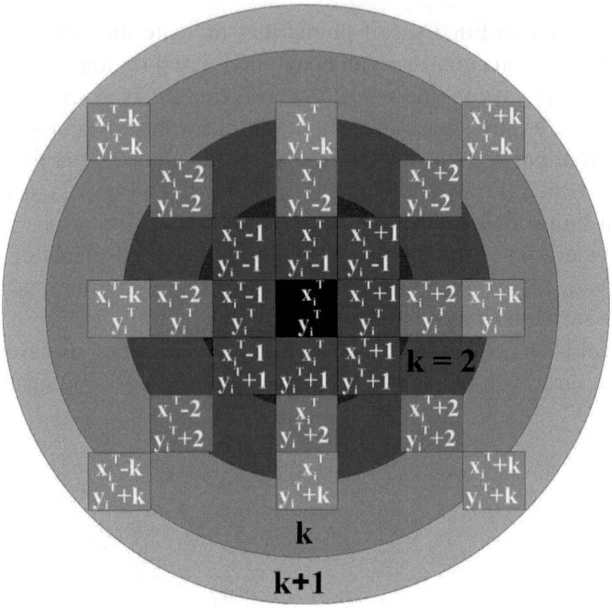

Figure 5.43 Pick-point algorithm used to determine the target point

able. This fact poses a challenge when transforming image point T from image space to world space. To solve this problem, the *pick-point algorithm* of Figure 5.43 was developed. This procedure searches in the neighborhood of point T until pixels carrying disparity information are found. It then averages their positions to find a suitable target point. The algorithm works on expanding rings originating from position T, also called index 0 or i[0]. Every new layer explores eight surrounding pixels, as graphically represented in Figure 5.43 and mathematically defined in Equation 5.39. The maximum number of layers k is set by the user, and the positions of the eight pixels within each layer are each labeled with a unique number between 0 and $(H \cdot V - 1)$, as established by Equation 5.39, where H is the horizontal resolution of the disparity image and V is its vertical resolution. If a valid disparity is not found after exploring the k layers, the search is terminated and an error message is issued. When the initial pixel at point T with an assigned index i[0] has a valid disparity, there is no need to execute the pick-point algorithm; however, to enhance robustness, the first layer ($k = 1$) is scanned and the detected pixels are averaged with point T. The output of the pick-point algorithm is the position of the target point T in both image and world coordinates. The world coordinates are primarily expressed as camera coordinates, but they can easily be transformed to ground coordinates through Equation 5.26. Once expressed in ground coordinates, the relationship between target point location and front wheel steering angle can be established by Equation 4.14 or a similar geometrical relation.

$$\begin{bmatrix} i\,[0] \\ i\,[1] \\ i\,[2] \\ i\,[3] \\ i\,[4] \\ i\,[5] \\ i\,[6] \\ i\,[7] \\ i\,[8] \end{bmatrix} = y_i^{\mathrm{T}} \cdot H \cdot \begin{bmatrix} 1 \\ 1 \\ 1 \\ 1 \\ 1 \\ 1 \\ 1 \\ 1 \\ 1 \end{bmatrix} + k \cdot H \cdot \begin{bmatrix} 0 \\ -1 \\ -1 \\ -1 \\ 0 \\ 0 \\ 1 \\ 1 \\ 1 \end{bmatrix} + x_i^{\mathrm{T}} \cdot \begin{bmatrix} 1 \\ 1 \\ 1 \\ 1 \\ 1 \\ 1 \\ 1 \\ 1 \\ 1 \end{bmatrix} + k \cdot \begin{bmatrix} 0 \\ -1 \\ 0 \\ 1 \\ -1 \\ 1 \\ -1 \\ 0 \\ 1 \end{bmatrix} \tag{5.39}$$

Conventional linear regression methods are known to find it difficult to process steep lines with slopes close to 90°. Very often, the row detected by a central region of interest – such as the one shown in Figure 5.42 – is close to vertical, and so can be a serious source of errors. A *shifted line regression* technique has been developed to cope with near-vertical lines. The midpoint encoded points of problematic lines are shifted by 90° to lie in a horizontal position before the regression line is calculated. The linear equation of the shifted (horizontal) line produces correct fits as its slope is very far from being vertical. A 90° back-transformation of the correct line yields the true direction of the crop row. Vertical lines are expected in central regions of interest, and low correlation coefficients are typically a sign of upright point positions. The suggested procedure for dealing with this situation is as follows. The fit line for the quasi-vertical set of points is incorrect. However, the fit line found after shifting the x and y coordinates must be correct. This 90°-shifted line is characterized by Equation 5.40, and can be calculated using the modified slope m_j^S of Equation 5.41 and the new y-intercept b_j^S of Equation 5.42.

$$x_j = m_j^S \cdot y_j + b_j^S \tag{5.40}$$

$$m_j^s = \frac{n_j \cdot \sum_{i=1}^{n_j} x_{ij} \cdot y_{ij} - \sum_{i=1}^{n_j} x_{ij} \cdot \sum_{i=1}^{n_j} y_{ij}}{n_j \cdot \sum_{i=1}^{n_j} y_{ij}^2 - \left[\sum_{i=1}^{n_j} y_{ij}\right]^2} \tag{5.41}$$

$$b_j^S = \frac{\sum_{i=1}^{n_j} x_{ij}}{n_j} - m_j^S \cdot \frac{\sum_{i=1}^{n_j} y_{ij}}{n_j} \tag{5.42}$$

The line defined by Equations 5.40–5.42 is horizontal and needs to be shifted back to its original vertical position. The equation for the line that will determine the vehicle's path is given in Equation 5.43, and is completely defined by the slope m_j^* of Equation 5.44 and the y-intercept b_j^* of Equation 5.45. Note that the perpendicularity condition stated in Equation 5.46 is met. Figure 5.44 shows the results of line shifting for the central region of interest: (a) plots the results of conventional re-

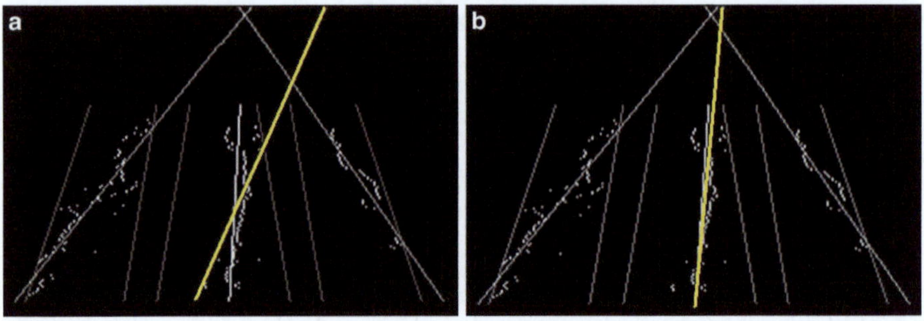

Figure 5.44 Results of 90° line shifting for regression on vertical lines: (**a**) original; (**b**) shifted

gression, and (b) gives the correction carried out after applying the proposed *shifted line regression* algorithm. Note the accuracy of the fit for the shifted line, which coincides with the *central path* resulting from averaging the three lines. Interestingly enough, the correlation coefficient r^2 does not vary, and although it represents a good fit, it will remain as low as it was before the transformation.

$$y_j = m_j^* \cdot x_j + b_j^* \tag{5.43}$$

$$m_j^* = \frac{n_j \cdot \sum_{i=1}^{n_j} y_{ij}^2 - \left[\sum_{i=1}^{n_j} y_{ij}\right]^2}{n_j \cdot \sum_{i=1}^{n_j} x_{ij} \cdot y_{ij} - \sum_{i=1}^{n_j} x_{ij} \cdot \sum_{i=1}^{n_j} y_{ij}} \tag{5.44}$$

$$b_j^* = \frac{\sum_{i=1}^{n_j} y_{ij}}{n_j} - m_j^* \cdot \frac{\sum_{i=1}^{n_j} x_{ij}}{n_j} \tag{5.45}$$

$$m_j^S \cdot m_j^* = 1 \tag{5.46}$$

The navigation engine based on the analysis of disparity images was implemented on a medium-sized tractor for field testing in Urbana (Illinois, USA) in October 2003. Visual features were provided by 100 m long rows of soybeans. The stereo camera (Mega-D, Videre Design, Menlo Park, CA, USA) had a 9 cm baseline and infrared filters with a 880 nm center wavelength and a 40 nm bandwidth. The initial lenses had a focal length of 7.5 mm, but on-site tuning revealed that the outcomes improved with 12.5 mm lenses. This simple modification left only two complete rows within the image (Figure 5.45a), and resulted in a new definition of the MROI; specifically, the initial three windows were substituted for a two-window approach, as shown in Figure 5.45b. The location of the target point was still set at the intersection of the guidance directrix and the top limit of the regions of interest, as pictured in Figure 5.45c. A comparison of the trajectory followed by an operator and the path traveled autonomously by the tractor, both recorded by an RTK-GPS

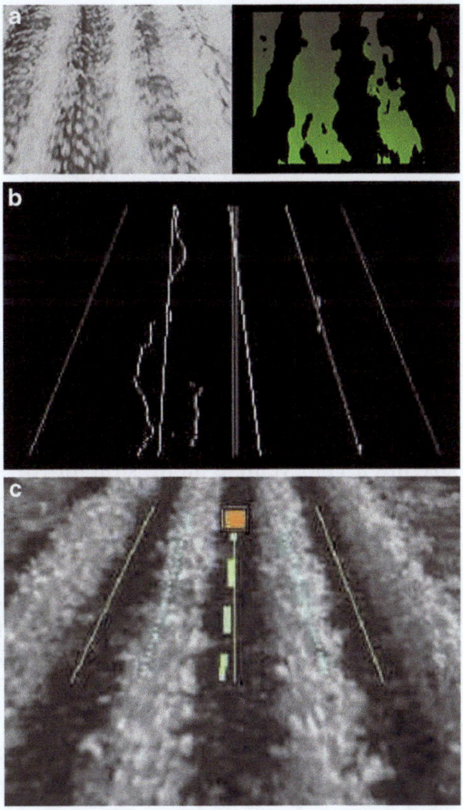

Figure 5.45 Stereo-based guidance system adapted to a soybean field

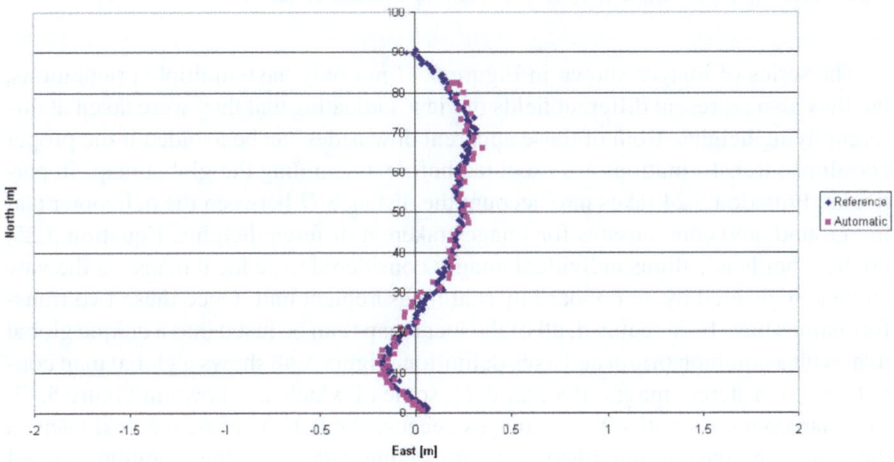

Figure 5.46 System evaluation achieved by comparing manual and automated trajectories

receiver, is provided in Figure 5.46. As shown in the figure, the trajectories of man and machine typically differed by less than 5 cm. More insights into this approach are available in [14].

5.10 *Case Study III*: 3D Terrain Mapping with Aerial and Ground Images

The fundamentals of global 3D mapping were described in Section 3.6. This case study shows how to apply that theory to real field scenes with both aerial and ground images. The aerial images used to construct the 3D global map developed in this section were acquired from the remote-controlled helicopter of Figure 5.28b, flying approximately ten meters over agricultural fields in Sapporo (Japan) in June 2002. Driving a land vehicle along straight rows is a much easier task than flying a remotely operated helicopter above crop rows, especially if the operator is located in a corner or at the side of a long field. In fact, once a set of aerial stereo images have been acquired, the first problem to address before they can be merged to construct the global map is their random orientation. Figure 5.47 illustrates this problem: all four aerial images of a corn field have different orientations. These images were captured with a 9 cm baseline stereo camera of resolution 320 × 240 supporting lenses of focal length 7.5 mm.

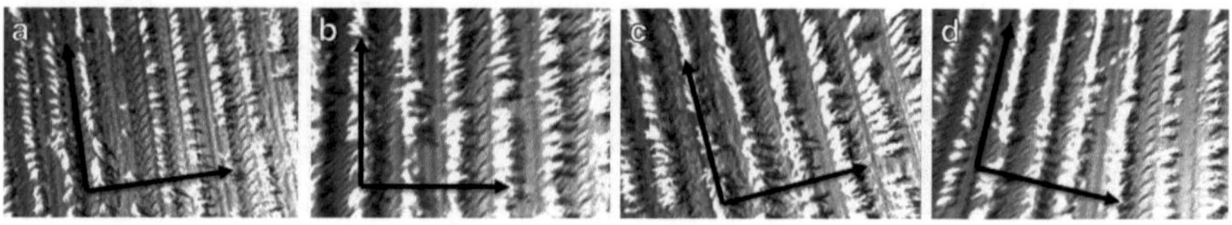

Figure 5.47 Airborne images of a cornfield showing multiple orientations

The series of images shown in Figure 5.47 not only have multiple orientations, but they also represent different fields of view, indicating that they were taken at different flying heights. Both of these apparent downsides can be avoided if the proper coordinate transformations are executed before assembling the global map. In particular, Equation 5.24 takes into account the distance D between the helicopter and the ground, and compensates for images taken at different heights. Equation 5.25, on the other hand, aligns individual images considered to be local maps via the yaw angle φ registered by an onboard inertial measurement unit. Once these two transformations have been realized, all of the local maps can be fused into a unique global map with a common origin and axes definition. Figure 5.48 shows a global map consisting of ten stereo images of a cornfield, some of which are shown in Figure 5.47. The map contains relatively few images because the helicopter flew quickly across the region and the computer had problems saving data due to the vibrations caused by the rotor. Nevertheless, it provides useful real-time production information, such

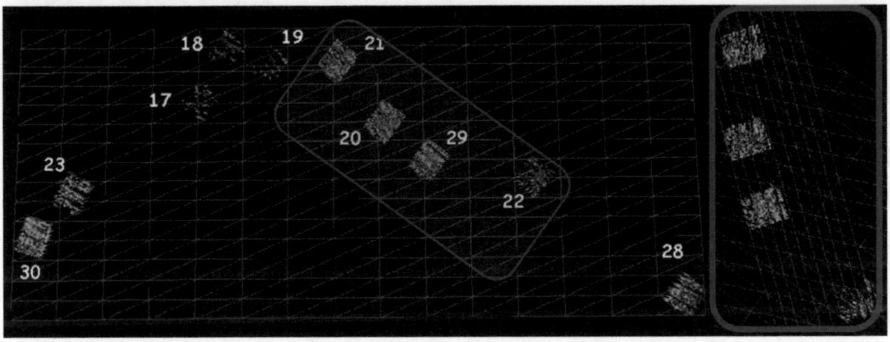

Figure 5.48 Virtual 3D map of a maize field generated from aerial stereoscopic images

as the distance between rows, the actual locations of the corn rows, the height of the crop, *etc*. A complete report of this project can be found in [10].

The applications of 3D terrain mapping are multiple: from the mere registration of field information as an aid to production tasks to planning paths taken during autonomous missions tackled by off-road vehicles. The difficulties involved in maintaining a constant camera height while capturing airborne imagery is not an issue for ground images acquired with a land vehicle; nonetheless, the perception quality of each stereo cloud and the proper arrangement of local maps are two fundamental matters that must be addressed. Unlike aerial images, where the electronic sensor array forming the image is parallel to the ground, cameras mounted on land vehicles tend to be tilted down for a better adjustment of the field of view to the field scene. The transformation of Equation 5.26 requires a good estimate of the inclination angle ϕ, and Equation 5.27 needs measurements for the pitch, roll, and yaw angles to correctly assemble the global map. However, when all of these parameters are

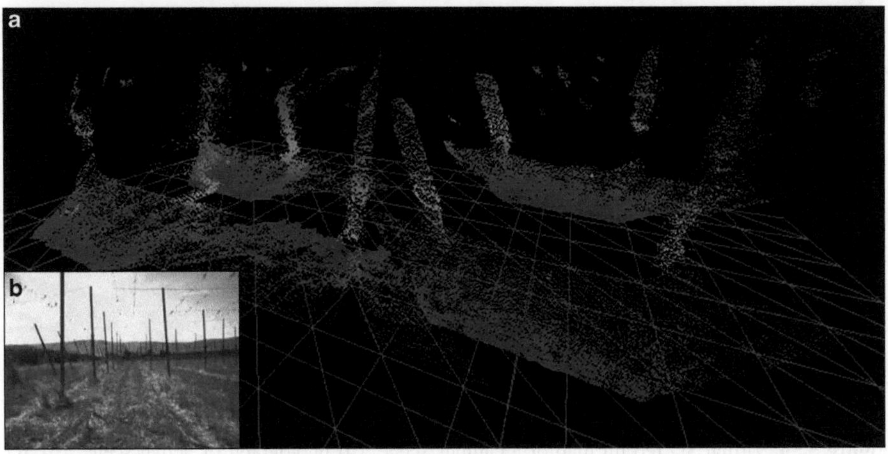

Figure 5.49 Virtual 3D global map (**a**) of a barren field with crop-supporting structures (**b**)

appropriately determined, the results are coherent. Figure 5.49 offers a perspective view of a global map (a) of a barren field with wooden posts that were prepared to hold hops, a twining vine (b). The spacing between the rows and between the posts within a row, the type of post (vertical or inclined), and the heights of the posts can be determined from the virtual 3D map. An intelligent planting system, for example, might utilize this information to determine where the plants should be planted.

Once a global map has been created, it can be reused for navigational purposes. This is not easy when only local references are available, but the introduction of global satellite positioning automatically allows the position of any feature in the field to be defined, as well as that of the mapping vehicle, in a global frame. A vehicle traversing unknown terrains can compose a map of the environment it perceives in a primary exploratory mission, but given the global properties of the map, it can be used and enhanced by other intelligent vehicles. The 3D global map of Figure 5.50b was generated by a utility vehicle driven along a sidewalk of a university campus. The photograph (a) of the scene shows the key features found by the mapping vehicles: the sidewalk and lateral rows of trees. The resulting global map indicates that the width and alignment of the concrete path were consistent in all of the images, which implies accurate behavior of the inertial measurement unit as well as the global positioning system. In addition to positioning information, this 3D map also codes the color components of the detected pixels such that the true color of the pathway is clearly distinguished from the surrounding trees and turf (represented by green hues). A navigation engine may associate pathway colors with traversable terrain, and other tones such as the intense red of a car with potential obstacles. The front view of the terrain map also provides the heights and exact positions of the side trees, indicating traversability from a vertical standpoint, which could force the vehicle to make a detour in order to avoid a protruding branch. The views portrayed in Figure 5.50b were chosen arbitrarily, but the information represented by this 3D map is a very powerful tool, and any conceivable vision, perspective, or point of view can be rendered. The most useful view will depend on the application devel-

Figure 5.50 Virtual 3D terrain map of a campus scene: (**a**) photograph of actual scene; (**b**) 3D global map; (**c**) field map

oped. The field map of Figure 5.50c consists of 20 images and a total of 492,428 points. Handling such a massive set of data in real time can be difficult if the perception engine processor is not very powerful. However, with optimized tools, the potential is almost unlimited. This map was rendered in the Immersive Visualization Laboratory at the John Deere Technology Center (Moline, Illinois, USA) where people wearing 3D vision glasses could actually walk through the virtual sidewalk and contemplate the trees at the sides. More information regarding ground 3D mapping is available in [11].

5.11 *Case Study IV*: Obstacle Detection and Avoidance

A natural consequence of detecting ranges in real time without the need to scan a beam is a preference for stereo vision as an obstacle-detection sensor, but it is one thing to capture obstacles in an image, and a very different one to understand that an object is interfering with the trajectory of the vehicle. Therefore, it is essential to explore the capacity of stereo to safely identify obstacles. Indeed, there are many kinds of obstacles, and not all of them have the same importance. The most important obstacles to be aware of are, obviously, people. Although intelligent vehicles are now being constructed, it is still unusual to find them outside of the lab, so people naturally assume that any moving machine they encounter is being controlled by a human operator, meaning that they expect the machine to react correspondingly. How is a person defined in a 3D point cloud? It is evident that choosing the camera parameters appropriately is vital for successful perception, as discussed in Section 5.4. Figure 5.51 shows a person standing in an orchard lane between two rows of trees (a), and its corresponding 3D virtual representation (b). Any clue that can aid the identification of the obstacle is welcome, and the real colors of the scene are useful for discriminating the background (the trees and the lane) from the stand-

Figure 5.51 Person detection in a conventional field scene: (**a**) actual scene; (**b**) corresponding 3D virtual representation

Figure 5.52 Perception of a person's movement within the field of view of a vehicle

ing person. A front view of this scene reveals that there is "something" in the middle of the lane, and it is high enough that it cannot be overcome by the moving vehicle.

The scene shown in Figure 5.51 corresponds to a unique shot: the camera captured an image in which an obstacle has been detected, but what happens next? The stereo camera can calculate the approximate distance between the vehicle and the obstacle; that is, the range. The vehicle is then likely to cease its motion as a safety precaution, but the camera can track the movement of the person to infer if they are approaching or moving away from the vehicle. The former case will probably cause the vehicle to halt; the latter might allow it to continue its mission at a slower pace. This way of actuating requires a higher degree of intelligence, and should constitute the reasoning rules and embedded behaviors of future intelligent mobile machines. Figure 5.52 is the composite of four pairs of stereo images that show the course traced by a person overtaking the vehicle carrying the camera. The person appearing in the lower left corner of the image is probably too close to the vehicle for safe operation; however, by the time the person is at the furthest position from the vehicle, motion could be resumed at a moderate speed. The unambiguous identification of the path due to its color is also an advantage when proceeding with a "smart move."

Apart from detecting people, it is also important to be aware of any obstacles interfering with the vehicle's planned path. The size of the minimum detectable obstacle determines the characteristics of the grid. Figure 5.53c shows a photograph of a common situation found in an orchard: during picking season, boxes are piled up in the lanes between tree rows, which are precisely the paths that vehicles must follow. Consequently, vehicles, pickers, ladders, boxes, and so on coexist in the same environment. The 3D virtual representation of the scene (Figure 5.53a) clearly identifies the positions, shapes, and sizes of the fruit boxes. The navigation engine must adjust its behavior to the new situation and employ the free space left between piled boxes for traveling.

Figure 5.53 Detection of a static obstacle in an orchard: (**a**) 3D virtual representation of scene; (**b**) field map; (**c**) actual scene

The first part of the objective, *detecting the obstacle*, can be successfully accomplished with only the stereo system, as demonstrated in Figures 5.51–5.53. The second part, *avoiding the obstacle*, also requires the integration of a *path planner* to determine the optimum vehicle trajectory (covered in Section 9.6) and the design of a control unit to execute navigation commands. Figure 5.54 illustrates the successful implementation of an obstacle avoidance algorithm in an intelligent utility vehicle. A 22 cm baseline camera was attached to the front bumper of the vehicle at 1.5 m above the ground, resulting in a field of view of approximately 8 m width and 20 m depth.

The stereo perception engine implemented in the vehicle of Figure 5.54 was based on the approaches described in Section 5.5: 3D density and density grids. Due to a severe drop in 3D density for ranges of more than 8 m, a compensation function similar to that shown in Equation 5.23 was also incorporated into the stereo algorithm. The stereo images captured by the camera had a resolution of 400 × 300, and the resulting point clouds were handled through top-view configured density grids that processed all of the data in the slice defined from 30 cm to 2 m. Cell sizes of be-

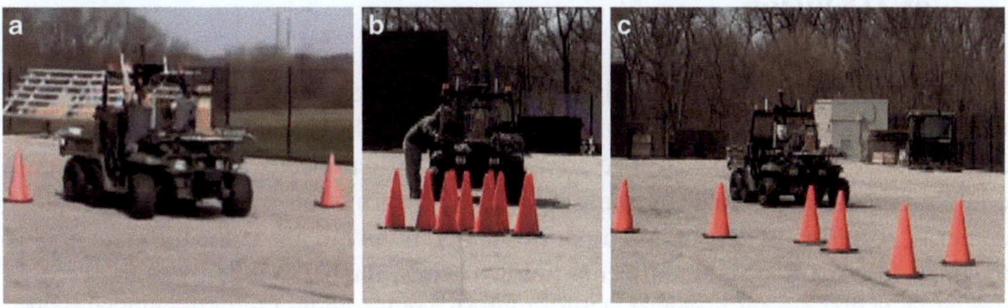

Figure 5.54 Intelligent vehicle performing obstacle avoidance tests

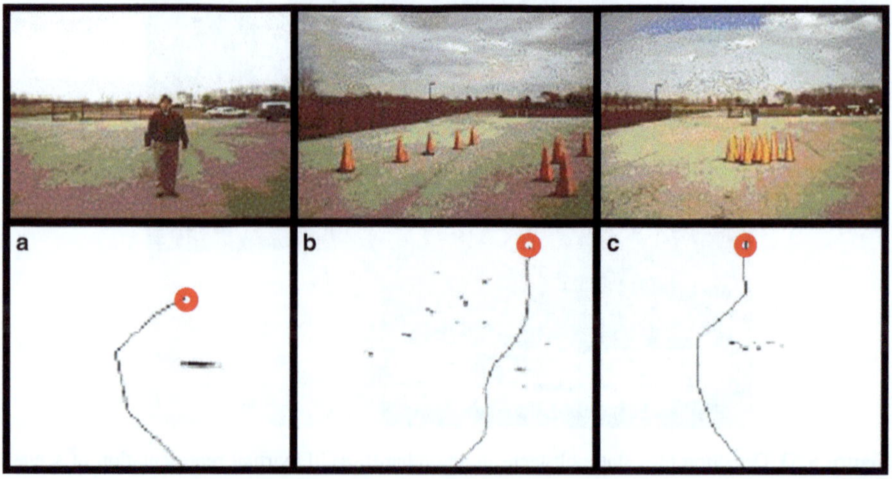

Figure 5.55 Obstacle avoidance tests for a utility intelligent vehicle

tween 125 and 150 mm were adequate to detect people and 1 m high plastic cones, as illustrated in the images of Figure 5.55, taken during the tests. The objective of the first run (Figure 5.55a) was to drive the vehicle autonomously to the target point situated 10 m ahead of the camera while avoiding a person standing in the middle of the trajectory. The vehicle arrived safely after following the trajectory drawn in Figure 5.55a, using a grid span of $6 \times 20\,\mathrm{m}^2$ and 15 cm cells. In the second test, the vehicle had to navigate along a corridor defined by plastic cones and reach a destination point located at ground coordinates (2, 16, 0) m. The course traced by the vehicle to meet the mission goal is shown in the central image (b). The third obstacle avoidance challenge (c) placed the objective approximately 18 m from the vehicle, with its trajectory blocked by a set of plastic cones grouped together. As shown in the figure, the vehicle deviated around the obstacle and resumed its direction straight to the objective set beforehand.

5.12 *Case Study V*: Bifocal Perception – Expanding the Scope of 3D Vision

The goal set for *Case Study V* is to demonstrate that *bifocal stereo heads* (Section 5.2, Figure 5.12) provide a richer and more robust level of perception than conventional stereo cameras. The advantage of bifocal heads arises from the data complementation obtained by merging the 3D information from both constituting cameras, resulting in denser and more populated point clouds. If the ultimate goal is to sense at two different but adjacent fields of view, the first challenge is to perceive two objects that are set far apart. This situation can be studied with the experimental layout displayed in Figure 5.56. As depicted in the figure, a person stood in front

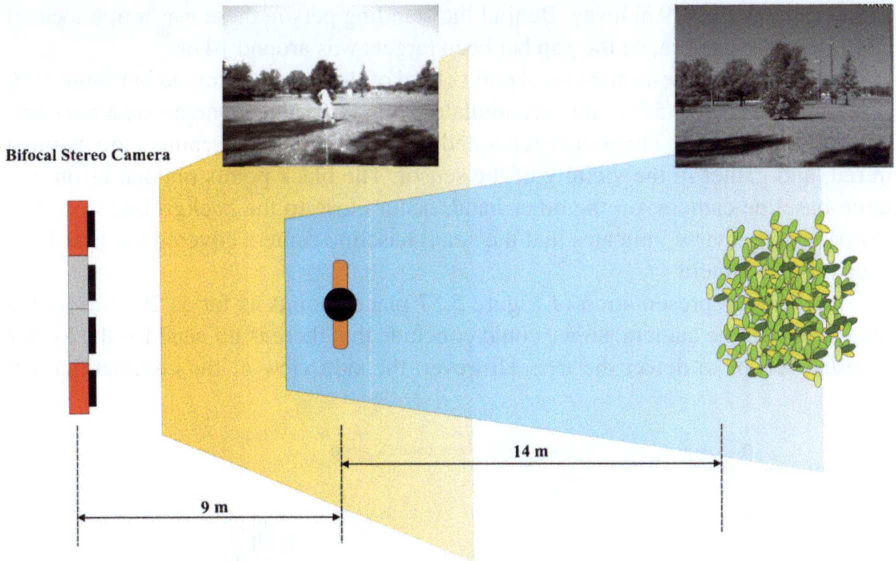

Figure 5.56 Simultaneous object detection at two separated locations with bifocal stereo

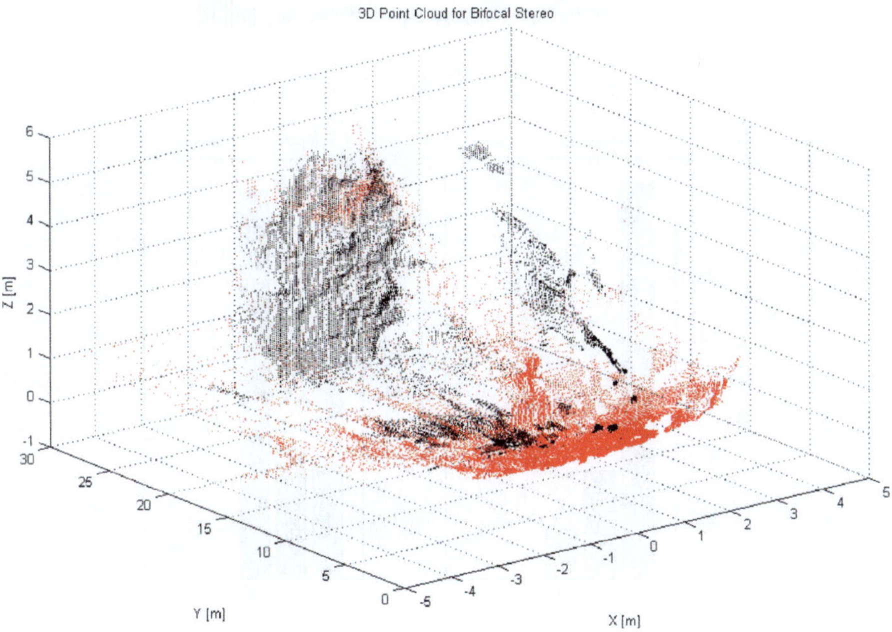

Figure 5.57 Perspective view of the merged point cloud for the scene represented in Figure 5.56

of the camera about 9 m away. Behind the standing person there was a tree located 23 m from the camera, so the gap between targets was around 14 m.

The resulting three-dimensional point cloud of the scene portrayed in Figure 5.56 is plotted in Figure 5.57. Points accumulate around two general areas: near the camera and near the tree. The points generated from the short-range camera are mapped in red, and gather in the vicinity of the sensor. The black points obtained from the large-baseline camera, on the other hand, occur close to the background tree. The 3D perspective view indicates that the small baseline camera covered the first 15 m ahead of the camera.

The virtual representation of Figure 5.57 places points as far as 20 m away for the short baseline camera, so we could conclude that there is no need for the longer baseline sensor to detect the tree. However, the side view of the complete cloud,

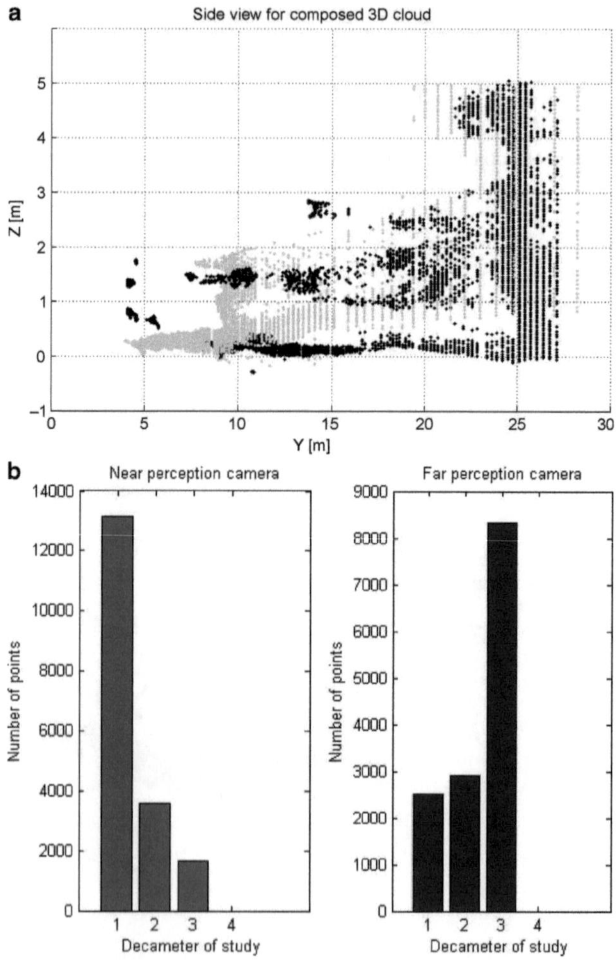

Figure 5.58 Side view (**a**) and distribution of points (**b**) for a bifocal camera 3D cloud

represented in Figure 5.58a, reveals a hard reality: the space between both objects (empty space in actuality) is full of noisy and misleading points. In other words, this result implies that – as anticipated – the short-range camera yields unacceptable points for distances of >10 m, whereas the other camera generates erroneous readings at near ranges. In conclusion, each camera arrangement should be utilized for a particular recommended range, beyond which poor perception occurs. The numerical justification for how well both camera assemblies complement each other can be found in the bar chart of Figure 5.58b. The points found in each of the first three decameters from the camera are separated out according to the camera that originated them. As expected, the first decameter mostly contains points from the near-perception camera, but those in the third decameter were usually perceived with the far-perception camera. The total number of points in the cloud constitutes the "critical mass" for the scene, from which features can be extracted and decisions made. What is important, though, is the source of each point; in other words, whether each matched point was detected with the most reliable and appropriate sensor available.

The scene studied in the previous case focused on two objects separated by a distance of approximately 14 m; what lay between them was of no particular interest. This objective could mask a lack of continuity in the composed point cloud for the medium ranges located between both objects. Good perception implies a smooth and consistent transition across the boundary between the fields of view of both

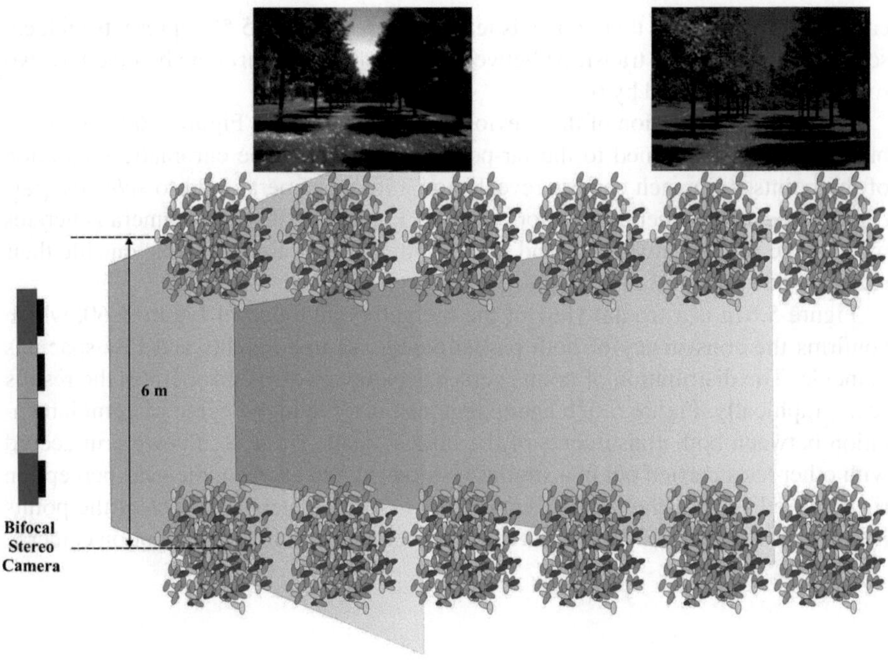

Figure 5.59 Field experiment focusing on the transition between fields of view

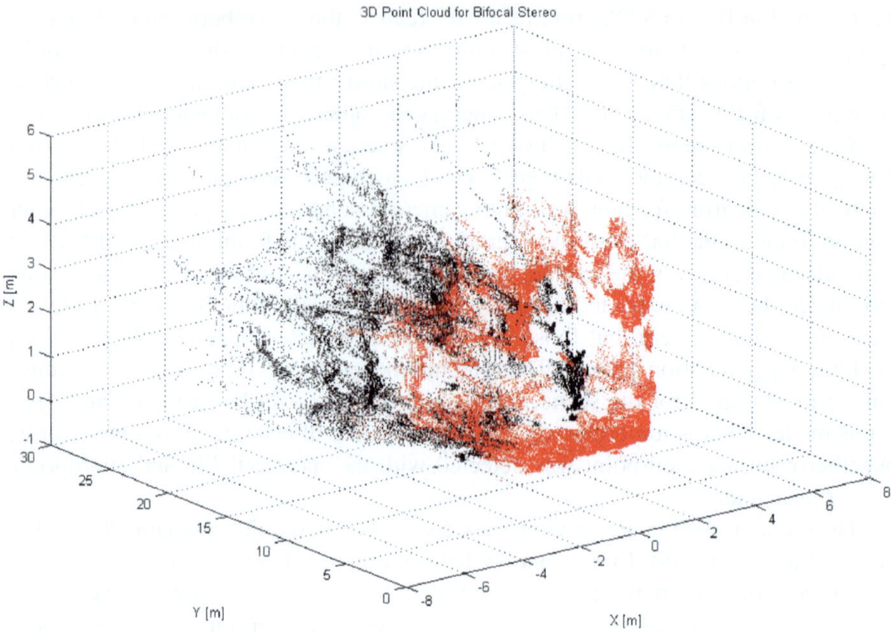

Figure 5.60 Perspective view of the merged point cloud for the scene represented in Figure 5.59

cameras in such a way that no gap is left uncovered. Figure 5.59 represents an ideal scene for checking the transition between fields of view: a turf lane bounded by two rows of trees separated by 6 m.

The 3D representation of the previous scene is plotted in Figure 5.60, where the black points are assigned to the far-perception camera. The chromatic separation of the points from each camera reveals how bifocal cameras lead to *selective perception*: the near-perception sensor covers 5–12 m while the other camera generates most of the points between 12 and 20 m. Both cameras are unreliable outside their recommended interval of perception.

Figure 5.61a is a frontal view of the merged point cloud of Figure 5.60, which confirms the consistency of both partial clouds, as tree heights and row spacings coincide. The distribution of points versus decameters also corroborates the results seen graphically. Figure 5.61b again demonstrates the high degree of complementation between both constituents of the bifocal head. These results were in accord with other tests carried out in a similar fashion [5]. On average, the near-perception rig acquired 74% of the points found in the first ten meters, but 78% of the points located between 10 and 30 m from the head belonged to the far-perception camera.

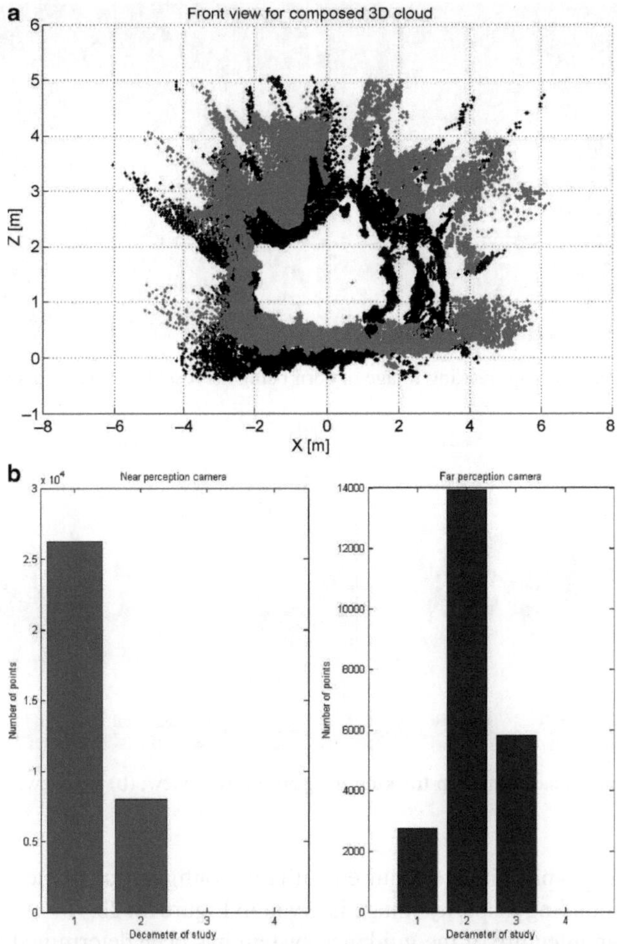

Figure 5.61 Front view (**a**) and distribution of points (**b**) for a bifocal camera 3D cloud

5.13 *Case Study VI*: Crop-tracking Harvester Guidance with Stereo Vision

The methodology discussed in *Case Study I* focuses on the use of the cut–uncut edge of the crops being harvested as a guidance feature to steer a combine. *Case Study VI*, on the other hand, involves guiding a harvester by detecting the rows ahead of the vehicle; that is, by means of a *crop-tracking algorithm*. The advantage of placing the sensor at the center of the combine is that it is independent of the particular side being cut, which alternates according to the direction of travel. In this study, the stereo camera was mounted at the center of the harvester cabin, where it looked forward and was tilted down to focus on the corn rows ahead of the combine.

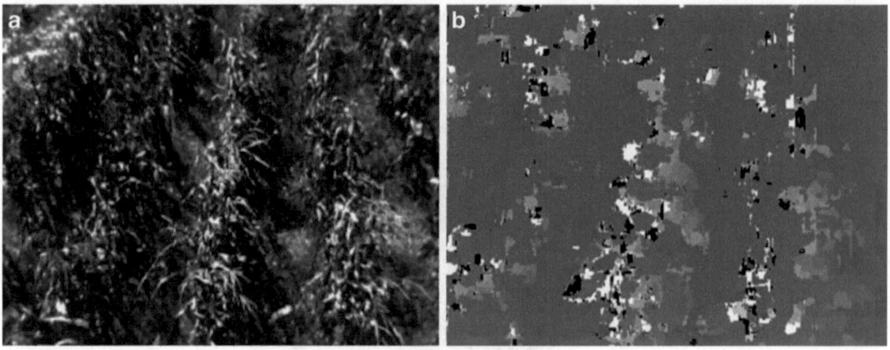

Figure 5.62 Standard crop-tracking image of corn being harvested (**a**) and its disparity map (**b**)

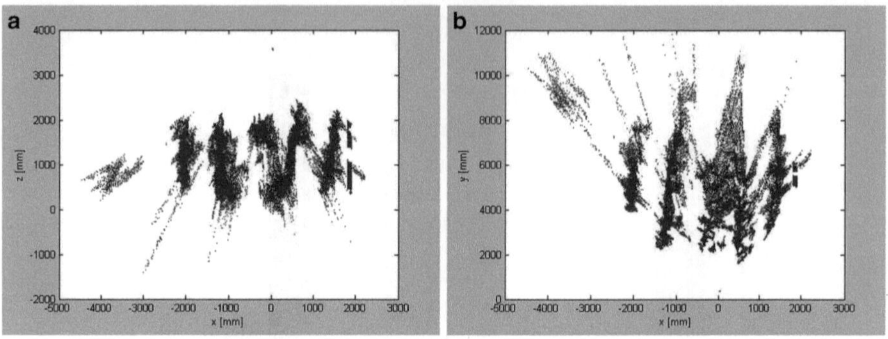

Figure 5.63 Point cloud from crop-tracking images: (**a**) front view; (**b**) top view

Figure 5.62a is a typical image acquired with this configuration of the stereo system, and its corresponding disparity image is given in Figure 5.62b.

Once the architecture of the guidance system has been determined, the processing algorithm initiates its task by acquiring stereo images of the crop-tracking scene and calculating the disparity image (Figure 5.62) that will lead to the 3D point cloud. Figure 5.63 plots front (a) and top (b) views of the point cloud generated from the image shown above. This is the first check on the perceptive quality of the system, and the scene portrayed should faithfully represent reality. The point clouds reveal that there are five rows of corn ready for tracking and that the field of view is approximately five meters wide and covers ranges of between two and ten meters. According to the front view of Figure 5.63a, the corn height was about two meters, which is a reasonable estimate for corn at harvesting time.

The unstructured discrete data of Figure 5.63 needs to be processed in order to find the optimum *guidance directrix*. The concept of *3D density* and *density grids* developed in Section 5.5 is useful for handling the sparse set of points found in the calculated cloud. An approximate area of 100 m^2 in front of the vehicle was processed through a density grid with a resolution of 35×75, square cells 20 cm on a side and 1.5 m deep, enclosing a volume of 0.06 m^3. The grid held 41,700 points

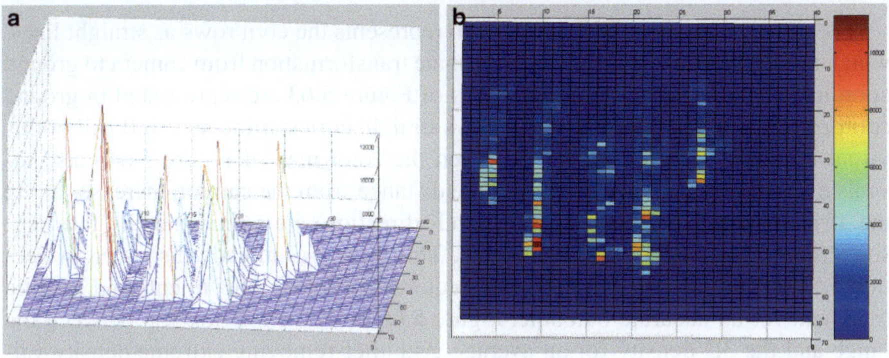

Figure 5.64 Density grid for crop-tracking images: (**a**) 3D perspective view; (**b**) top view

after removing outliers, and the maximum density was 9749 points/m^3. Figure 5.64a shows a 3D view of the calculated density grid, where only the five crop rows within the field of view of the camera have noticeable density. The top view of the grid, shown in Figure 5.64b, indicates that five straight rows were detected ahead of the harvester.

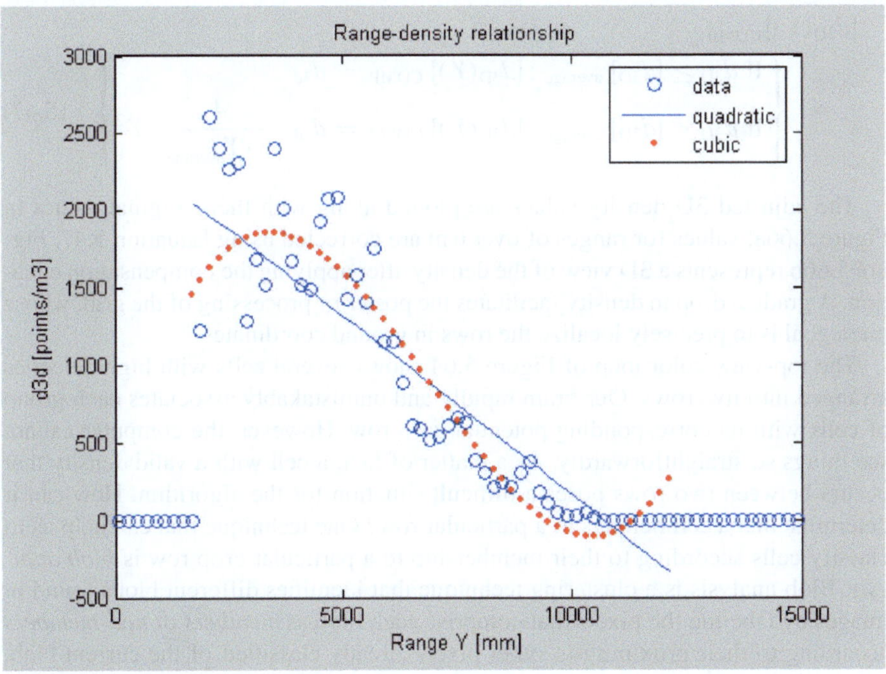

Figure 5.65 Distribution of 3D densities according to range for crop-tracking images

The top view depicted in Figure 5.64b represents the corn rows as straight lines, which is an indication of a correct coordinate transformation from camera to ground coordinates. Notice that the point clouds of Figure 5.63 are represented in ground coordinates, and that the front view shows straight corn starting at $z = 0$. All of this visual information makes sense; however, the color map of Figure 5.64b also reveals a severe decay in 3D density as the distance from the camera increases along the 15 m represented. The rear 30 cells (Y direction) are represented by low densities (dark blue, under 2000 points/m^3). This issue was already discussed in Section 5.5 (Equation 5.23), but each case requires particular adjustments and specific treatment for an accurate outcome. Figure 5.65 plots the relationship between the range and the 3D density for an average grid after removing extreme density values. A quadratic fit to the data resulted in a linear fit, as the quadratic coefficient was null. Both fitting curves underestimate the 3D densities for ranges close to the camera, which are the most important as they represent the best-determined data. Figure 5.65 proves that a simple trend curve, either quadratic or cubic, is not satisfactory for compensating for the drop in density as ranges grow. Equation 5.47 gives the compensating expression implemented in this case study, where Y_{average} is the average range in the grid and $[d_{3D}]_{\text{average}}$ is the mean 3D density. Note that these two parameters will change with every image captured. For the example used here, the average density was 1201 points/m^3 and the average range was 7.7 m.

$$[d_{3D}(Y)]_{\text{COMP}} \rightarrow \left\{ \begin{array}{l} \text{If } d_{3D} \geq [d_{3D}]_{\text{average}} : [d_{3D}(Y)]_{\text{COMP}} = d_{3d} \\ \text{If } d_{3D} < [d_{3D}]_{\text{average}} : [d_{3D}(Y)]_{\text{COMP}} = d_{3d} \cdot \dfrac{4}{3 \cdot Y^2_{\text{average}}} \cdot Y^2 \end{array} \right\} \quad (5.47)$$

The adjusted 3D density values are plotted along with their original values in Figure 5.66a; values for ranges of over 6 m are corrected using Equation 5.47. Figure 5.66b represents a 3D view of the density after applying the compensation equation. A gradual drop in density facilitates the posterior processing of the grid, whose main goal is to precisely localize the rows in ground coordinates.

The top-view color map of Figure 5.64 shows several cells with high densities arranged into five rows. Our brain rapidly and unmistakably associates each group of cells with its corresponding potential corn row. However, the computer cannot see things so straightforwardly. As a matter of fact, a cell with a valid density that occurs between two rows poses a difficult situation for the algorithm. How can it determine that a cell belongs to a particular row? One technique that can help us to classify cells according to their membership to a particular crop row is *blob analysis*. Blob analysis is a clustering technique that identifies different blobs found in images by labeling the pixels that comprise each blob as *members* or *non-members* according to their proximity to other pixels already classified of the current blob. This operation can be carried out considering either four-connectivity, *i.e.*, considering only top, bottom, left and right neighbors; or eight-connectivity, which also includes diagonally neighboring pixels. This analysis is frequently used in image

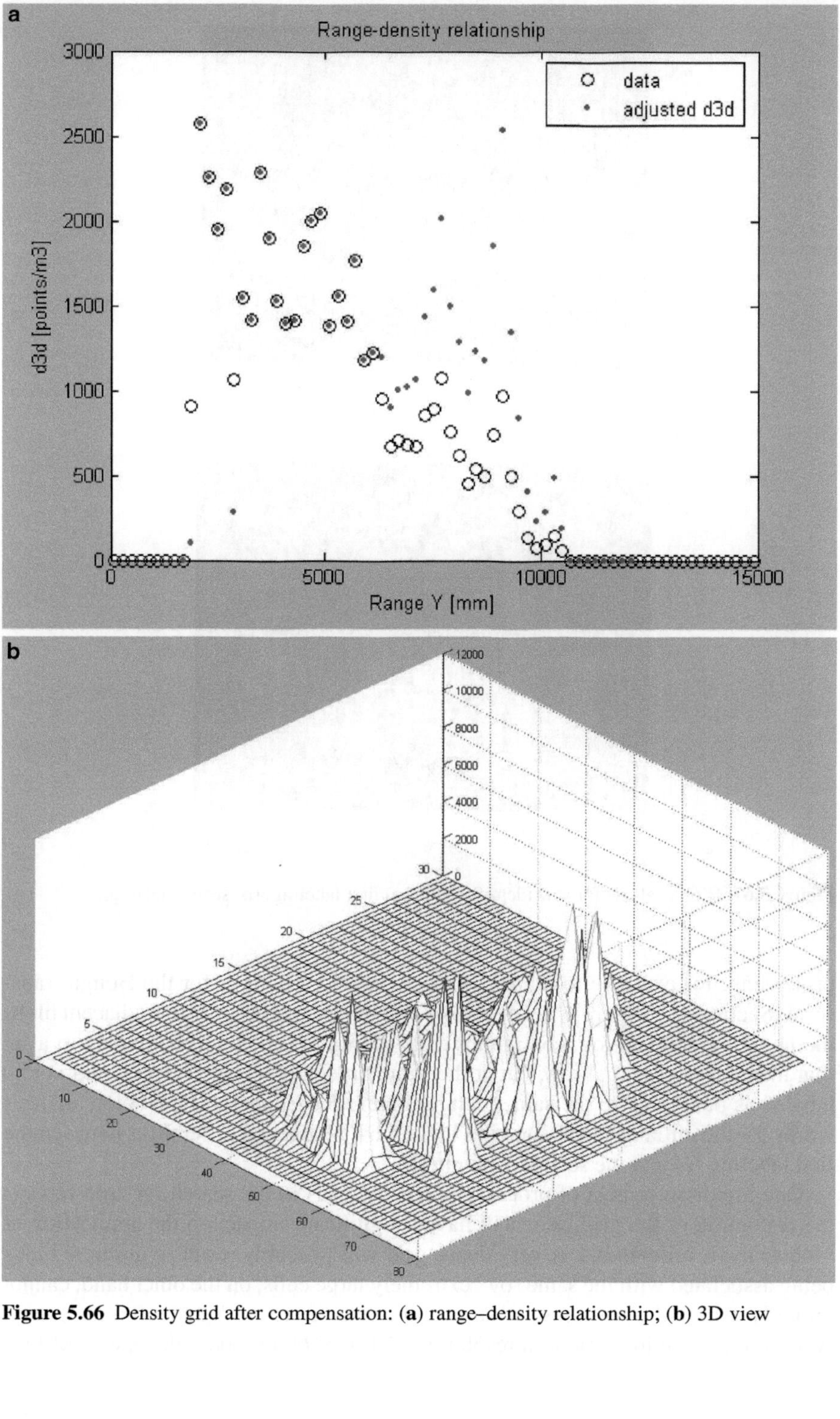

Figure 5.66 Density grid after compensation: (**a**) range–density relationship; (**b**) 3D view

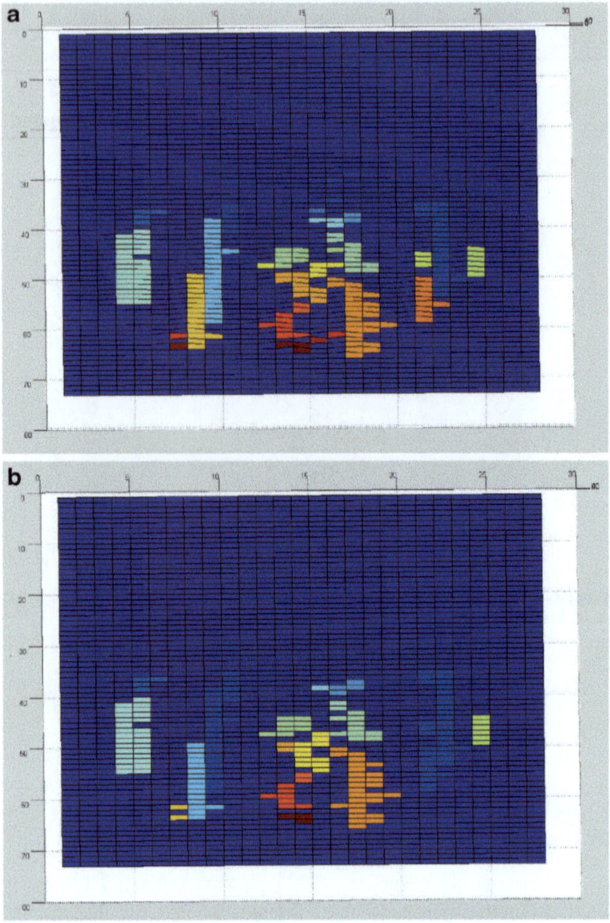

Figure 5.67 Blob analysis for row identification: (a) first labeling; (b) second labeling

processing, for example to reliably locate sinusoid crossings for the Hough transform (Section 4.3; Figure 4.17). It involves an iterative process where adjacent blobs merge into a unique blob; contiguous cells should be labeled as belonging to a set that indicates a potential row, but at the same time neighboring rows should not be labeled as belonging to the same feature. Figure 5.67 illustrates how blob analysis works for the initial grid depicted in Figure 5.64. Notice the differences between the first labeling (a) and the second labeling (b).

One important aspect of processing density grids in the search for crop rows is the resolution of the grid, as it will have a significant impact on the application of blob analysis reiterations. A very dense grid will probably result in multiple blobs being associated with the same row; extremely large cells, on the other hand, cannot properly delimit the positions and sizes of the detected rows. A tradeoff is necessary. Amplifying the initial grid to a resolution of 150 × 60 produces the results of Fig-

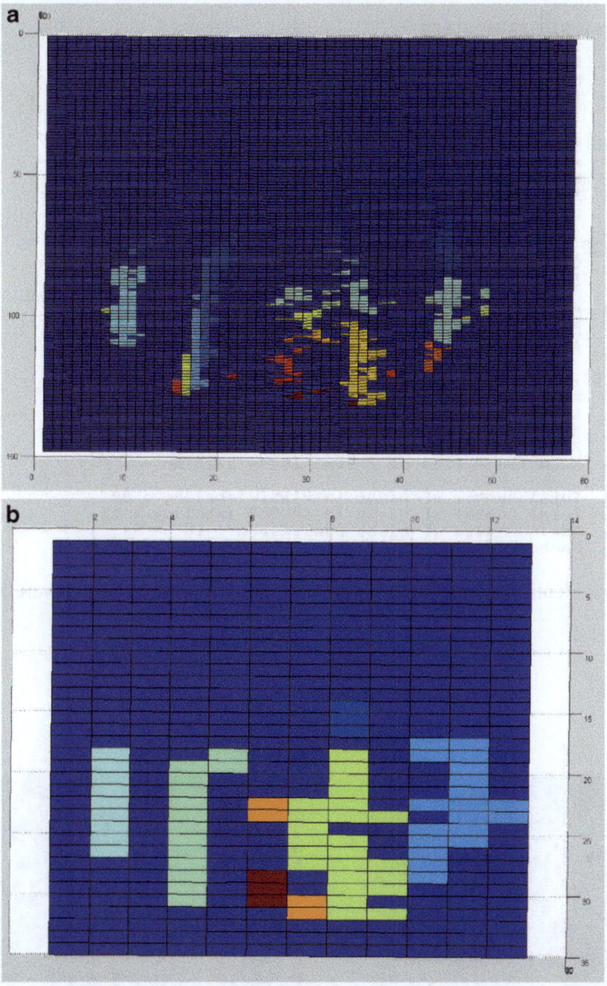

Figure 5.68 Effect of grid resolution on blob analysis: (**a**) high resolution; (**b**) low resolution

ure 5.68a, and simplifying it to 35 × 12 leads to Figure 5.68b. The former generates a multiplicity of blobs per row, as expected, but the latter introduces some confusion in the central rows. There is always a loss of information with low-resolution grids; therefore, a medium-high resolution is probably the right choice.

The next step in the sequence of operations after optimal blob estimation is to localize the *centroid* for each *blob*. A high-resolution grid, such as that shown in Figure 5.68a, yielded the best solution for the original image given in Figure 5.62. However, some of the multiplicities present in Figure 5.68a need to be eradicated. Changing the properties of the grid automatically implies modifying some of the parameters in the blob analysis routine. Doubling the resolution of the initial grid meant that the cells quadrupled in number, and the new cell size was 100 × 100

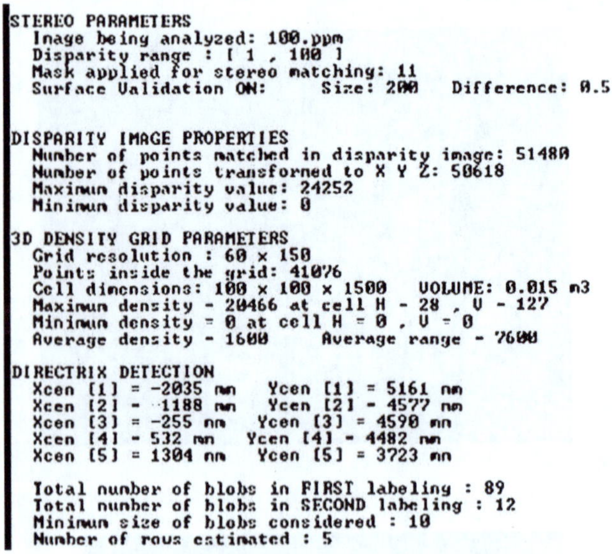

Figure 5.69 Row identification and localization: output screen from the software application

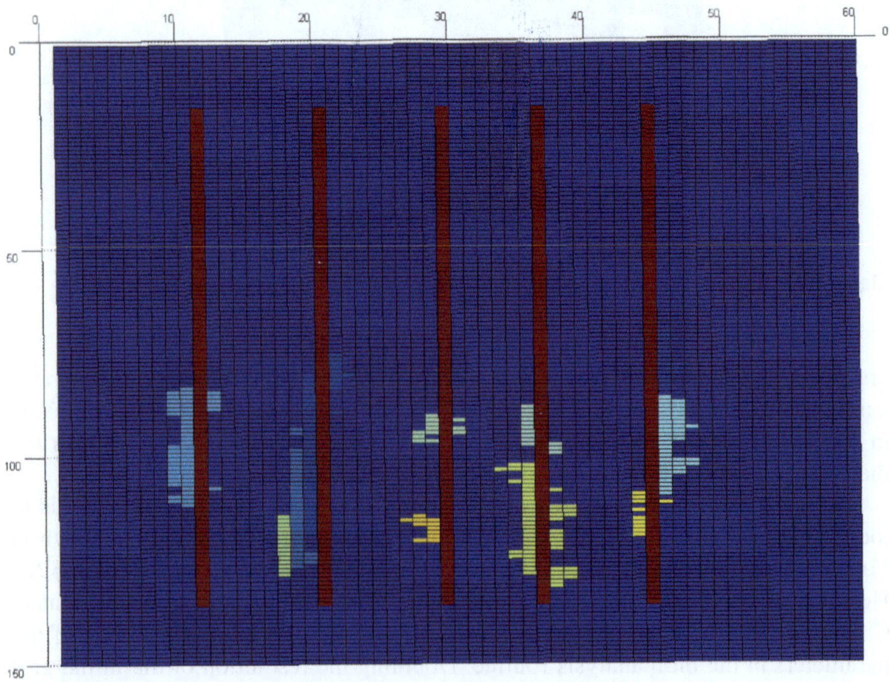

Figure 5.70 Row identification and localization: graphical representation

(mm^2) instead of 200×200 (mm^2). The minimum allowable blob size, for instance, must be modified too. In the example provided here, the minimum size was five cells per blob, but this was augmented to ten cells per blob to get rid of small, poorly defined blobs. When this change was introduced in the high-resolution grid of Figure 5.68a, five rows were detected, with the following separations between them: 85, 93, 79, and 77 cm. The average separation was 83 cm for rows spaced $\sim$ 75 cm apart, which is quite reasonable for the cell size considered. The exact positions of the blob centroids are specified in the output screen of Figure 5.69, and they are represented graphically in Figure 5.70.

The last step, once the positions of the rows in the field have been estimated, is to find the offset of the harvester; that is, how far the vehicle is from the optimum path determined by the centers of the rows. This is the *output* of the system: a recommended *directrix* and the *offset* needed to follow it. Since the row spacing is

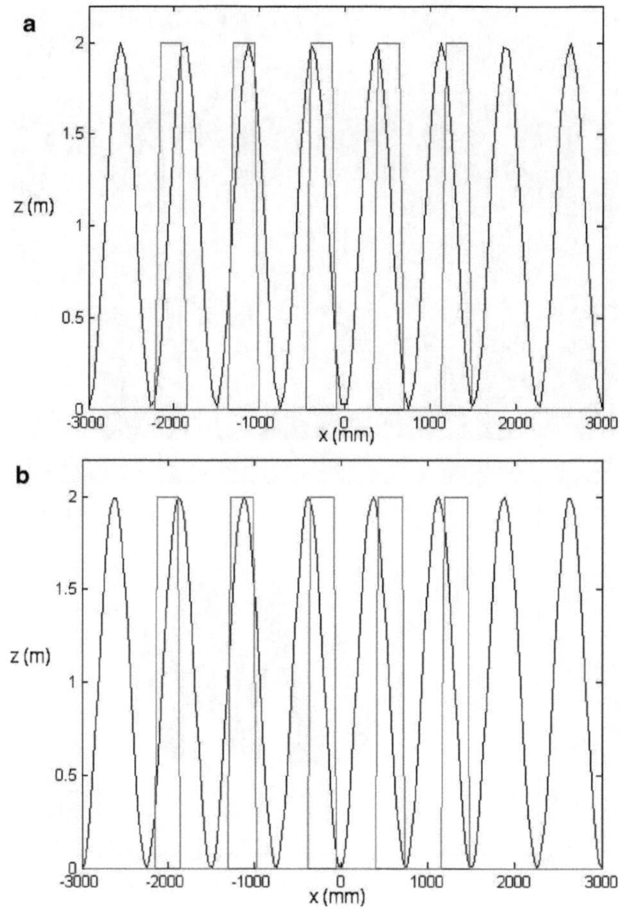

Figure 5.71 Cross-correlation results: (**a**) before correcting the offset; (**b**) after offset correction

known and constant, it is possible to establish a correlation between the theoretical positions of the rows and their actual placement as determined by the blob analysis plotted in Figure 5.70. *Cross-correlation* is a standard method used to find the best match between two series. In this case, one series is the positions of the rows as given in the results screen of Figure 5.69, and the other series is a sinusoidal theoretical function that models equally spaced rows. The formula used to generate the theoretical outline of the rows is given in Equation 5.48, where the actual (theoretical) spacing between rows is represented by S_R. Figure 5.71a shows how both series overlap before the application of the offset, and Figure 5.71b shows the overlap after the best correlation has been found, which in this case was only +30 mm. Such a tiny offset is difficult to perceive in the corrected graph (b), and that indicates that the harvester is fairly well centered, as can be seen in Figure 5.62a. In fact, these

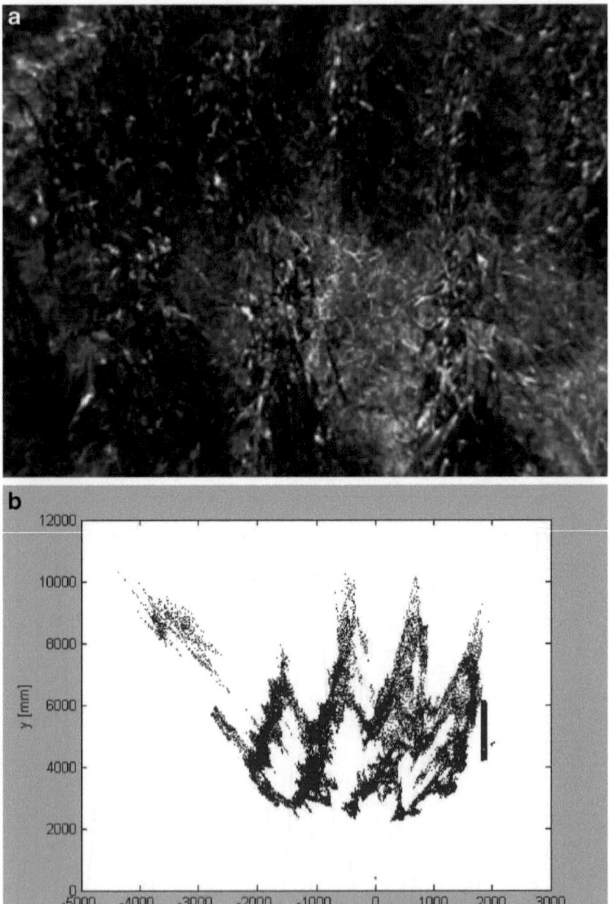

Figure 5.72 Tracking challenging (patchy) rows: (**a**) monochrome image; (**b**) top view of point cloud

images were acquired while driving the combine along a path that was as centered as possible in order to avoid damaging the crops, which were going to be harvested soon afterwards.

$$F_{\text{crop}}(x) = 1 + \sin\left(-\frac{\pi}{2} + \frac{2\cdot\pi\cdot x}{S_{\text{R}}}\right) \tag{5.48}$$

There are many circumstances that can challenge this system. One particularly important case is an absence of tracking features caused by naked patches in the field ahead of the stereo camera. This is the case portrayed in the stereo image of Figure 5.72a. The top view of the resulting point cloud is plotted in Figure 5.72b, where four skewed rows have been detected.

The final positions of the rows are shown in the grid of Figure 5.73. Note that the algorithm is expecting straight rows, and the fact that these rows appear at a certain angle of inclination has a negative effect on the estimation of their locations. However, the system identified the four rows that can be discerned in the original monochrome image. The output data screen is included in Figure 5.74a, where the exact positions of the rows are given in ground coordinates. Note that the calculated offset is zero, which implies that the harvester is well positioned. The results of this example therefore show that the harvester was well placed but probably erroneously orientated, which is a typical example of heading error without offset error. The results from the cross-correlation algorithm are graphed in Figure 5.74b.

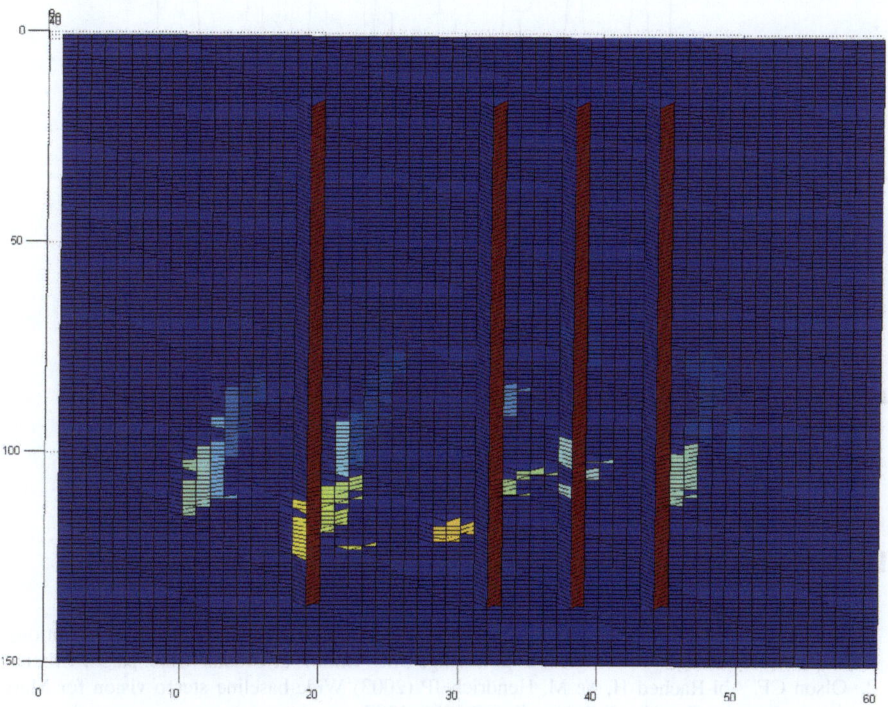

Figure 5.73 Tracking challenging rows: identification and localization of rows

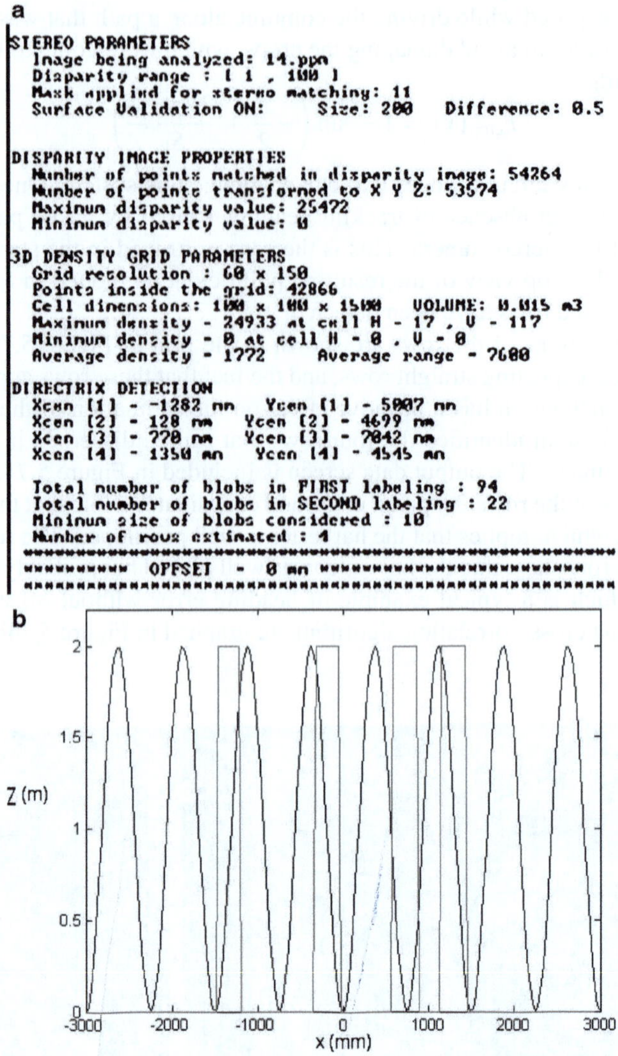

Figure 5.74 Tracking challenging rows: (**a**) data screen; (**b**) cross-correlation results

References

1. Rovira-Más F (2003) Applications of stereoscopic vision to agriculture (Ph.D. dissertation). Dept. Agricultural and Biological Engineering, Univ. Illinois at Urbana–Champaign, Urbana
2. Olson CF, Abi-Rached H, Ye M, Hendrich JP (2003) Wide-baseline stereo vision for Mars rovers. Proc Int Conf Intell Robots Syst 2:1302–1307

3. Zabih R, Woodfill J (1994) Non-parametric local transform for computing visual correspondence. Proc Eur Conf Comput Vis Stockholm 3:151–158
4. Konolige K (1997) Small vision systems: hardware and implementation. Int Symp Robot Res 8:111–116
5. Rovira-Más F, Wang Q, Zhang Q (2009) Bifocal stereoscopic vision for intelligent vehicles. Int J Veh Technol 123231
6. Rovira-Más F, Wang Q, Zhang Q (2009) Noise reduction in stereo disparity images based on spectral analysis (ASABE Paper no. 096258). ASABE, St. Joseph
7. Rovira-Más F, Wang Q, Zhang Q (2008) Configuration of stereo systems for off-road intelligent vehicles (ASABE Paper no. 083541). ASABE, St. Joseph
8. Martin MC, Moravec HP (1996) Robot evidence grids (Tech. Rep. CMU-RI-TR-96-06). Robot. Inst., Carnegie Mellon Univ., Pittsburgh
9. Rovira-Más F, Reid JF, Zhang Q (2006) Stereovision data processing with 3D density maps for agricultural vehicles. Trans ASABE 49(4):1213–1222
10. Rovira-Más F, Zhang Q, Reid JF (2005) Creation of three-dimensional crop maps based on aerial stereoimages. Biosystems Eng 90(3):251–259
11. Rovira-Más F, Zhang Q, Reid JF (2008) Stereo vision three-dimensional terrain maps for precision agriculture. Comput Electron Agric 60:133–143
12. Yokota M, Mizushima A, Ishii K, Noguchi N (2004) 3-D map generation by a robot tractor equipped with a laser range finder (ASABE Publ. no. 701P1004). ASABE, St. Joseph
13. Rovira-Más F, Han S, Wei J, Reid JF (2007) Autonomous guidance of a corn harvester using stereo vision. Agric Eng Int (CIGR Ejournal) XI:ATOE-07-013
14. Rovira-Más F, Zhang Q, Reid JF (2004) Automated agricultural equipment navigation using stereo disparity images. Trans ASABE 47(4):1289–1300

Chapter 6
Communication Systems for Intelligent Off-road Vehicles

6.1 Onboard Processing Computers

Intelligent vehicles require, at the very least, the presence of a main computer to acquire data from multiple sensors at the desired sampling rates, to save, manipulate and display key data, and, when required, to make use of the processed information in order to support automatic control functions. Although most sensors output an analog voltage signal after signal conditioning, some sensors produce digital outputs. Analog signals normally pass through an analog multiplexer and an analog-to-digital converter before being fed into the computer, but digital signals are usually passed to the computer directly via a serial port. Onboard computers can also be used to control vehicle processes, either directly with digital signals or with analog signals that are then transformed to digital signals by digital-to-analog converters. These computers typically include the following major components: a central processing unit, software programs, a random-access memory, mass storage systems, and data input/output (I/O) devices. The *central processing unit* (*CPU*) controls the operations of the computer system and performs all of the arithmetic processes. The CPU follows instructions from the computer's operating system, its built-in programs, and user-provided software. Most computers have a single CPU, which is often used to control several peripheral devices, such as displays, joysticks, or keyboards. Because there is, generally speaking, only one CPU per computer, instructions tend to be executed in a sequential manner; that is, only one instruction at a given time. The *random-access memory* (*RAM*) is a temporary information storage subsystem in the computer. It can store both program instructions and numerical data when the computer is being operated. The RAM consists of electronic components with no moving mechanical parts; therefore, information can be retrieved from or saved to (*i.e.*, stored in) the RAM at a very high rate. As a result, the data saved in the RAM can easily be changed. However, the RAM is volatile too, and all of the information stored in it is lost when the power supplied to the computer is interrupted. Thus, there is also another kind of memory called the *read-only memory* (*ROM*), which is used for the permanent storage of information.

The values measured by sensors are often represented by levels of electrical variables, such as voltages and currents, and are normally represented in base 10 (decimal numbers). However, computers represent numbers in base 2 (binary) by adopting the "on–off" feature of electronics. The state of "on" is assigned to a numerical value of 0, and the state of "off" is defined as a numerical value of 1. To represent a decimal number, a series of 0/1 digits are required. For example, the four digits 1001 are needed to represent the decimal number 9 using the binary numbering system of computers. Each of these digits represents a *bit* of the binary number, where bit stands for *binary digit*. A number represented using four binary digits is called a 4-bit number. The leftmost 1 in the binary number 1001 is the *most significant bit* (*MSB*). The rightmost 1 is the *least significant bit* (*LSB*). It is common practice in computer science to break long binary numbers into segments of 8 bits, which are known as *bytes*. There is a one-to-one correspondence between binary numbers and decimal numbers. For example, a 4-bit binary number can be used to represent all the positive decimal integers from 0 (represented by 0000) to 15 (represented by 1111). This procedure of transforming binary numbers to decimal numbers and *vice versa* is used to represent positive decimal integers. However, it is also necessary to represent negative and floating-point numbers, such as -3.56×10^3. Special techniques are used to handle these numbers. Negative numbers are commonly represented by a technique known as *two's complement*. As noted above, 4 bits can be used to represent the decimal integers between 0 and 15. The same 4 bits can also be used to represent numbers from –8 to 7 via the two's complement technique. According to this method, the MSB is always 0 for positive numbers and 1 for negative numbers. Thus, positive numbers from 0 to 7 are represented by varying the three LSBs appropriately (from 0000 to 0111); similarly, negative numbers from –8 to –1 are represented by the binary numbers ranging from 1000 to 1111. Floating-point numbers are handled by keeping track of both parts of the number (the mantissa and the exponent). Separate arithmetic operations are performed on the two parts in a similar manner to hand calculations.

The simplest way of connecting an external device to a microprocessor is to make the added device look like a memory location to the microprocessor. In such a memory-mapped system, the input/output (I/O) ports of the computer operate as if they were dealing with regular internal memories, with each port being assigned one or more addresses; as a result, microprocessors generally use the same buses for both memory and I/O transfers. An alternative to this memory mapping is *isolated input/output*, in which memory and I/O addresses are separately decoded, and *attached input/output*, in which the I/O ports are activated by special instructions. The memory-mapped system is probably the most common method used in practice, and it can be implemented with any microprocessor. The isolated input/output approach, on the other hand, can only be employed with microprocessors that are specifically designed for it and maintain separate in and out instructions. Robust design of the core systems that comprise an intelligent vehicle demands perfect compatibility between computers and sensors, which very often requires some of the following operations. *Electrical buffering* and *isolation* is needed when the connected sensor operates at a different voltage or current to that of the microprocessor bus system, or when there are different ground references. The term *buffer* is used for a device that provides isolation and current or voltage amplification. A common amplifier is therefore frequently employed as a buffer in electronic circuitry. *Timing control* is needed when the data transfer rates of the sensor and microprocessor are different. This synchronization can be achieved by introducing special connecting lines between them to control the timing of data transfer. Such lines are referred to as *handshake lines*, and the process is known as *handshaking*. The external device sends a "data ready" signal to the I/O section. The CPU then determines that the "data ready" signal is active, reads the information through the I/O section, and sends an "input acknowledged" signal to the sensor. This signal indicates that the transfer has been completed and thus more data can be sent. For an output, the device sends an "output request" or "peripheral ready" signal to the I/O section, which makes the CPU send the data to the sensor. The next "peripheral ready" signal may be used to inform the CPU that the transfer has been completed. *Code conversion* is necessary when the code used by the sensors differs from that utilized by the microprocessor. Microprocessors operate on a fixed word length of 4 bits, 8 bits, or 16 bits. This determines the number of lines in the microprocessor data bus. External devices may have a different number of lines, perhaps requiring a longer word than that of the microprocessor. Within an 8-bit microprocessor, data is generally manipulated eight bits at a time. Therefore, to transfer eight bits simultaneously to a device, eight independent data paths are required. Such a form of transfer is termed *parallel data transfer*. It is not, however, always possible to transfer data in this way, and a sequential procedure involving one bit at a time is the only means of communication. This form of transfer is called *serial data transfer*. Serial data transfer is a slower method of exchanging information than parallel data transfer. Thus, if serial data transfer is the method used by the sensor, there will be a need to convert incoming serial data into parallel data for the microprocessor, and *vice versa* when outputting data from the microprocessor. With the parallel transmission of data, one line is used for each bit; serial systems, in contrast, use a single line to transmit data in sequential bits. There are two basic types of serial data transfer: asynchronous and synchronous. With *asynchronous transmission*, the receiver and the transmitter each use their own clock signals, so it is not possible for a receiver to know when a word starts or stops. This condition forces each transmitted word to carry its own "start" and "stop" bits so that it is possible for the receiver to tell where one word stops and another starts. This mode of transmission is normally applied when the transmitter and the receiver are remote. With *synchronous transmission*, the transmitter and the receiver have a common clock signal and thus transmission and reception can be synchronized.

6.2 Parallel Digital Interfaces

In parallel communication, n lines of an n-bit computer are available for communication between the computer and a multiplicity of external devices. The voltage levels on these lines are typically either 0 or 5 V, and are sensed simultaneously. Another important method that the computer uses to communicate with the external world is through *interrupts*; when an interrupt is received from a sensor, the program on which the computer is operating is suspended to shift to a special subroutine. The implementation of parallel I/O and interrupts is achieved through special chips. It is crucial that n lines of the data bus of the computer are connected to the chip so that the microprocessor can send or receive data from the port. Other basic connections to the parallel I/O chip are the power and ground lines, and some means of selecting and/or enabling the chip when it needs to be activated.

The most commonly used parallel interfaces are the *Standard Parallel Port* (SPP) and the *Enhanced Parallel Port* (EPP). The SPP supports an 8-bit, unidirectional interface to transmit data from the computer to the peripheral, transferring information as either ASCII code or graphical data through a 25-pin connector (DB-25). Apart from the eight unidirectional data lines, a typical SPP connector has three primary control and handshake lines ("busy," "ack," and "strobe"). There are other six control lines for various statuses and control functions and also eight ground lines. The SPP has a low data transfer speed of about 100 kbytes/s and is unidirectional, which motivated the development of the *Enhanced Parallel Port* (EPP) to provide bidirectional data transfer with a higher data rate of up to 2 Mbytes/s. The EPP also uses a 25-pin connector. The main difference between SPP and EPP connectors is that the EPP has only six control and handshake lines, whereas the SPP has nine. In 1994, the Institute of Electrical and Electronics Engineers (IEEE) approved the parallel port standard 1284. This standard classifies parallel ports by the transfer mode used, and covers different types of connectors in terms of their pin assignments, cables and the electrical operations required by each interface. When a parallel port is IEEE 1284 compliant, it supports the EPP mode at data rates of up to 2 Mbytes/s over cables as long as 10 m. The *General-Purpose Interface Bus* (GPIB), defined in the IEEE 488 Standard, is designed to connect multiple devices to a central computer. It transfers data asynchronously via eight parallel data lines plus several control and handshaking lines. In many ways, the GPIB acts like a conventional computer bus or network, allowing one computer to uniquely communicate with up to 15 individually-addressed devices. In general, the maximum length of a cable linking two devices is 2 m, and the total cable length of the entire network must be no longer than 20 m. The standard GPIB connector consists of 24 pins divided into four groups: a group of eight bidirectional lines for digital data transfer, a group of three handshaking lines for data transfer control, a group of five interface management lines for bus control and status identification, and a group of eight ground lines.

6.3 Serial Data Transmission

As previously mentioned, interfacing among sensors, computers, and other electronic devices usually involves electrical buffering, timing and handshaking, analog-to-digital conversions (A/D or D/A), and data transfer. Physical and messaging compatibility must be assured before exchanging information, and this can be achieved through two different approaches: parallel or serial. Handshaking signals are used to control the flow of serial data, and timing is necessary to pass clock signals between transmitters and receivers. *Parallel communication* systems send each data bit of a message simultaneously along parallel data lines. *Serial communication* systems, in contrast, transmit data in a chain-like process along a single pathway [1]. Parallel communication is recommended for high speed and short distances, whereas serial ports can connect over much longer distances. Serial data transmission is extensively used in off-road vehicles; for example, to reference the vehicle trajectory or to localize digital images with GPS receivers (see Section 6.6). Transmitted data will generally contain two types of information: one is the data that one device wishes to send to another; and the other is the information termed the *protocol data*, which is used by the interface to control the transfer of the data. A *protocol* is a formal set of rules governing data format, timing, sequencing, access control and error control.

Some terms frequently used in serial data transfers include mark, space, character, and cluster. A *mark* is a logical 1 represented by 5 V, and a *space* is a logical 0 that is always represented by 0 V. Marks and spaces are grouped into a *character*, often 1 byte of data consisting of eight marks and spaces. A character is normally separated from others by a *start bit* (a space), a *stop bit* (a mark), or a *parity bit*. The assembly of the start bit, the characters, the stop bit, and sometimes the parity bit is defined as a *cluster*. The rate of data communication is measured by the *baud rate*, defined as the number of bits that can be transmitted per second. The typical rates for serial digital interfaces are 100, 300, 600, 1200, 2400, 4800, 9600, and 19 200 baud. If the transmitter sends bits of information when the receiver is not expecting them, interpretation errors may occur. For this reason, some kind of synchronization must be established between the computer and the external device. There are two ways of timing serial sequences: synchronous and asynchronous. In both methods, transmitter and receiver are regulated by clock pulses sent through the communication line by the transmitter according to its clock. In *synchronous communication*, once the transmitter begins sending data, it continues at a regular pace until all the data have been sent, but the transmitter and the receiver must be timed by the same clock or by clocks of identical frequency. In *asynchronous communication*, data flow along the transmission line one cluster at a time, as illustrated in Figure 6.1. The transmitter sends a character of 8 bits (for example), bounded by the start and stop bits, and then may wait a period of time at its own discretion. The usual convention establishes a mark for the stop bit and a space for the start bit. After a cluster has been transmitted, the final stop bit leaves the line in a mark status, remaining in this condition until the next start bit switches the line to a space status. If the waiting time is zero, the clusters are transmitted continuously. An important feature of asynchronous communication is that, even though transmitter and receiver need to share the same clock frequency, there can be a slight deviation between their two frequencies without disturbing the accuracy of transmission, since transmitter and receiver always synchronize at the start of each new cluster.

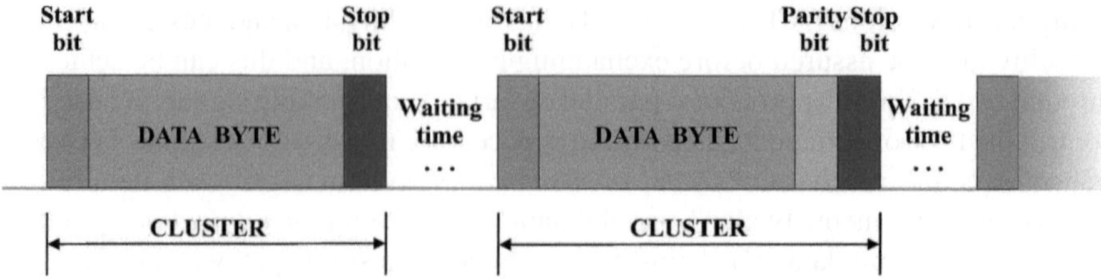

Figure 6.1 Asynchronous data transmission for serial communication

If we compare both transmission methods, synchronous transmission demands extra provisions to set identical timing for both the transmitter and the receiver, while in asynchronous mode the two clock rates only need to be very close to each other. Synchronous communication, on the other hand, does not require the start and stop bits, since both transmitter and receiver are in agreement on where the characters start and end once the transmission begins. An advantage of synchronous transmission is therefore that more data can be transmitted at the same bit rate. Nevertheless, asynchronous communication is more widely used because it is more convenient, and will be the only type considered hereafter.

The serial transmission of data is not always flawless, and a bit in a character may be distorted such that a 0 is received when a 1 was sent, and *vice versa*. One procedure used to detect this type of error is the application of parity bits. A *parity bit*, which can have either *odd parity* or *even parity*, is an extra bit added to a cluster as represented in Figure 6.1. If odd parity is chosen, the sum of the 1s in the combination of data and parity bit is an odd number. Likewise, if even parity is chosen, the sum of the 1s is an even number. It should be noted that the use of parity will only detect an error in 1 bit of data. The serial data internally handled by the computer belongs to the *transistor–transistor logic* (TTL) level. In this procedure, the voltage is interpreted as a logical 0 if it is below a certain level (below 0.8 V for inputs and 0.4 V for outputs), and decoded as a logical 1 if it is above another threshold level (above 2.4 V for inputs and 2.8 V for outputs). Between the voltage ranges representing logical 0 and logical 1, there is a band of voltage that provides a margin of safety between these two logical levels. The conventional interval 0–5 V is reliable for short distance communication, but for long distances the signals are susceptible to distortions caused by electrical noise, differences in ground potential, and physical damage from short circuits. To deal with these problems, some interfaces such as the RS-232C provide greater voltage differences between marks and spaces.

The most popular serial interface for robotic ground mobile platforms has been the *RS-232C*, which supports transmission rates of up to 20 kbps over distances of up to 15 m, which covers the dimensions of the majority of off-road vehicles considered in this book. The RS-232C protocol was established by the Electronic Industries Association (EIA, currently Electronic Industries Alliance) in 1969. Because of its widespread use, the RS-232C has become one of the most "nonstandard standards" available, with the capacity to withstand open circuits and shorts without damage, as well as signals of up to ± 25 V. The RS-232C uses positive voltages of between

12 and 15 V to represent spaces, and negative voltages between of –12 and –15 V to represent marks. The difference in voltage between mark and space is thus 24–30 V. The RS-232C standard uses either a 25-pin D-shell connector or a 9-pin connector; therefore, when linking devices that integrate connectors of different type, a cable adapter is required to establish a correspondence between a DB-25 and a DB-9 connector. The RS-232C incorporates two data lines (pins 2 and 3) to support full-duplex operations, allowing the connected devices to simultaneously transmit and receive data asynchronously.

In practice, RS-232C interfaces are used to connect many different types of sensors and electronic devices with computers. The serial port in a computer is normally set up as the data terminal equipment (DTE). The meaning of handshaking lines is software dependent and they may not need to be used. If required, only three lines ("TXD," "RXD," and signal ground) are necessary to transmit data without handshaking. Overall, there are two approaches for implementing handshaking: full handshake support with seven wires, and a simple no-handshake connection with three wires. With the simpler connection that employs only three wires, the handshake lines are permanently enabled and cannot be used to control the data flow on the interface, although this can still be controlled with special data characters in a software handshaking protocol. These data characters are expressed in ASCII (American Standard Code for Information Interchange), the most widely used computer code for handling alphanumeric (text) data, and the one usually employed for data transfer between electronic equipment over standard interfaces. It is a 7-bit code consisting of printable (alphanumeric) and control characters. As mentioned above, the RS-232C standard does not specify the protocol used for data transmission. The vast majority of RS-232C interfaces use an asynchronous protocol. When no data is being transmitted, the line is at the marking level (logic 1). At the beginning of the transmission, a start bit is sent, causing a line transition to the spacing level (logic 0), which tells the receiver that data is coming. Next, the information bits – usually 7 or 8 bits – are sent, one at a time, where a bit value of 1 is at the marking level and a bit value of 0 is at the spacing level. These data are followed by an operational *parity bit* for error detection. Finally, one or more stop bits at the marking level are sent to indicate the end of the data byte. Since RS-232C line drivers and receivers are inverters, the marking level (logic 1) corresponds to a negative voltage (−3 V to –15 V), and the spacing level (logic 0) corresponds to a positive voltage (+3 V to +15 V) on the interface line. To set up an asynchronous RS-232C communication line, both devices at the two ends of the line must be set to the same data rate (also called the baud rate). In addition, the number of data bits must be known. This can vary from 5 to 8 bits, with 7 or 8 bits being the most common. The next parameter needed is the parity bit, used as a simple error-detection scheme to determine if a character was incorrectly received. The number of logic 1s in the transmitted character is then summed, including the parity bit. For even parity, the parity bit is chosen to make the number of 1s an even number, and for odd parity the total number of 1s is made odd. For example, the ASCII character "a" is 61h, or 01100001 in binary. For even parity, the parity bit would be 1 (summing up four 1s, which is an even number), whereas for odd parity, the parity bit would be 0 (leaving three 1s to make an odd number). When a parity bit is used, typically with 7-bit data characters, the transmitting end determines the correct parity bit value, as just described, and incorporates it into the character sent. The receiving end calculates the expected value of the parity bit from the character obtained, and compares it to the parity bit actually received; if these values are not the same, an error is assumed. The final communication parameter in asynchronous mode is the number of stop bits, which can be set to 1, 1½, or 2 stop bits, although 1 bit is the most commonly used. Male 9-pin or 25-pin connectors are typically attached to cables, whereas female sockets are mounted on the communicated devices. There are convenient commercially available software applications that facilitate the connection of electronic devices to the main onboard computer via the RS-232C interface.

Besides RS-232C, several other serial communication interfaces are generally used, such as the standard *RS-423A*, which can be considered an enhanced version of RS-232C with a number of notable advantages. RS-423A has a lower driver voltage range (between ± 3.6 and ± 6.0 V); however, it possesses a much higher allowable data rate of up to 100 kbps, and a maximum cable length of over 1 km. Another important difference is that RS-423A can support multiple receivers on the same line, up to a maximum of 10, which is very useful for unidirectional data transfers in a broadcast mode, such as those used to update multiple displays with the same information.

Another popular EIA serial transmission standard is the *RS-422A* interface, which uses differential data transmission on a balanced line. A differential signal requires two wires, one for non-inverted data and the other for inverted data. It is transmitted over a balanced line, usually a twisted-pair wire with a termination resistor at the receiver end. The received data is the difference between the non-inverted data and the inverted data, and no ground wire is required between receiver and transmitter because the two signal lines are referenced to each other. However, there is a maximum common-mode (ground-referenced) voltage range on either line of -0.25 V to $+6$ V, because most RS-422A drivers and receivers are powered by the same $+5$ V power supply. This signal ground is usually connected between the transmitter and the receiver to keep the signals within this common-mode range. This technique that is based on differential signals makes it possible to use of high data rates (10 Mbps) over long cable lengths because of its immunity to noise. If external noise induces a corrupted signal on the transmission line, it will be the same on both conductors, and the receiver will cancel out the common-mode noise by taking the difference between the two lines. With single-ended transmission lines, in contrast, noise spikes will probably show up as false data at the receiver end. As with RS-423A, RS-422A can also have up to ten receivers on the same line with a single transmitter.

The EIA *RS-485* interface is basically a superset of the RS-422A standard, with electrical specifications similar to those of RS-422A. RS-485 also features a differential transmission scheme with balanced lines that can operate at speeds of up to 10 Mbps over cable lengths up to 4000 feet long. It uses different output voltages, including a popular mode range of between -7 V and $+12$ V. The most significant difference is that RS-485 interfaces can support as many as 32 drivers and 32 receivers on the same line. To allow for this capability, RS-485 drivers must be able to switch into a high-impedance (tri-state) mode, so that only one driver is transmitting data at any given time. As with RS-422A, all receivers can be active at the same time. RS-485 interface cards are readily available and typically use the same connector (DB-9) and pin designations as RS-422A interface cards. The RS-485 driver output can be tri-state using a control signal on the card.

The limitations of serial interfaces when large amounts of information need to be transferred in real time among vehicle components have resulted in a progressive shift of this technology towards Ethernet communication systems. *Ethernet* was originally developed for local area networks (LAN), and has been standardized as IEEE 802.3. Ethernet is made up of four basic elements: the physical medium, the signaling components, the media access control protocol, and the frame. The physical medium encompasses the cables and other components used to carry the signals for the sensor network. The most popular medium for Ethernet systems is twisted-pair wiring terminated with 8-pin RJ-45 (telephone style) connectors, although coaxial and fiber optic cables are also used. There are cable length limitations based on signaling speed and media type to ensure that the maximum round-trip time is not exceeded. This timing limitation can be overcome by dividing a large LAN into multiple, separated, LANs using switching hubs. The signaling components are the electronic devices that transmit and receive data via the physical medium. These components include signal line transceivers and a computer Ethernet interface, often residing on a network interface card (NIC) or motherboard. Ethernet has been used to allow multiple computers to communicate in complex autonomous vehicles, such as those participating in the Grand Challenge competition. The utility vehicle used in *Case Study IV* of Chapter 5 required two different computers, one to host the steering controller, and the other to process 3D data captured with a stereo camera. Both computers were connected via an Ethernet cable.

6.4 Video Streaming: Frame Grabbers, Universal Serial Bus (USB), I^2C Bus, and FireWire (IEEE 1394)

Machine vision is probably the most valuable method of perceiving the local vicinity of an intelligent vehicle. Throughout this book we have seen a multiplicity of applications regarding visual sensors, such as monocular vision, multispectral analysis, thermographic mapping, and stereoscopic vision. While each technique possesses distinct features that make it ideal for a particular situation, all of them share a common property: their outputs are images, either analog or digital, that need to migrate from the sensing array located inside the camera – *i.e.*, the imager – to the processing engine (generally a remote computer). This transfer can be achieved by following different communication protocols, but the ever-increasing resolution of commercial cameras means that this technology is in constant flux. Even so, the most common interfaces that are currently in use are analog video, USB, I^2C, and FireWire. The need to handle digital images in real-time is unavoidable for many navigation functions such as steering or obstacle avoidance, and complicates the design of perceptive units. A telephone line with the ability to transmit at 9.6 kbps, for instance, would need over 10 min to transmit an 8-bit megapixel image.

Even though digital cameras are increasing in acceptance and popularity, it is possible to use an analog camera to capture a scene of interest. The project presented in *Case Study I* of Chapter 4, for example, features a conventional tractor guided via an analog monocular monochrome camera. When the camera is analog, given that computers can only deal with digital data, the video input from the camera must be digitized with a *frame grabber*: an electronic card that is usually installed in the main processor or in the image-processing computer. Before selecting an off-the-shelf frame grabber, we must ensure that capture rate restrictions are easily met, and that the firmware provided incorporates the proper libraries needed to integrate the digitizer into the whole vision engine.

The unlimited possibilities that digital cameras offer at present have led to the almost complete dominance of this kind of sensor over traditional analog cameras. Various digital cameras from low-cost webcams to sophisticated high-quality visual sensors are currently being implemented in robotic applications. Generally speaking, low-profile imaging devices tend to use the Universal Serial Bus (USB), due to its widespread use; in fact, last-generation computers typically included several USB ports. However, higher demands may overrun the capabilities offered by USB connections, so an alternative protocol with a notably higher bandwidth could be necessary. This is provided by the FireWire and I^2C protocols. Because of the USB's limited bus speed, most products work at low sampling rates. Stereoscopic vision, for example, has been using FireWire and I^2C communication alternatives for a long time. The red binocular camera of Figure 5.6a uses two FireWire connectors, whereas the camera portrayed in Figure 5.6b features the I^2C communication protocol.

The *Inter-IC* communication bus (I^2C) is a data bus designed by Philips that allows the exchange of data between devices when the cable linking them is relatively short. The I^2C cable comprises two lines, a bidirectional serial data line and a serial clock line, which are both connected to the positive 5 V power supply. The device that controls bus operations is the master, and the devices it controls are the slaves [1].

The *Universal Serial Bus* (*USB*) is the result of the efforts of an industrial consortium. It combines all of the advantages of a multiplatform industry standard, including low cost, extended compatibility, and a large number of available peripherals. The USB was developed as a replacement for standard serial and parallel ports on personal computers, and it is a high-speed, multi-drop serial bus with data rates as high as 12 Mbps. It is a true bus that can support as many as 127 devices with one host controller, typically a computer. The USB interface uses a strictly controlled wiring system that prevents erroneous connections. In addition, it can provide DC power to peripheral devices (5 V at up to 5 A) and is hot-swappable; that is, you can safely connect or disconnect USB devices from the bus without powering down or rebooting the host controller. The USB uses a special four-conductor cable that is up to 5 m long. Two of the data wires form a twisted pair that carry a differential data signal; the other two wires provide optional +5 V power to the peripherals. Because it is designed for a single host device, you cannot normally use it to connect one computer to another, as opposed to an IEEE-1248 parallel port. However, some manufacturers produce special USB cables with custom software for transferring data between computers. The USB interface implements a sophisticated communication protocol based on three types of packets: token, data, and handshake. The host computer initiates a transaction by sending out a 7-bit token packet that addresses the desired device, with a limit of 127 devices (*i.e.*, 128 unique addresses) on the bus. Next, data is exchanged via a data packet containing up to 1023 bits of data along with an error checking bit. Finally, a handshake packet is transmitted to finish the transaction. As with most technologies related to computers, the USB standard continues to evolve. The first generally used USB standard was version 1.1. A few years later, USB 2.0 was developed, with a 40 × improvement in speed that boosted the transmission rate up to 480 Mbps, although it is backward compatible with the original 12 Mbps USB devices and cables. USB devices typically negotiate with the host to run at the highest speed allowed on the bus.

The *FireWire* communication system, technically known as the *IEEE 1394* protocol, was originally developed by Apple Computer, Inc. in 1995, and is a high-speed serial bus for connecting peripheral devices to a computer or to each other. FireWire is a peer-to-peer system, in contrast to the USB host-based protocol, and therefore is one of the fastest data communication standards ever developed. The industry standard IEEE 1394 is a very high speed bus, with original data rates of 100, 200, 400, and even 800 Mbps. Transfer occurs thorough a simple 6-pin connector, and the system is very easy to use. Up to 16 devices or 64 nodes can be connected to a single IEEE 1394 bus, with individual cable lengths of up to 4.5 m. As with USB, it is also hot-swappable and provides power to the connected devices. IEEE 1394 was originally designed with high-bandwidth applications in mind, such as those requiring digital video streaming. The FireWire cables consist of six conductors: two twisted pairs and two power wires. Since there is no default host node, any 1394 device can supply power. The cable power is between +8 V and +40 V relative to the ground cable. The power provided by a device is limited to a maximum of 1.5 A, and a power-consuming device initially cannot draw more than 1 W. Similarly to the USB interface, the IEEE 1394 uses differential signals to transmit high-speed data reliably, but it incorporates two signals instead of the unique signal used in USB connections. In order to improve high-speed capabilities, there are low-voltage differential signals (LVDS) with amplitudes of only about 200 mV. The IEEE 1394 standard supports two types of data transfers: isochronous and asynchronous. A simple *isochronous* transfer, which has the highest priority, can use up to 80% of the available bus bandwidth (or cycle time). This transfer could be as long as 5 kbytes in one cycle if no other device is requesting an isochronous transfer for the same cycle. Isochronous transfers are suitable for time-critical and high-bandwidth data, such as digital video, as these transfers are fast and virtually real-time, although they do not contain error correction data. The isochronous philosophy rests on the idea that it is better to drop a few pixels in a video frame than to corrupt the frame timing and, consequently, get a distorted image. Here, speed is more important than data quality. *Asynchronous* transfers are not guaranteed a certain share of bus bandwidth, but they are given a fair chance of bus access when they are allowed after isochronous transfers. The maximum size of an asynchronous data block depends on the transfer rate of the system, with a maximum block of 2048 bytes for a 400 Mbps bus. Since an asynchronous block can get sent each cycle (that is, every 125 μs), a maximum asynchronous rate of about 16 Mbps can be achieved. Asynchronous transfers use error checking and handshakes. They can be slower than isochronous transfers but are better suited for applications where errors cannot be tolerated. Additionally, the IEEE 1394 bus uses an arbitration system that ensures that all devices on the bus have an opportunity to transfer data and are not locked out by high-priority devices. IEEE 1394 devices are fairly complex to design, and this partly accounts for their higher cost than USB peripherals. As with USB, IEEE 1394 continues to evolve through faster implementations. The newer standard *IEEE 1394b* is compatible with existing hardware with data rates up to 400 Mbps, but adds higher rates of 800, 1600, and 3200 Mbps, which keeps it well ahead of USB 2.0 (with a maximum rate of 480 Mbps). The standard IEEE 1394b also supports long transmission lengths of up to 100 m, using twisted-pair cables at a data rate of 100 Mbps. This is still an order of magnitude faster than RS-422 or RS-485 transmission. Data rates of 3200 Mbps can be supported on glass optical fiber cables up to 100 m long. FireWire connectors can be either *4-circuit* or *6-circuit*; therefore, it is crucial to check what kind of connector each sensor accepts before selecting the linking cables of the vehicle.

6.5 The Controller Area Network (CAN) Bus for Off-road Vehicles

The term *network* is used for a system that allows two or more computers and/or microprocessors to be linked in order to exchange data. The logical form of the links is known as the network *topology*. The term *node* is used for a point in a network where communication lines or units are connected to the network lines. Commonly used topologies include data bus, star, hierarchy or tree, ring, and mesh networks. In the past few decades, the need for improvements in automotive technology has increased the usage of electronic control systems for basic functions such as engine timing, anti-lock braking systems, and enhanced ignition without the traditional distributor. With conventional wiring networks, data can only be exchanged individually through dedicated signal lines. As the complexity and number of devices in the vehicle increases, using one independent signal line per sensor has gradually become more difficult and expensive.

In order to overcome the limitations of conventional automotive wiring, Bosch developed the *controller area network* (*CAN*) bus in the mid-1980s. The CAN is a high-performance serial data communication protocol that efficiently supports distributed real-time control under very high levels of security. It allows controllers, sensors, and actuators to be connected on a common serial bus. This network of devices can be assimilated into a scaled-down, real-time, low-cost version of the networks typically used to connect computers. Any device on a CAN network can communicate with any other device using a common pair of wires. The data transfer protocol of a CAN strictly relies on a simplified *open system interconnection* model (OSI). In addition, depending on the field of application, the CAN standard also incorporates a number of application-based protocols. This reason was strong enough for the International Standardization Organization (ISO) and the Society of Automotive Engineers (SAE) to standardize the CAN specifications internationally as *ISO 11898*. Due to its growing popularity in automotive and industrial applications, the CAN standard has been increasingly used in a wide variety of applications, such as agricultural equipment, nautical machinery, medical instrumentation, semiconductor manufacturing, and the like.

When a CAN device transmits data onto the network, an identifier that is unique throughout the network precedes the data. This identifier defines not only the content of the data but also the priority privileges. A CAN identifier, along with its associated data, is often referred to as a *CAN object*. When more than one CAN device transmit messages simultaneously, the identifier is used to determine the priority with which each device accesses the network. The lower the numerical value of the identifier, the higher the priority. If an identifier collision is detected, the losing devices immediately cease their transmission and wait for the highest priority message to complete before automatically retrying. Because the highest-priority identifier continues its transmission without interruption, this scheme is termed nondestructive bitwise arbitration, and each CAN identifier is often referred to as an arbitration ID. This ability to resolve collisions and continue with high-priority transmissions makes CAN buses ideal for real-time applications.

In a CAN network, the messages transferred across the network are called *frames*. The CAN protocol supports two frame formats, with the essential difference between them being the length of the arbitration ID. In the *standard frame format,* also known as CAN 2.0A, the length of the ID is 11 bits. In the *extended frame format,* also known as CAN 2.0B, the length of the ID is 29 bits. ISO 11898 supports only the standard frame format, but standard SAE J1939 also includes CAN 2.0B. Figure 6.2 shows the fundamental fields of the standard and extended frame

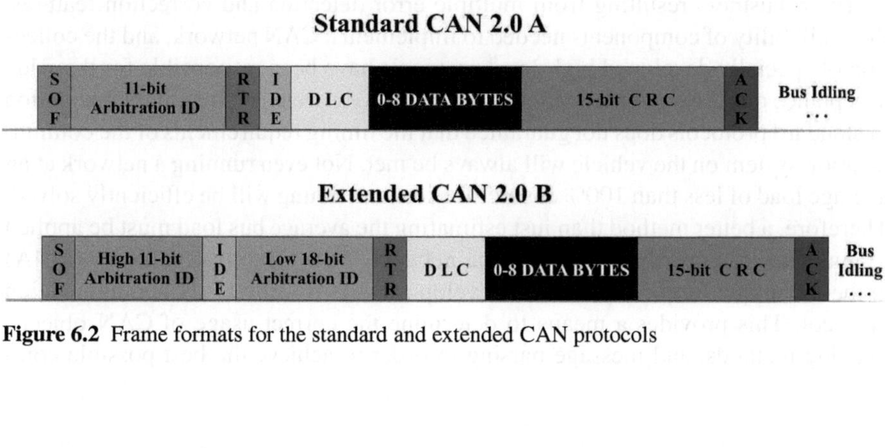

Figure 6.2 Frame formats for the standard and extended CAN protocols

formats. The *start of frame* (SOF) is a single bit (0) that marks the beginning of a CAN frame. Following the SOF are the *arbitration ID* fields containing the identifier for the CAN frame. The standard format 2.0A has one field of 11 bits, while the extended format 2.0B has two fields 11 and 18 bits in length. In both formats, the bits of the arbitration ID are transmitted from high to low order. The *remote transmit request* (RTR) bit is labeled "dominant" (0) for data frames and "recessive" (1) for remote frames. The *data frames* are the fundamental pieces of data transferred on a CAN network, and are used to transmit information from one device to one or more receivers. The devices on the network transmit a *remote frame* warning in order to request the transmission of a data frame for a given arbitration ID. The remote frame is used to request data from a source device, rather than waiting for the data source to transmit the data on its own. The *identifier extension* (IDE) bit differentiates between standard and extended frames. Because the IDE bit is dominant (0) for standard frames and recessive (1) for extended frames, standard frames always have a higher priority than extended frames. The *data length code* (DLC) is a 4-bit field that indicates the number of data bytes in a data frame. In a remote frame, the DLC indicates the number of data bytes in the requested data frame. Valid data length codes range from 0 to 8 bytes. The *data bytes* (DB) field contains the critical information circulated on the network and has a length of between 0 and 8 bytes. The remote CAN frames always contain 0 data bytes. The 15-bit *cyclic redundancy check* (CRC) detects erroneous bits in frames. To do so, the transmitter calculates the CRC based on the preceding bits of the frame, and all receivers recalculate it for comparison. If the CRC calculated by a receiver differs from the CRC in the frame, the receiver detects an error. All receivers use an *acknowledgement* (ACK) bit to recognize successful receipt of the frame. If the frame has been received satisfactorily, the ACK bit is transmitted as recessive (1), and is overwritten as dominant (0) by all of the devices that have received the frame successfully. The receivers acknowledge correct frames regardless of the acceptance test performed on the arbitration ID. If the transmitter of the frame detects no acknowledgement, the receivers may have detected an error (such as the CRC error), the ACK bit could be corrupted, or there may be no receivers available on the network. After the required number of recessive bits has been sent, the CAN bus is idle, and the next frame transmission can begin.

The robustness resulting from multiple error detection and correction features, the availability of components needed to implement a CAN network, and the collection of specially developed high-level protocols have been responsible for the wide acceptance of CAN communication buses. However, even the direct implementation of standard protocols does not guarantee that the timing requirements of the communication system on the vehicle will always be met. Not even running a network at an average load of less than 100% ensures that critical timing will be efficiently solved. Therefore, a better method than just estimating the average bus load must be applied to guarantee the proper behavior of the network. *Rate monotonic analysis* (RMA) is the technique frequently used to develop a model for the CAN communication protocol. This provides a means to determine the correct usage of CAN objects, queuing methods, and message parsing in order to achieve the best possible communication performance. This model facilitates the identification of three important parameters: (1) the average bus loading; (2) individual message latencies; and (3) the system's spare capacity. The correct determination of these factors allows users to guarantee stable and robust operations. The RMA is a collection of quantitative methods that enable real-time system developers to understand, analyze, and predict the timing behavior of many systems performing in real time. It applies to all types of real-time systems and can be used for hardware, software, or both [2]. In practice, it is a collection of simple mathematical methods that allow system designers to determine if a set of tasks can be guaranteed to meet their deadlines. This ability to meet deadlines is referred to as *schedulability*, in such a way that a task is *schedulable* if it is assured to occur before its deadline. For a periodic task, the deadline is usually the period; that is, the task must finish before the next time that is scheduled to run. When analyzing a real-time system, we must determine whether one or more resources can be shared effectively by accessing users within the time constraints of the system. In the case of software, the main resource is the processor time, but in the case of a communication network, the main resource is bandwidth. In general, the recommended steps to carry out an RMA analysis are: (1) selection of scheduling algorithms; (2) construction of message models; (3) schedulability study to check if all the messages are schedulable; and (4) parameter adjustment when needed.

In a CAN network, the message packets are considered to be the tasks of RMA analysis, and they compete for time in the network. When multiple nodes are ready to transmit a message at the same time, the CAN protocol determines their priority in a similar manner to when the tasks have different but fixed priorities. Each CAN message identifier always contains the message priority. Once a node has won arbitration, it is free to transmit the entire message without interruption; that is, the tasks are non-preemptive and cannot be interrupted once started. Each task, or message, must be guaranteed to meet its deadline; therefore, system-level timing considerations (such as end-to-end deadlines across nodes) must be satisfied. The following steps trace the procedure applied: (1) calculation of the CAN bus usage; (2) prediction of the worst-case latency of a set of messages; (3) verification of the system schedulability; and (4) estimation of the breakdown utilization as an indication of the system's spare capacity.

The *bus utilization* is the sum of the transmission times of the individual messages. In particular, the message utilization C_{m} for message "m" is the sum of the time required to transmit the entire message, including overhead bits, data bits and stuff bits, as detailed in Equation 6.1, where s_{m} is the number of data bytes in the message, s_1 is the number of stuff bits for the first 39 bits, and τ_{bit} is the bit time of 4 μs for 250 kbps. According to Equation 6.1, the number of overhead bits for a 29-bit identifier is 67, and there are 8 bits per data byte. The *stuff bits* are added automatically by the CAN controller to guarantee proper synchronization between nodes. The number of stuff bits for the first 39 bits can be computed exactly in the model, since the identifier will be known. The remaining bits for the data and the CRC checksum will be different for each unique set of data bytes; therefore, the worst case is considered to be when all the bits are either 1 or 0. In that case, the number of stuff bits is the number of CRC and data bits divided by five, since a stuff bit is inserted for every 5 bits of the same value. The bus utilization U_t is then given by Equation 6.2, where n represents the number of messages in the system. The parameter U_t gives an accurate indication of average bus loading, which is helpful from a relative perspective; for example, when measuring the relative impact of adding or deleting particular messages or changing their transmission rates. By itself, however, U_t does not guarantee the schedulability of the entire message set; a system with a low U_t could in fact be unschedulable if the timing deadlines are notably severe.

$$C_{\mathrm{m}} = \tau_{\mathrm{bit}} \cdot \left[67 + 8 \cdot s_{\mathrm{m}} + s_1 + \left(\frac{15 + 8 \cdot s_{\mathrm{m}}}{5} \right) \right] \tag{6.1}$$

$$U_t = \sum_{m=1}^{n} C_m \tag{6.2}$$

The next step in the design of CAN networks is to estimate the response time of each message. This requires a knowledge of the message utilization time computed above, the worst-case waiting time W_{m} (defined as the time that the message must wait for higher-priority messages), and the blocking time of a lower-priority message that just started transmitting before message "m" became ready to for transmission. These parameters can be determined from Equations 6.3–6.6 from Tindell [3]. A message "m" is schedulable if its worst case response time is less than or equal to the deadline for that message minus the jitter of the message, as expressed in Equation 6.3, where R_{m} is the worst-case response time of the message, J_{m} is the message jitter of the task relative to the beginning of the sending task, and D_{m} is the deadline of message "m," which is less than or equal to its period T_{m}. The worst-case response time of message "m" is finally given in Equation 6.4, where C_{m} is the longest time taken to send the message in the CAN bus found with Equation 6.1, and W_{m} is the worst-case queuing delay that can be computed recursively by Equation 6.5 (where B_{m} is the longest time that message "m" can be delayed by a lower-priority message as determined by Equation 6.6, with a period T_j for message "j"). The error function $E(t)$ indicates the longest time spent recovering from errors in any interval of duration t, as developed in [3].

$$R_{\mathrm{m}} \leq D_{\mathrm{m}} - J_{\mathrm{m}} \tag{6.3}$$

$$R_{\mathrm{m}} = W_{\mathrm{m}} + C_{\mathrm{m}} \tag{6.4}$$

$$W_{\mathrm{m}}^{n+1} = B_{\mathrm{m}} + \sum_{\forall j} C_j \cdot \left\lceil \frac{W_{\mathrm{m}}^n + \tau_{\mathrm{bit}} + J_j}{T_j} \right\rceil + E_{\mathrm{m}} \left(W_{\mathrm{m}}^n + C_{\mathrm{m}} \right) \tag{6.5}$$

$$B_{\mathrm{m}} = \max_{\forall k} (C_k) \tag{6.6}$$

One of the most important and useful features of the CAN bus is its high reliability, even in extremely noisy environments. This protocol provides a variety of mechanisms to detect errors in frames and retransmit the frames until they are successfully received. In addition, CAN networks also provide an error confinement mechanism that is used to remove a malfunctioning device from the network when a high percentage of its frames are defective, thus preventing faulty devices from disturbing the overall network traffic.

When a CAN device detects a problem in a frame, the device transmits an error flag consisting of a special sequence of bits. When the transmitting device detects this error flag, it retransmits the data frame to correct the error. The common CAN error forms include *bit error*, *stuff error*, *CRC error*, *form error*, and *acknowledge error*. During frame transmission, CAN devices monitor the bus on a bit-by-bit basis. If the level of the bit monitored is different from the transmitted bit, a bit error is detected. This bit error check applies only to the DLC, DB, and CRC fields of the transmitted frame. Whenever a transmitting device detects five consecutive bits of equal value, it automatically inserts a complemented bit, called a stuff bit, into the transmitted bit stream. This stuff bit is automatically removed by all receiving devices. The bit stuffing framework is used to guarantee enough edges in the bit stream and thus maintain synchronization. A stuff error occurs whenever six consecutive bits of equal value are detected on the bus. A CRC error is detected by a receiving device if the calculated CRC differs from the actual CRC in the frame. A form error occurs when a violation of the fundamental CAN frame encoding is detected; for example, if a CAN device begins transmitting the SOF bit for a new frame before the EOF sequence has been completed for a previous frame, meaning that it has not waited for the bus to become idle. An acknowledgement error is detected by a transmitting device every time it does not detect a dominant ACK bit.

To provide for *error confinement*, each CAN device must implement a transmission error counter and a reception error counter. The transmission error counter is incremented when errors are detected for transmitted frames, and decremented when a frame is transmitted successfully. The receiver error counter is used for received frames in the same way. The error counters increment more for the errors tracked than they are decremented for successful reception or transmission. Using this arrangement of the error counters, the CAN protocol can generally distinguish temporary errors, such as those caused by external noise, from permanent failures, such as a broken cable. With regard to error confinement, each CAN device can be in one of three states: *error active*, *error passive*, and *bus off*. When a CAN device is powered on, it begins in the error active state. A device in the error active state can normally take part in communication, and transmits an active error flag when an error is found. This active error flag (a sequence of dominant 0 bits) causes the current frame transmission to terminate, resulting in a subsequent retransmission. A CAN device remains in the error active state as long as the transmission and reception error counters are both below 128, so in a normally functioning CAN network, all devices are in the error active state. If either the transmission error counter or the reception error counter surpasses 127, the CAN device transitions into the error passive state. A device in the error passive state can still communicate, but it transmits a passive error flag when an error is detected. This passive error flag (a sequence of recessive 1 bits) generally does not cancel frames transmitted by other devices. Since passive error flags are not able to prevail over any activity on the bus line, they are noticed only when the error passive device is transmitting a frame. Thus, if an error passive device detects a reception error on a frame which has been received successfully by other devices, the frame is not retransmitted. One special rule to keep in mind is that when an error passive device detects an ACK error, it does not increment the transmission error counter. Thus, if a CAN network consists of only one device, and that device attempts to transmit a frame, it retransmits continuously but never goes into the "bus off" state. If the transmission error counter increments above 255, the CAN device transitions into the bus off state. A device in the bus off state does not transmit or receive any frame, and consequently cannot have any influence on the bus. The bus off state is used to disable a malfunctioning CAN device that frequently transmits invalid frames, so that the device does not adversely impact other devices on the network. When a CAN device has transitioned to bus off, it can only be placed back into error active state with both counters reset to zero by manual intervention. For sensors or actuators, this often involves powering the devices off and then on before resuming normal functioning.

6.6 The NMEA Code for GPS Messages

Regardless of the architecture devised for the intelligent vehicle under design, the availability of global positioning will never hurt; rather, it will very likely be beneficial to the vehicle. Even if global coordinates are only used to register the trajectory of the vehicle or as a tracking tool to know its location, GPS is becoming standard equipment for off-road automation. Since the prices of receivers are dropping fast, satellite signals are becoming ever more reliable and accurate, and commercial solutions are flourishing everywhere, GPS is becoming a common component of intelligent vehicles operating outdoors. As a matter of fact, the majority of the vehicles (if not all) used in the examples and case studies presented throughout this book incorporate some type of global navigation and localization system. All of these reasons compel system designers to understand and become literate in the integration of GNSS receivers with the other vehicle systems.

Most GPS receivers output data electronically through a serial port, typically referred to as COM port. However, the mere presence of a serial port in the receiving computer does not guarantee the direct and easy transfer of significant information from the GPS receiver; on the contrary: users need to cautiously configure the receiving computer or processor according to the manufacturer's instructions to ensure smooth communication between receiver and computer. When configuring a GNSS sensor, it is good practice to install a test application, usually provided by the manufacturer of the receiver as a firmware, to verify that data are being sent correctly and at the expected rate. Some receivers output data in customized formats developed by the manufacturer, but they still tend to follow the basic procedure of serial com-

Table 6.1 GPS NMEA messages commonly used for intelligent vehicles

Identifier	Specifications
GGA	Time, position, and fix data
GSA	GPS receiver operation mode, satellite information, and DOP values
VTG	Course and speed relative to the ground
ZDA	Time and date
RMC	Time, date, position, course, and speed data provided by a GPS or Transit navigation receiver

munication described in Section 6.3; that is, messages consist of a number of bytes bounded by start and stop bits called clusters. Fortunately, however, there is a standard format that has gained universal acceptance among practitioners, especially in real-time applications involving ground, marine and aerial vehicles: NMEA 0183.

The *NMEA 0183 Interface Standard* was created by the US National Marine Electronics Association (NMEA), and consists of GPS messages in text (ASCII) format that include information about time, position, velocity, and signal precision transmitted through a serial interface. There are multiple bit rates that can be used to transfer NMEA codes, but data bits are typically 8 bits in size, and have only 1 stop bit. This protocol uses neither handshaking nor parity bits. Messages are initiated by a six-character identifier starting with a dollar sign ($) and data fields are separated by commas. The next two characters after the dollar sign identify the sender (GP for GPS; GL for GLONASS), and the following three characters indicate the type of message (Table 6.1). If information on a field is not available, it will be represented by null bytes, which appear in the data string as two adjacent commas with nothing between them. The first character straight after the final character of the last data field is an asterisk, which is immediately followed by a two-digit checksum expressed as a hex number. GPS receivers operate at output rates of either 1 or 5 Hz; however, guidance applications require frequencies of 5 Hz.

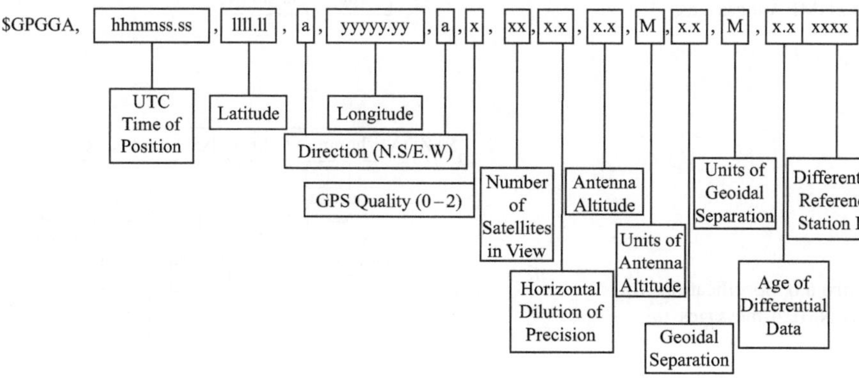

Figure 6.3 Specifications for GGA NMEA messages

Before integrating a GPS receiver into the intelligent system of a vehicle, it is essential to become familiar with NMEA code strings, so that the key information can be extracted from the original messages and the appropriate parsing can be carried out. To do this, it is necessary to know the meaning and position of every field in the expected strings. As mentioned above, the second part of the identifier – more precisely the last three characters – indicates the type of message sent by the GPS receiver. Table 6.1 reproduces the most common messages used in off-road vehicle automation, as described by the NMEA [4], whose detailed explanation is provided below in Figures 6.3–6.7. The most common messages used for off-road applications are GGA and VTG; especially the latter, as it includes velocity estimates relative to the ground.

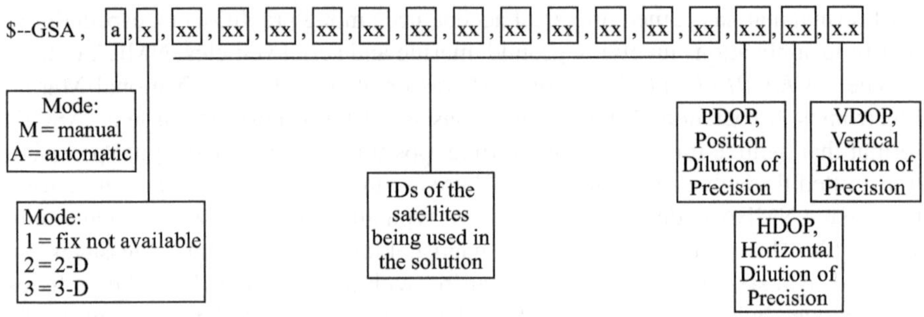

Figure 6.4 Specifications for GSA NMEA messages

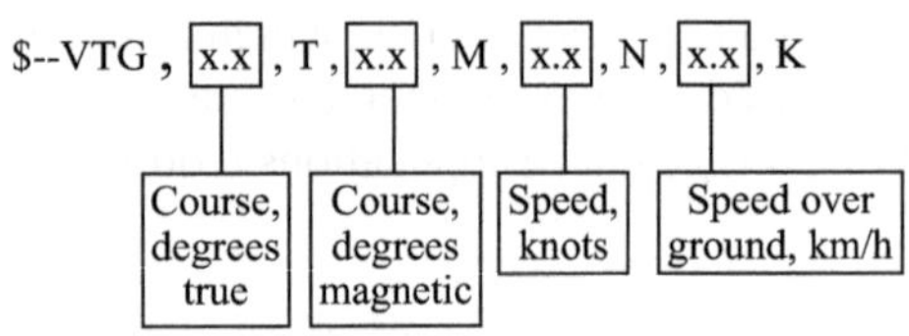

Figure 6.5 Specifications for VTG NMEA messages

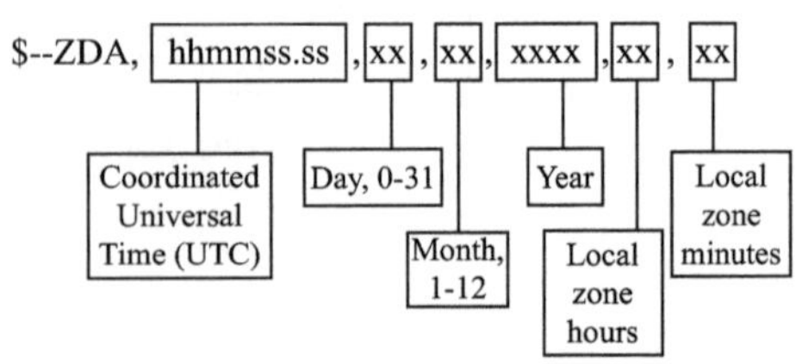

Figure 6.6 Specifications for ZDA NMEA messages

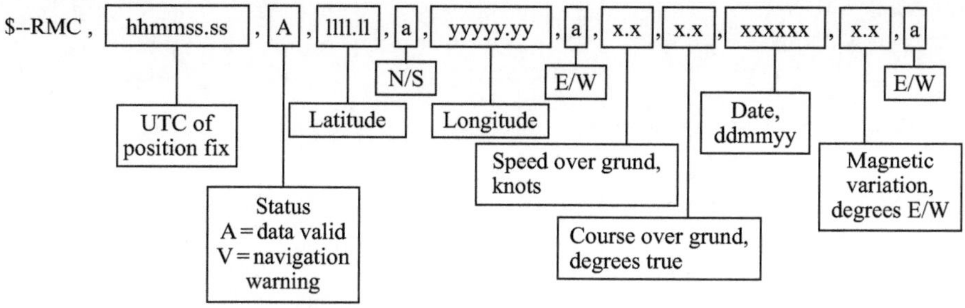

Figure 6.7 Specifications for RMC NMEA messages

6.7 Wireless Sensor Networks

One of the automatic modes described in Chapter 1 is teleoperation, which requires wireless communication between the vehicle and the center of operations, either to guide the vehicle or to actuate any of its attachments, such as a harvester downloading pipe. Some precision agriculture applications can benefit from remote data acquisition or transmission, so that while the vehicle is performing a task in the field, data is being downloaded to the farm computer station or received by the vehicle in real time. Nevertheless, in spite of this potential use of wireless technology by off-road equipment, it is not used very often within intelligent vehicles because most of the sensors needed for normal operations are onboard, and reliability is higher when physical connections are realized through protected wires. The large number of electronic devices on modern vehicles can generate harmful interference between radiofrequency emitters and receivers, which can result in erroneous data and even risky situations.

A *wireless sensor network* (WSN) is a set of autonomous sensors that is used to transmit or receive data for sensing and control. Each sensor on the network is typically coupled to a radio transceiver, a microcontroller, and a battery. Generally speaking, data rates and power consumptions are relatively low, with top bit rates of 250 kbps and power consumptions of < 1 W, depending on the duty cycle. The IEEE 802.15.4 standard is devoted to low-rate wireless personal area networks, and it defines three different frequencies according to the bit rates 20, 40, and 250 kbps. Normal distances for indoor applications are between 30 and 100 m, but outdoor distances can reach up to 1 mile with a free line of sight.

References

1. Bolton W (2000) Instrumentation and measurement. Newnes, Woburn
2. Klein MH, Ralya T, Pollak B, Obenza R, Harbour MG (1993) A practitioner's handbook for real-time analysis: guide to rate monotonic analysis for real-time systems. Kluwer, Boston
3. Tindell K, Burns A, Wellings A (1994) Calculating controller area network (CAN) message response times. Dept. Computer Science, Univ. of York, York
4. NMEA (2002) NMEA 0183: Standard for interfacing marine electronic devices. NMEA, New Bern

Chapter 7
Electrohydraulic Steering Control

7.1 Calibration of Wheel Sensors to Measure Steering Angles

The feedback control system that is used to autosteer an intelligent vehicle requires the real-time estimation of the angle turned by the front wheels if the machine uses Ackerman steering, or the angle turned by the rear wheels for vehicles that use inverse Ackerman steering. Articulated vehicles, on the other hand, feature non-turning wheels, as steering is achieved by turning one part of the vehicle's structure with respect to the other around an articulation joint. Therefore, these vehicles do not use sensors to measure the angle turned by the wheels, so they will not be discussed in this chapter. The calibration procedure described in the following paragraphs is appropriate for the most common types of agricultural vehicles: front wheel steered tractors and sprayers, harvesters and pickers, and undersized utility vehicles. However, any other off-road vehicles that steer by turning their front or rear wheels, such as lawn mowers and domestic robotic platforms, can also use it to automate navigation.

The closed control loop that governs autosteering requires an update frequency of at least 10 Hz, which imposes a similar refresh rate on the wheel angle sensor. In order to integrate the measurement of angles into the control loop, the output of the wheel sensor needs to be transformed to an electronic signal, typically a voltage in the range 0–5 V. Sensor calibration is simply the set of operations that are required to establish a *relationship between the angle turned by the wheel (in degrees) and the sensor output (in volts)*. There are three widely used wheel-angle sensors: hydraulic flow meters, linear potentiometers, and optical encoders. Flow meters measure the flow of hydraulic oil to and from the steering cylinder to estimate the angle turned by the wheels. Oil flow, conveniently converted to volts, is therefore correlated to the steering angle. Linear potentiometers are variable resistors that are fixed to the steering cylinder rod; a displacement of the rod that is actuating the wheels implies a change in resistance that can easily be related to the wheel angle. Optical encoders, unlike flow meters and linear potentiometers, provide a direct measurement of the angle turned by the wheel. A sensing shaft is typically attached to the king pin of the wheel being monitored. The shaft spins a disc with a distinctive pattern that is read by a light beam, and a voltage that is proportional to the angle turned by the shaft is output. The encoders can be relative or absolute, but either type is adequate for steering feedback as long as it is properly calibrated and initialized, since the shaft will never turn over a complete circle. Flow meters are notorious for their lack of accuracy, and even though they provide compact solutions for wheel-angle estimation, other alternatives are much more attractive for precise steering. Linear potentiometers have the advantage of directly outputting electrical commands,

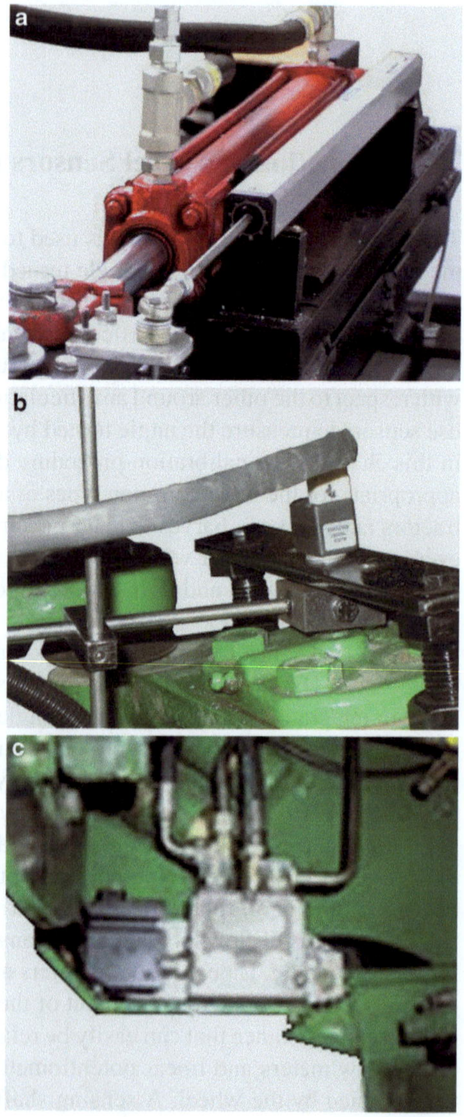

Figure 7.1 Sensors for wheel-angle estimation: (**a**) linear potentiometer; (**b**) optical encoder; (**c**) hydraulic flow meter (courtesy of Deere & Co.)

although an equation expressing the correlation between rod extension and angle turned needs to be found through calibration. Their main disadvantage is the attachment of the potentiometer to the cylinder rod, where crops, weeds, or branches can interfere with and damage the sensor during normal operation. Optical encoders have the benefit of directly measuring the angle turned by the wheel, given that the encoder's shaft is linked to the king pin of the wheel, which is the axis that the wheel pivots around. Assembling the encoders on the king pin is probably the most delicate stage in the determination of the steering angle. Figure 1.7 provides two examples of optical encoders mounted on large (a) and mid-sized (b) tractors. Illustrations of the three most common wheel angle sensors currently used in off-road equipment are provided in Figure 7.1.

The procedure used to calibrate wheel-angle sensors requires the acquisition of data from two different sources: the electrical signal from the sensor and the actual angle turned by the wheel. Voltage is easy to measure and record, but the precise estimation of the real angle turned by the wheel is not a straightforward task. The more accurate the calibration, the better behaved the vehicle. Figure 7.2 illustrates the entire process of wheel-angle calibration for the encoder shown in Figure 7.1b. The mid-sized tractor of Figure 7.2 incorporates two equal optical encoders, one for each front wheel. Both encoders are fed by an independent battery through a voltage reducer device. The output signal for each encoder is recorded by a portable computer located in the cabin. The calibration test consists of turning the steering wheel

Figure 7.2 Calibration procedure for optical encoders mounted on the front wheels of a tractor

slowly to cover the whole range from maximum left turn to maximum right turn, acquiring data from as many intermediate points as necessary. The voltages recorded by the onboard computer have to be reliably matched to the actual angles turned by the wheels. The operator in charge of steering the wheels at small intervals is very unlikely to be able to assess the angle being turned by the wheels from the driver's seat. In order to facilitate this task, a camcorder connected to an output monitor allows the driver to turn the wheels by the desired amount. The angle turned by the wheel is indicated by a goniometer on a rotating scale placed under each wheel, and is monitored by the camcorder. A second operator supports the calibration test from the ground, recording the angles (directly read from the scales) for both wheels.

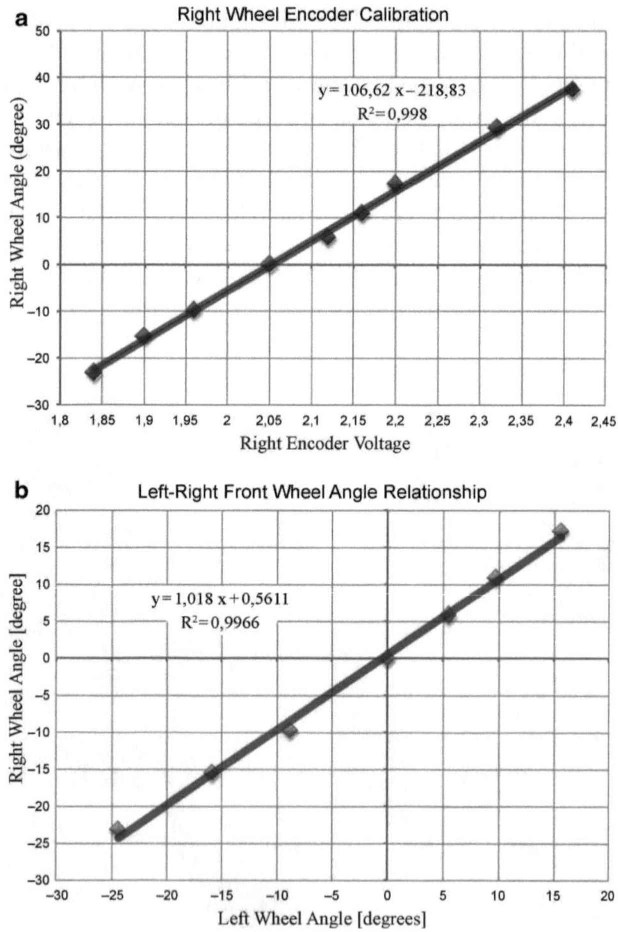

Figure 7.3 Calibration results for the encoder (installed in right wheel) shown in Figure 1.7a: (**a**) relationship between the voltage readings of the encoder and the corresponding angles turned by the right wheel; (**b**) angles turned by both front wheels for moderate angles

The results of the calibration test are a set of equations that relate the voltage output of each encoder to its corresponding wheel steering angle. When only one encoder is used, the left and right wheel angles must also be mathematically related. Given that a mechanical linkage couples both wheels, the relationship found will always be valid unless the steering mechanism is physically changed. Figure 7.3 represents the calibration equation for the absolute encoder of Figure 1.7a installed in the tractor of Figure 1.6, which was automatically guided along corn rows. Figure 7.3a shows the relationship between the voltage readings of the encoder and the corresponding angles turned by the right wheel, where the encoder was installed. Figure 7.3b relates the angles turned by both front wheels for moderate angles; it is clear that a linear relationship can be assumed in this case, which applies to the tracking of straight crop rows.

7.2 The Hydraulic Circuit for Power Steering

Hydraulic power circuits are designed to transmit energy from a generating source to the place where mechanical work needs to be performed using pressurized oil. Fluid power systems perform this energy transfer in three steps: conversion from kinetic energy to potential energy, delivery of the potential energy to where is needed, and conversion from potential energy back to kinetic energy. Figure 7.4 illustrates the energy conversion process in a typical fluid power system. As shown in this schematic, fluid power systems use a *hydraulic pump* to convert the mechanical power provided by a prime mover (usually a diesel engine on mobile equipment) into hydraulic power by raising the energy level of the *pressurized oil* from zero to a maximum level. The potential energy carried by the pressurized fluid is delivered to a control valve through a flexible hose. The *control valve* regulates the direction and amount of oil through different ports connected to more hoses, which are then used to deliver the fluid to actuators that actually perform the useful work, such

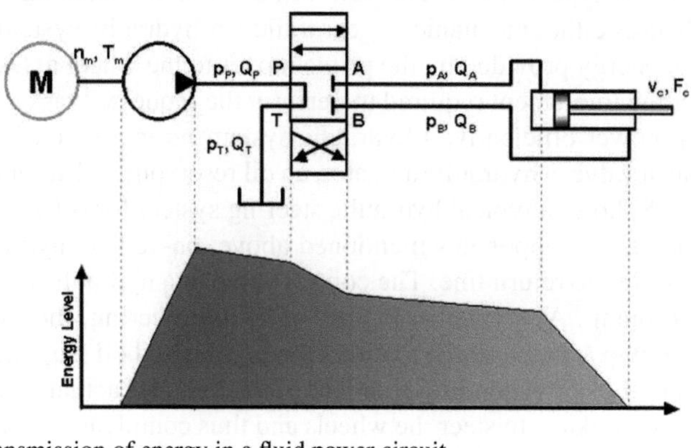

Figure 7.4 Transmission of energy in a fluid power circuit

as vehicle steering. It should be noted, however, that some hydraulic energy will always be lost during the transmission process due to friction as the fluid passes through the hoses and valves, as well as fluid leaks.

A fluid power system can be divided into the following subsystems based on function: power generation, distribution, deployment, and accessories. The *power generation* subsystem consists of the prime mover, say a diesel engine, and the hydraulic pump, and its main function is to convert the chemical energy present in the fuel to the hydraulic energy present in the pressurized oil. The *power distribution* subsystem often employs a variety of valves to control and regulate the pressure, flow rate and direction of the fluid in its way to the hydraulic actuators. The *power deployment* subsystem uses some hydraulic actuators – commonly hydraulic cylinders or hydraulic motors – to use the potential energy carried by the pressurized liquid to move the targeted loads. The *accessory* subsystem includes the reservoir (often called the tank), hydraulic hoses, filters, accumulators, and other ancillary components necessary for the proper functioning of the circuit.

Off-road vehicles are mainly designed to perform numerous tasks on the move, which means that they frequently need to relocate. Consequently, they are widely used in agriculture, construction, mining, and military operations. Agricultural tractors, hydraulic excavators, bulldozers, backhoe loaders, and tunnel drills are some common examples of mobile off-road equipment. Different types of mobile equipment are designed to perform different types of work, and so their appearances and structures can differ markedly. For example, an agricultural tractor and a hydraulic excavator are quite different in both appearance and structure because the tractor is mostly designed to pull and actuate farm implements in motion, while the hydraulic excavator is designed to perform static earthmoving tasks (although it does require frequent relocation). However, these two types of machines do have something in common: they include a prime mover, a power transmission system, and an implement. The most common prime movers used in off-road vehicles are diesel engines, which convert the chemical energy of diesel fuel into mechanical energy in the form of torque and angular speed to drive implements and power the vehicle via the mechanical transmission. Implements are, generally speaking, specific tools that perform a particular type of work. The power transmission system on mobile equipment normally uses either mechanical (gear trains) or hydraulic systems to transmit the mechanical energy provided by the prime mover to the wheel axles (usually the differential) or the implement required to perform the requested task.

To transmit power efficiently, a hydraulic system normally consists of at least a pump, a control valve, a hydraulic actuator, an oil reservoir, and several connecting hoses. Figure 7.5 shows a typical hydraulic steering system for off-road equipment. Apart from the basic components mentioned above, on-vehicle hydraulic systems also need a filter in the return line. The control valve in a hydraulic steering system is often a hand pump. As shown in Figure 7.5, while steering, the steering wheel turns the hand pump (via the shaft) to direct the pressurized oil (supplied by the hydraulic pump) to the corresponding chamber of the cylinder actuator, which in turn drives the steering linkage to steer the wheels and thus complete the turning maneuver. The characteristics of hydraulic power systems are heavily determined by the

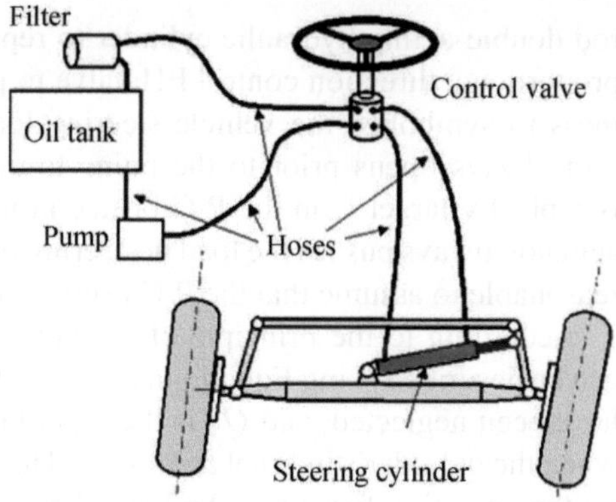

Figure 7.5 Basic hydraulic steering system used for off-road vehicles

physical properties of the pressurized fluids they use, but they offer many attractive features for mobile applications, and have therefore been widely utilized in mobile equipment to the point that they are the most common power delivery method for actuation systems. One of the most remarkable features they possess is an unrestricted geometric design, which allows the hydraulic energy to be delivered to all directions at the same capacity. Another attractive feature for mobile applications is their high power-to-weight ratio, which facilitates the design of compact power transmission systems that can deliver sufficient power to drive heavy loads. Other important features are their fast response to control inputs, their ability to instantly start or stop actuator motion, their capacity to supply a constant force or torque at infinitely variable speeds in either direction with smooth reverses, and their ability to stall without causing any damage to the system, thus providing very reliable overload protection. The major shortcomings of hydraulic systems are the difficulty involved in achieving accurate speed conversion ratios (mainly due to fluid leaking and compressibility), and their lower power transmission efficiencies compared to other means of power transmission.

An electrohydraulic (EH) steering system can be modeled – without any loss of generality – by assuming that an EH valve controls a hydraulic cylinder attached to a mass, which allows us to simplify the actual power-steering system of the vehicle. As shown in Figure 7.6, this simplified model of a steering system uses a single-

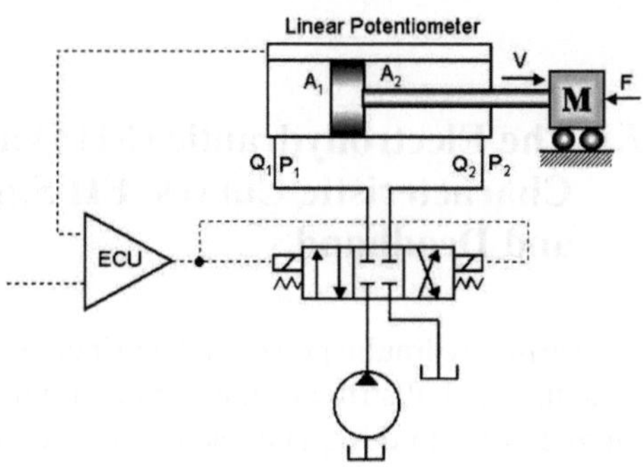

Figure 7.6 Simplified model used to design an electrohydraulic steering system

rod double-acting hydraulic cylinder to represent the steering actuator, a four-way proportional direction control EH valve to provide direction control, and a moving mass to symbolize the vehicle steering load. Because the cylinder-to-tank (C-T) port always opens prior to the pump-to-cylinder (P-C) port, and the C-T orifice is typically larger than the P-C orifice in this proposed control valve, the steering actuator always pushes the load in steering operations. In this situation, it is therefore reasonable to assume that the P-C orifice controls the rate of steering.

According to the principle of flow continuity, the actuator's extending motion can be described using Equation 7.1, where friction, oil compressibility, and leakage have been neglected, and Q_1 is the input flow that makes the rod move at speed V when the cylinder's internal area is A_1. The system's momentum can be determined by the actuating force and the load. The back-pressure on the steering cylinder is always much lower than the system pressure, and the friction associated with the cylinder is always much smaller than the steering load. Therefore, to simplify calculations for the EH steering system, it is realistic to ignore these two parameters during the development of the model.

$$Q_1 = A_1 \cdot V \tag{7.1}$$

Directional control valves regulate the oil flow rate by introducing flow restrictions to the internal pathways of the valve through adjustable orifices. Therefore, the flow rate supplied to the head-end chamber of the cylinder can be determined by the *orifice equation* given by Equation 7.2, where Q is the flow rate through the orifice, r is the fluid density, and DP is the pressure drop. Note that both the orifice coefficient C_d (0.6∼0.8) and the orifice area A_o are functions of the spool displacement, which is a function of the control signal. The behavior of an EH steering system is affected by the dynamics of the valve spools and hydraulic steering actuators, as well as the tire–ground interaction. Because an off-road vehicle often drives on changing terrain with a varying load, it is difficult to analytically determine the natural frequency and damping ratio of a particular system. As a result, it is highly recommended that system identification tests should obtain reliable system parameters under different operating conditions to facilitate the design of effective steering controllers.

$$Q = C_d \cdot A_o \cdot \sqrt{\frac{2}{\rho} \cdot \Delta P} \tag{7.2}$$

7.3 The Electrohydraulic (EH) Valve for Steering Automation: Characteristic Curves, EH Simulators, Saturation, and Deadband

Delivering hydraulic power is the principal function of the fluid power transmission system, which distributes the right amount of hydraulic power (carried by the pressurized flow) to designated actuators according to pre-designed strategies. Among the core components employed to construct power delivery systems, control valves demand special attention. Hydraulic control valves are used to regulate the pressure, the flow, and the direction of the oil transported within the enclosed system, and are therefore classified into three categories: pressure control, flow control, and direction control. However, this classification does not necessarily result in significant differences in physical structure; in other words, some valves can be used as a pressure control, a flow control, or a direction control valve for different applications with only minor structural modifications, or even without the need for any alteration.

An alternative to classifying valves according to their structural properties, as discussed in the previous paragraph, is to characterize them in relation to their control functions (such as pressure, flow and direction control valves) or based on their control mechanisms (such as on–off, servo, and proportional solenoid valves). Regardless of the classification system used, a hydraulic control valve controls the oil passing through it by adjusting the cross-sectional area of the channel that the fluid flows through in the valve. This adjustable area in an EH valve is technically named the *orifice area* in engineering practice. Physically, from a fluid-mechanical standpoint, the orifice is a controllable hydraulic resistance. Under steady-state conditions, a hydraulic resistance can be expressed as the ratio of the pressure drop across the valve to the passing flow rate, numerically given by Equation 7.3, where R_h is the hydraulic resistance, ΔP is the pressure drop, and Q is the flow rate traversing the valve.

$$R_h = \frac{\partial \Delta P}{\partial Q} \tag{7.3}$$

Control valves can use many orifice configurations to create multiple hydraulic resistances for different applications. Therefore, it is essential to determine the relationship between the pressure drop and the flow rate across the orifice. The orifice equation that is often used to describe this relationship is Equation 7.2, where C_d is the orifice efficiency, A_o is the orifice area, and r is the density of the fluid. The pressure drop across the orifice, ΔP, is equivalent to the pressure loss across the valve. The orifice coefficient, C_d, plays an important role in determining the amount of flow passing through the orifice, and tends to be estimated experimentally. In general, the orifice coefficient C_d varies greatly with spool position, but does not

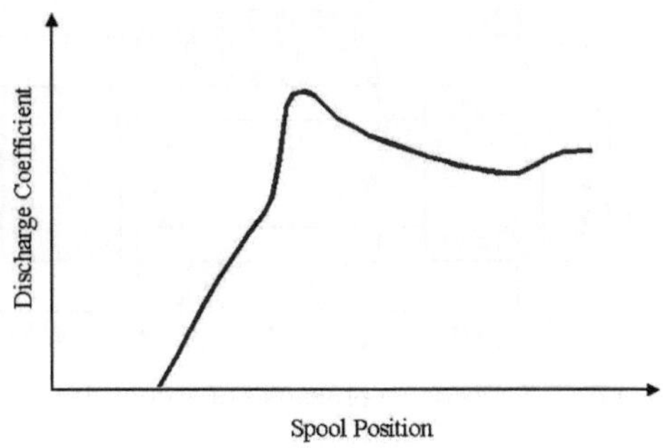

Figure 7.7 Relationship between discharge coefficient and spool position for a spool valve

appear to vary much with respect to the pressure drop across the orifices of spool valves, as illustrated in Figure 7.7. Analytical results obtained from computational fluid dynamics simulations show that valve spools and sleeve geometries have little effect on the orifice coefficient of the valve for large spool displacements. Even though it has been shown experimentally that the orifice coefficient varies with the spool position, it is a common practice to consider this coefficient to be constant (typically at around 0.65) to simplify design calculations, because the actual value of the coefficient barely changes after the orifice area surpasses a critical value. Additional studies showed that, in most cases, the orifice opening of a spool valve is greater than this critical value, so there is no great benefit from considering C_d to be variable. The orifice area for a spool valve A_o can be calculated using Equation 7.4, where D_{sp} is the spool diameter, x is the total spool displacement, and x_o is the spool displacement corresponding to the spool deadzone.

$$A_o = \pi \cdot D_{sp} \cdot (x - x_o) \tag{7.4}$$

The flow control characteristics of spool valves are similar to those of cartridge valves, as spools are also subjected to forces induced by the preloaded spring, the oil pressure, and the flow passing through the valve. The pressure force acting on a spool can either be balanced in direct-actuating valves by the symmetric structure of the spool, or unbalanced in pilot-actuating valves. In the former, an additional mechanical force such as the actuating force exerted by a solenoid driver or a manual lever is used to drive the spool to the desired position to control the orifice opening. In the latter, the unbalanced pressure force is used to push the spool away from its present position to a new one. As illustrated in Figure 7.8, the sliding spool is often centered by two preloaded springs, whose pressing forces tend to keep the spool in the central (neutral) position. The spring force F_s can be estimated with Equation 7.5, where k_1 and k_2 are the spring constants of the left and the right springs, x_o is the initial compression of both springs, and x is the spool displacement. We can very often assume that the left and right springs are identical, so they have the same spring constant k. In that situation, Equation 7.5 can be redefined as Equation 7.6.

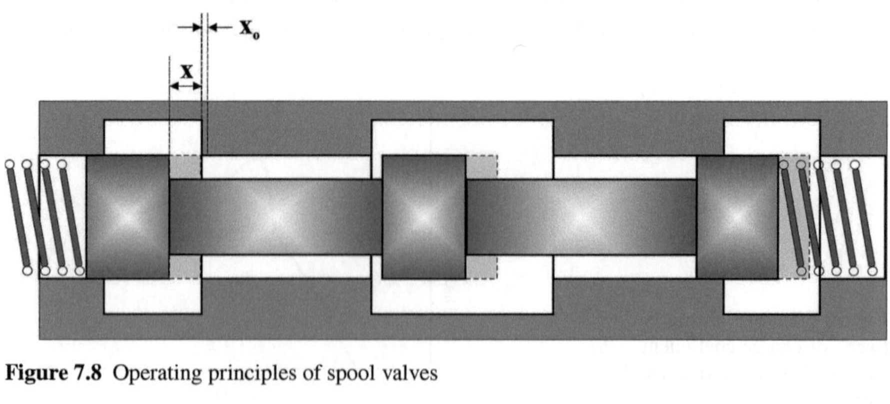

Figure 7.8 Operating principles of spool valves

It can always be assumed that the total pressure force acting on a directly actuated spool is zero because of the symmetric distribution of the pressure on a spool. The flow forces acting on the spool can be calculated using Equation 7.7, where F_F is the flow force, Q is the flow rate, V is the flow velocity, and β is the flow velocity angle, normally taken to be 69° [1].

$$F_S = k_1 \cdot (x + x_o) - k_2 \cdot (x_o - x) \tag{7.5}$$

$$F_S = 2 \cdot k \cdot x \tag{7.6}$$

$$F_F = \rho \cdot Q \cdot V \cdot \cos\beta \tag{7.7}$$

Valve transform curves, which often indicate the flow that passes through a spool valve at different spool positions, are commonly used to study the behavior pattern of a EH valve. As depicted in Figure 7.9, the typical valve transform curve of an overlapped spool valve normally consists of five characteristic zones classified into three categories: a central *dead zone*, two *modulating zones,* and two *saturation zones*. This archetypical transform curve reveals several key operating features related to the width of the dead zone, the width of the active zones, the flow gains and their linearity, and the saturation flow on both directions of the spool stroke. The dead zone is bracketed by the two *cracking points* of the valve, which are defined as the points in the curve where the valve just begins to open its flow passage. Inside this dead zone, spool shifting will result in very little flow passing through the valve due to the overlapping design of the valve. This fact implies, *a priori*, that a zero-lapped valve or an underlapped valve has a minimum dead zone, or even no dead zone. The dead zone is a vital parameter of EH valves because a significant dead zone can have a major impact on the performance of the hydraulic system in terms of both flow and pressure control. Consequently, dead-zone compensation techniques are often used in electrohydraulic systems.

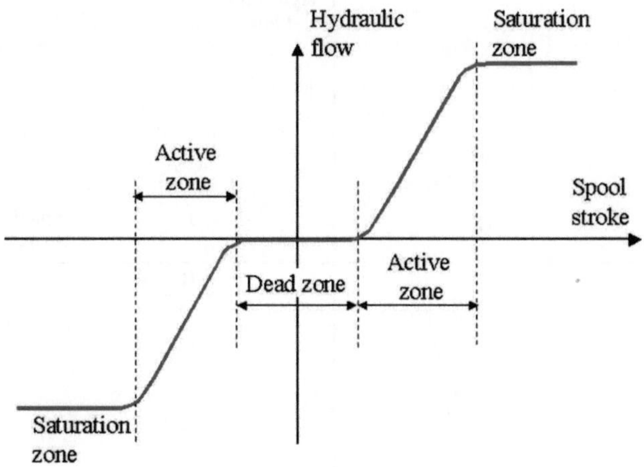

Figure 7.9 Typical valve transform curve for a general spool valve

The active zones are the actual operating ranges of a valve. One fundamental parameter that is often used to describe the control characteristics of a valve is the flow gain of the valve. A definition for the *flow gain* for a spool valve is the change in output flow with respect to the change in spool stroke, as expressed in Equation 7.8, where G_Q is the flow gain, Q is the flow rate passing through the valve, and x is the spool stroke measured from its neutral position. Ideally, a hydraulic valve should have a constant flow gain over the entire active range of the valve. In that case, the valve is called a *linear valve*. In fact, most hydraulic valves do not have a linear gain across their active ranges, so they are termed *nonlinear valves*. Fortunately, the flow gain is only a fundamental parameter for describing the control characteristics of the valve; it is not a particularly important parameter for actually controlling the valve because the flow gain is also affected by the pressure drop across the valve, which makes it variable with time. For this reason, the *average flow gain* is normally used as the design parameter for many hydraulic applications. Another important parameter that reveals some of the basic properties that are important for controlling a valve is the *hysteresis* of the valve. The hysteresis of a valve is estimated as the point of widest separation between the flow gain curves obtained with increasing inputs relative to that deduced with decreasing inputs, as measured along a horizontal line denoted H_v in Figure 7.10. Physically, this means that the capacity of the flow to pass through a certain valve opening is lower when the spool is opening the valve than when it is closing it. This phenomenon frequently affects the control performance of the flow if hysteresis is not properly compensated for. *Saturation* of a spool valve is habitually caused by its physical limitations, which mean that there is a maximum passage area for the flow. The flow saturation value in the valve transform curve can be used to estimate the *rated flow* that passes through the valve for a specific operating pressure. This parameter provides the information needed to

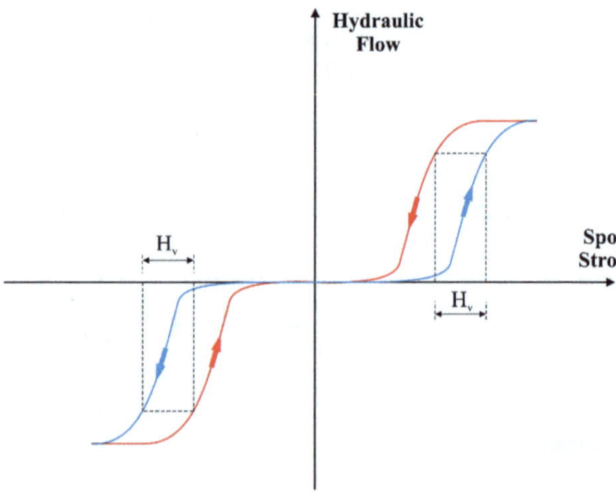

Figure 7.10 Spool valve hysteresis

determine the speed control capacity of the hydraulic system for the selected valve.

$$G_Q = \frac{\Delta Q}{\Delta x} \tag{7.8}$$

As previously discussed, a great number of off-road vehicles – mainly large machines working on harsh terrain – are equipped with a hydraulic steering system that can respond promptly and accurately to heavy steering demands. An efficient way of automating the steering mechanism of such a vehicle is to control the *electrohydraulic* (EH) *valve* that governs the steering cylinder. Electronically controlled hydraulic valves often behave nonlinearly, which makes it more complicated to study them, as the input signal, especially the frequency of actuation, has an important effect on the behavior of the valve. Autosteered off-road vehicles possess a distinct way of actuating in terms of the frequency and amplitude of the cylinder rod that produces the turns, so successful autosteering demands a deep understanding of the workings of EH valves at these excitation signals.

In order to design the complete steering system and a control strategy for it, it is necessary to deduce the mathematical model that best approximates the physical system, which is known in control terminology as the *plant* (Section 7.4). The model of the plant will typically be defined by a set of differential (or difference) equations or their corresponding transfer functions. For the specific case of EH valves, the elucidation of these model equations tends to be complicated by the high nonlinearity of the hydraulic system. However, an alternative solution based on the idea of *hardware in-the-loop simulators* is feasible. Such an apparatus features a hybrid assembly of hardware and software such that components that are difficult to model – the EH valve and cylinder in particular – are physically integrated into the simulator; the other operating functions (such as navigation commands, controller gains and algorithms) are all managed from the simulator computer. This method of connecting the mechanical components of the system with the control loop to simulate and develop control strategies is an efficient way of designing and evaluating a controller algorithm when the theoretical model of the plant is difficult or impossible

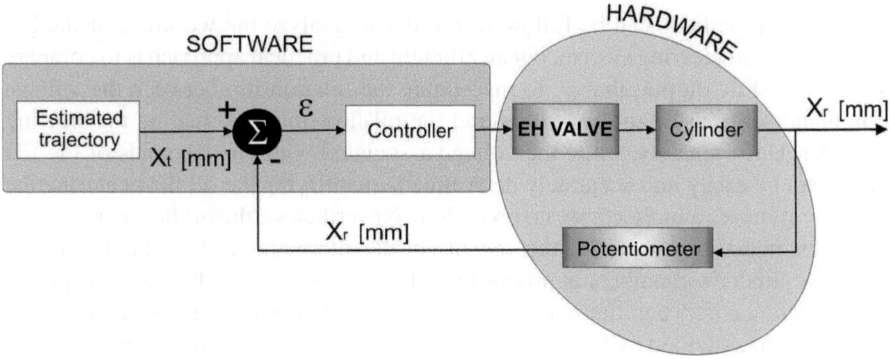

Figure 7.11 Block diagram for the control system of a hardware in-the-loop hydraulic simulator

to determine with acceptable precision. The diagram of Figure 7.11 illustrates the concept of hardware in-the-loop simulation, where the hardware elements are the EH valve, the steering cylinder, and the feedback sensor (linear potentiometer) that estimates the instantaneous position of the cylinder rod. This assembly reproduces the steering system of an off-road vehicle, but the conclusions extracted from it are scalable to real platforms. Controlling the position of the cylinder rod is equivalent to controlling the angle turned by the wheels, as rod and wheels are mechanically connected through the steering linkage.

The first step in steering automation is to calculate the turning angles that allow the vehicle to follow the desired trajectory. The onboard navigation engine, based upon information gathered from perception and localization sensors, estimates the theoretical angle θ_t. Using the calibration results, it is always possible to establish a three-way relationship between the position of the cylinder rod, the voltage output of the wheel-angle sensor, and the actual angle turned by the wheel. Even though an optical encoder mounted on the king pin of the wheel will directly provide an estimate for the angle turned, the intermediate calculation of the rod position is very helpful for assessing the workings of the EH valve, as an estimate for the oil flow traversing the valve is required to study its performance. In fact, once the steering mechanism has been properly analyzed, determining this relationship becomes a straightforward task. In this case, the initial angle θ_t inferred by the navigation computer corresponds to a particular extension of the cylinder rod X_t, thus setting the input signal to the steering control system. This angle, and its corresponding rod position, is termed *theoretical* because it is obtained from sensors external to the steering system, and the calculations of these angles depend on the desired and actual positions of the vehicle. Conversely, at a given time, the end of the cylinder rod will be located at a different position, denoted the *real* position X_r, which is determined by the wheel-angle sensor that provides feedback messages to the control loop. Therefore, at a given instant t_i in the recurrent control sequence, there will be a theoretical position and a real position of the steering cylinder rod; the difference between them will constitute the instantaneous *control error* ε, mathematically expressed by Equation 7.9.

$$\varepsilon_i = (X_t)_i - (X_r)_i \tag{7.9}$$

Different methods can be followed in order to analyze the workings of the EH valve of an autosteering system, but an efficient and practical approach is to compare its input and its output; that is, to investigate the relationship between the voltage command sent out by the controller and the oil flow that is actuating the steering cylinder. Unfortunately, while the voltage associated with the solenoids of the EH valve can be easily and accurately determined, quantifying the oil flow entering the cylinder requires a more elaborate procedure. Nevertheless, the oil flow can be indirectly estimated by measuring the velocity of the alternating rod, as the flow entering the cylinder's chambers is proportional to the extension and retraction speeds. The difference between the real positions of the cylinder rod for two consecutive measurements divided by the time interval between the measurements gives the approximate instantaneous speed of the rod, as explicitly indicated by Equation 7.10.

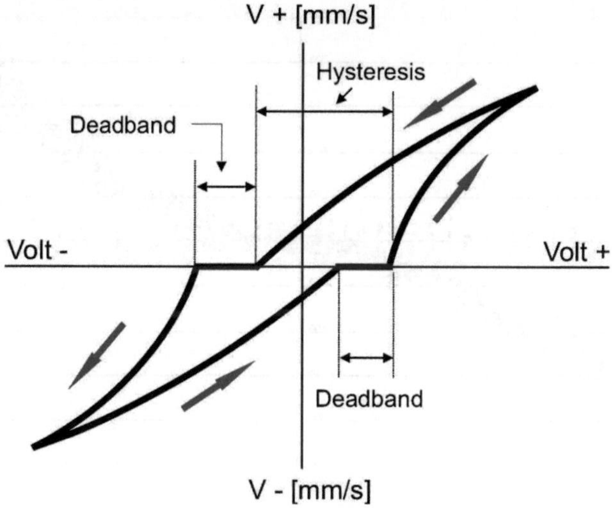

Figure 7.12 Characteristic curve of an electrohydraulic valve

A graphical representation of the voltage that actuates the EH valve (abscissa) versus the velocity reached by the rod (ordinate) is designated the *characteristic curve* of the EH valve. A stereotypical characteristic curve for an EH valve is displayed in Figure 7.12, where two fundamental phenomena are indicated: *hysteresis* and the *deadband*. The former quantifies differences between the extension and retraction strokes, while the latter indicates the presence of a range of voltages where changing the voltage does not result in any rod motion.

$$V_i = \frac{(X_r)_i - (X_r)_{i-1}}{t_i - t_{i-1}} \tag{7.10}$$

The main advantage of characteristic curves is that they can help us to understand how EH valves function while also providing a means to evaluate the whole control strategy. To do this, a sequence of input commands can be applied to the EH valve and the response of the cylinder can be assessed through its characteristic curve. The curve plotted in Figure 7.12 represents a theoretical all-purpose characteristic curve. However, autosteering poses quite specific demands on the frequency (f) and displacement (A) of the cylinder rod. A more realistic scenario would be represented by small amplitudes of 25 mm and frequencies of 1 Hz. The particular response of a hardware-in-the-loop simulated steering system that is subjected to such an input signal is depicted in Figure 7.13, where the phenomenon of hysteresis is clearly noticeable.

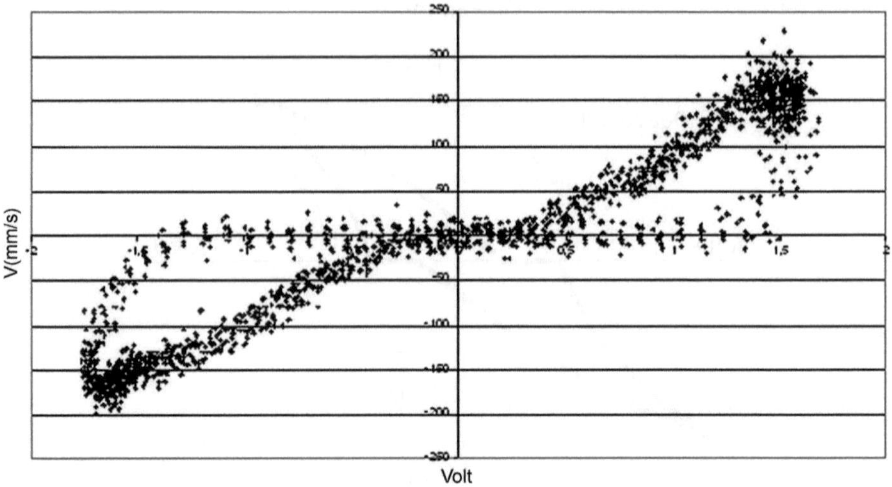

Figure 7.13 Characteristic curve for an EH valve tested at 1 Hz and for displacements of 25 mm

7.4 Steering Control Loops for Intelligent Vehicles

Intelligent off-road vehicles are designed to travel across and perform specific operations in a great variety of terrains. Automated path tracking is one of the fundamental operations required of such vehicles. In order to achieve accurate path tracking for intelligent ground vehicles, it is essential to design an appropriate steering controller. An *automatic controller* is an electronic device that is used to convert an operational command into an electrical control signal that drives an actuator; this actuator maintains or adjusts the status of the system (technically called the *plant*) in order to achieve the desired operational goal. In general, a typical control method works by modifying the actuator control signal according to the reference operational command and feedback signals. This can involve three possible paths: *forward*, *feedforward*, and *feedback*. The general control system of Figure 7.14 shows the three most common types of path used to automate ground vehicles. Control sys-

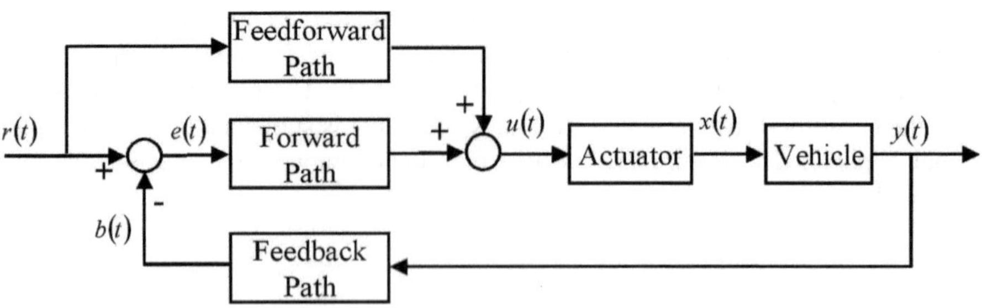

Figure 7.14 Block diagram of a generic steering controller

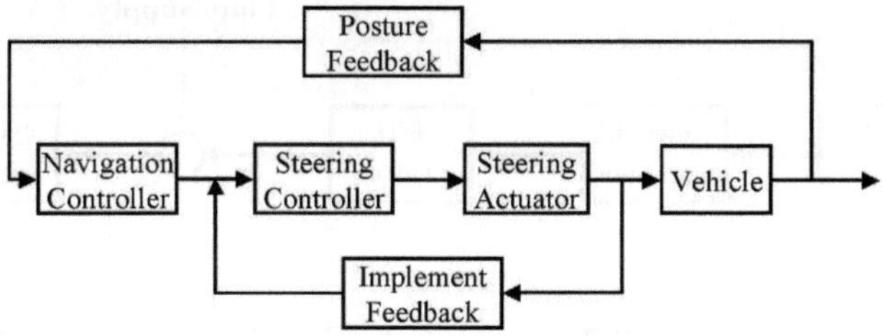

Figure 7.15 Basic components of intelligent steering controllers

tems mainly differ in the method used to modify the input commands and feedback signals.

Because of the wide variety of operating conditions that off-road vehicles need to cope with, the main challenge when designing a high-performance steering controller is to create appropriate steering commands to navigate the vehicle under a set of specific operating conditions. This problem can be solved in two steps: the design of the navigation controller algorithm, and its implementation in the steering mechanism of the vehicle. As illustrated by Figure 7.15, the basic idea of this approach is to separate the navigation controller from the steering controller in the design process. The result of this is the splitting of the control goal into two separate subgoals: searching for the optimum steering command in terms of vehicle posture, and accomplishing the desired steering actions promptly and accurately. This allows us to design a navigation controller independently of the steering actuating system, without needing to concerned ourselves with the turning dynamics. This procedure not only simplifies the design process, but also results in more robust and performant intelligent steering controllers.

Navigation control is a fundamental function of intelligent vehicles. There are a few obstacles to overcome in order to incorporate human maneuvering behavior into the design process of an industrial controller. The first obstacle is to provide a satisfactory design environment by integrating control dynamics and human behavior into the design process. The second is the need for a trainable real-time model that yields the right performance for different operators. Finally, the third difficulty is the need for an adaptive algorithm that adjusts the control system parameters automatically.

The major challenge when designing an *adaptive controller* is to find an effective means of tuning the steering controller parameters according to unpredicted changes in the system dynamics, also known as system disturbances. The rate of success of an adaptive control strategy can be estimated through improvements in steering performance, for example by reducing the deadband of the system or enhancing command modulation [2]. In a hydraulically actuated steering system, the deadband is determined by the command level corresponding to the first movements of the cylinder. The quality of command modulation includes gain variations and the linearity of velocity control under varying loads. When a *fuzzy-based adaptive* approach is applied, the navigation controller can modulate the input control signal in response to variations in the external loads by adjusting the valve transform curve.

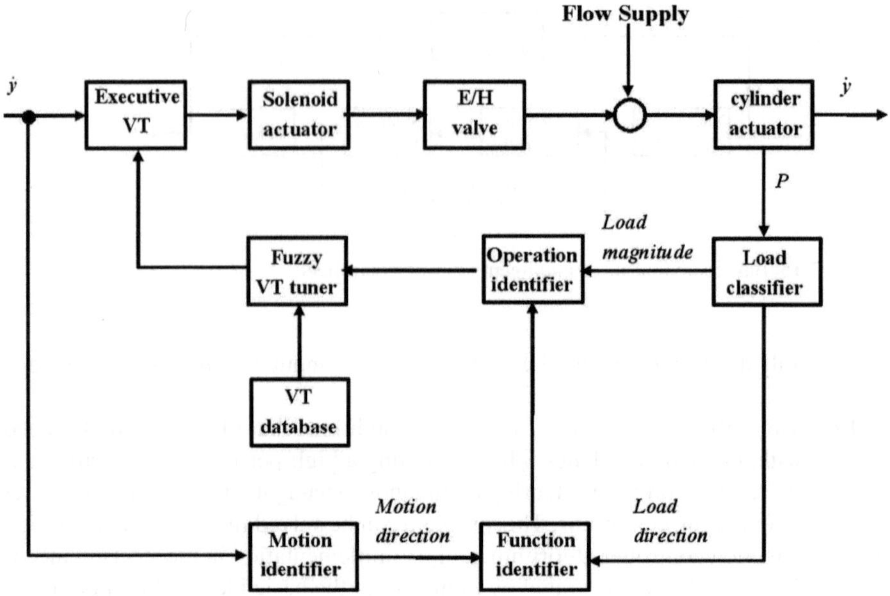

Figure 7.16 Working principle of a fuzzy adaptive navigation controller

As proposed in Figure 7.16, a fuzzy adaptive controller consists of an actuator operation identification function and an executive valve transform (VT) tuning function. The steering rate of a hydraulic actuator (cylinder) is closely interrelated to the magnitude and direction of the system load, which defines the operating conditions of the controller. Specifically, the fuzzy tuning algorithm needs to check the magnitude of the system load and the direction of motion of the actuator, select the appropriate valve transforms, and execute the best valve transform to achieve optimal control. The load quantifier is a key component in this fuzzy adaptive control scheme; this converts a real-valued system load into a couple of fuzzy-valued variables that provide the information fed into the fuzzy reasoning engine in charge of resolving the typical lack of linearity of the steering control. The database of valve transform curves stores curves for various potential operating conditions. A confidence factor is assigned to each selected valve transform. If two valve transforms are chosen in the fuzzy reasoning process, the controller applies a defuzzification method to both transforms to get an executive one. Compared to conventional control solutions, this fuzzy adaptive controller can tune the selected valve transform accurately for the operational conditions identified, effectively compensate for its high nonlinearity, and achieve accurate velocity control of a hydraulic steering actuator. Additionally, the fuzzy adaptive controller offers a general solution, as it is capable of controlling different steering systems that perform similar operations.

After separating navigation (or task) control from steering (or implement) control, the practical execution of steering control can be reduced to a simple steering actuator (cylinder–valve) control problem, the design objective of which is to real-

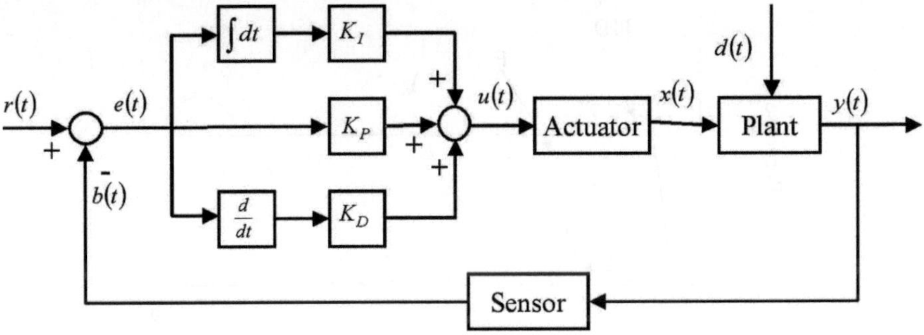

Figure 7.17 Implementation of a PID controller

ize the desired steering actions accurately and promptly according to the input commands. This simplification allows us to use most control methods to achieve a satisfactory steering performance for the controller. The following paragraphs introduce the main primary controllers, including PID, FPID, fuzzy, and neural-network controllers.

The acronym *PID* stands for *proportional–integral–derivative*, and it refers to a classical control method with well-developed controller design tools and tuning techniques. PID controllers are among the most commonly applied control methods in many fields of automation, including vehicle steering control [3]. In practice, a PID controller can be integrated with other control methods, such as fuzzy control, to minimize the negative effects of disturbances. As illustrated in Figure 7.17, a PID controller achieves its control goal by utilizing a *feedback signal* that reflects the actual operational state of the plant being controlled. When a PID controller receives a control command, the controller first compares it with the feedback signal to identify the difference between the desired setpoint and the actual state; it then introduces a correction to the plant with the controlled actuator. Due to its ability to make control adjustments based on actual plant states, a PID controller can also correct for undesirable behavior induced by external disturbances during plant operation. A PID controller uses three error correction methods, as shown in Figure 7.17, to achieve its control goals. Equation 7.11 mathematically describes how a control signal is calculated in terms of the identified error, where $u(t)$ is the control signal, $e(t)$ is the measured control error, and K_{P}, K_{I}, and K_{D} are the proportional, integral and derivative gains, respectively. In most autosteering applications, PID controllers are often implemented in a discrete form, as expressed in Equation 7.12, where $u(k)$ and $e(k)$ are sampled number sequences for the control signal and the error, respectively.

$$u(t) = K_{\mathrm{P}} \cdot e(t) + K_{\mathrm{I}} \cdot \int e(t) \cdot \mathrm{d}t + K_{\mathrm{D}} \cdot \frac{\mathrm{d}e(t)}{\mathrm{d}t} \tag{7.11}$$

$$\begin{aligned} u(k) - u(k-1) = {} & K_{\mathrm{P}}[e(k) - e(k-1)] + K_{\mathrm{I}} e(k) \\ & + K_{\mathrm{D}}[e(k) - 2e(k-1) + e(k-2)] \end{aligned} \tag{7.12}$$

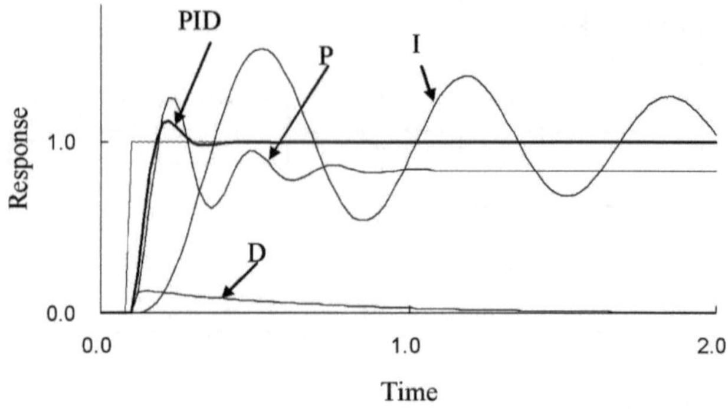

Figure 7.18 Contributions of the individual modes of PID controllers to the final response

Designing a PID controller mainly involves determining the appropriate control *gains* (K_P, K_I, and K_D) that will correct the error rapidly and accurately. Among the three control modes (proportional, integral, and derivative), the proportional control forms a control signal that reacts directly to the error in order to correct it. Thus, a proportional controller is simple to implement and simple to tune, but has an important performance limitation caused by its fast response to system errors: it can easily cause overcorrection and eventually result in an offset. The integral part of the control command composes a control signal in response to the integral (sum) of the error, and provides a means to perform effectively with zero-mean errors. However, its phase lag of 90° for the control signal may cause a reduction in the stability of the system. The differential control mode reacts to the rate of change in the controlled variable, measured by the derivative of the error, and is often used to stabilize the system. As plotted in Figure 7.18, when a PID controller receives a step command to increase its setpoint, the proportional segment of the controller reacts quickly to the deviation between the setpoint and the plant output feedback – *i.e.*, the error $e(t)$ – but often induces a steady-state error. The integral segment is sensitive to small errors and is used to eliminate the steady-state error provoked by the proportional mode, but its high sensitivity to small errors may cause unstable behavior. The differential segment is only sensitive to abrupt changes in control errors. Appropriately combining the P, I and D segments in a PID controller can yield satisfactory performance by compensating for the flaws of the individual segments, as illustrated in Figure 7.18. In practice, a PID controller can be designed with any combination of proportional, integral and derivative control modes in a forward path. This means that a PID-type controller can contain all three modes, it can contain any combination of two modes (PI, PD, ID), or it can have only a single mode (P, I, or D). The feedback path in a PID-type controller always employs proportional terms.

Different approaches are typically applied when designing PID controllers. One of the most common design methods identifies the transfer function of the plant being controlled, either theoretically or experimentally. This procedure is applicable to plants of a single variable that can be modeled as being linear time-invariant and lumped. A number of technical tools are available for the design of a PID controller based on this approach. However, a problem appears when there is a need to design a PID controller for plants that cannot be represented using a linear model, such as nonlinear plants. A classic solution for dealing with such plants is to *linearize* the plant model, and then design the PID compensator based on the linearized model, thus treating the plant as if it originally behaved as a linear model. Another approach that is often used in industry is to adjust the parameters of the PID controller in the hope that a satisfactory overall response will be obtained; however, there is no simple and general method for designing nonlinear PID controllers.

A well-designed controller must be carefully tuned to ensure the stability of the system and to meet other performance requirements. Normally, the tuning criteria include the following five performance requisites: stability, response rate, noise ratio, disturbance rejection capability, and parameter sensitivity. Many tuning methods have been developed since the birth of control theory. In general, the controller tuning process first searches for appropriate gains that will ensure the desired stability and response, and then it checks disturbance rejection and parameter sensitivities. As a practical rule, a complete tuning process is a two-step procedure comprising the identification of initial gains (based on the open-loop system) and the modification of those gains (based on the closed-loop system). There are several commercially available tools for tuning PID controllers, such as the control systems toolbox of Matlab. A traditional approach to tuning PID compensators is the Ziegler–Nichols method, which provides a way to find the PID gains by applying the following rationale. First, both the integral and the differential gains are set to zero, as if the controller were a proportional controller, and the proportional gain is increased slowly from a very low level until the system reaches instability. At this point, it is important to identify the two critical values of the proportional gain at which the system response starts to oscillate (the oscillation point) and becomes unstable (the instability point). Next, the integral and differential gains are determined based on the identified proportional gain. Because of the multiple loops needed to find a set of suitable PID gains, the Ziegler–Nichols method has both technical and economical disadvantages for tuning industrial controllers. There are also many self-tuning methods for automatically optimizing controller gains, which can be applied for initial setup or in response to changes under operational conditions.

A *feedforward-PID* (FPID) controller consists of the feedforward loop and the conventional PID loop drawn in Figure 7.19. The feedforward loop is used to create basic control signals in terms of an identified relationship between the control input

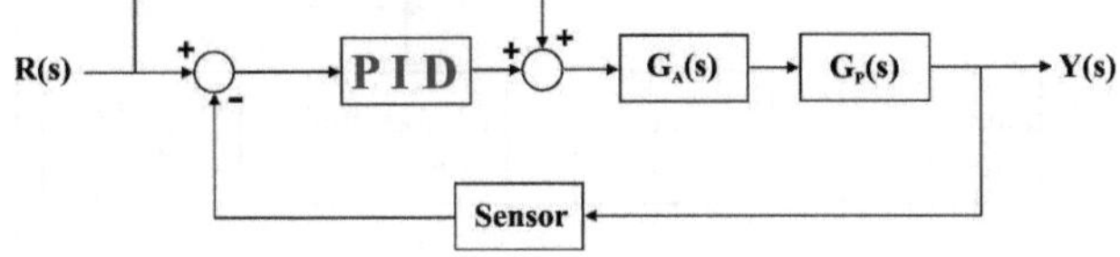

Figure 7.19 System diagram of a feedforward-PID control system

and the system response, which is often presented in the form of an inverse transform function of the plant being controlled. Based on the identified input–output relationship, it is possible to modify the inverse transform into a set of scheduled gains to compensate for the nonlinearity of the plant. In a similar manner to regular PID controllers, the PID loop is used to correct the control signal in response to the control error to improve tracking accuracy. FPID controller design can be accomplished in three steps: feedforward loop design, PID loop design, and finally F-PID integration. PID loop design follows the normal PID controller procedure discussed in the previous paragraphs, and may consist of different combinations of P, I and D segments according to specific applications. Feedforward loop design focuses on determining the feedforward function $G_F(s)$ and the gain K_F. In most cases, the inverse function of the plant dynamics, G_P^{-1} (s), is selected as the transfer function for the feedforward loop, and the feedforward gain K_F is often set to be equal to or less than 1 [4]. Given that the feedforward loop sends the command signal directly to the control actuator, it can avoid the delay caused by feeding the state variables of the plant back to a feedback loop, resulting in a faster response to control commands. In addition, it should be noted that the gains in a feedforward loop do not impair system stability, as they are determined independent of the plant dynamics by just inputting the control commands.

A *fuzzy controller* uses human heuristic knowledge to intuitively create control rules that allow the controller to emulate human behavior to a certain extent. A general scheme of a fuzzy controller is depicted in Figure 7.20. In general, a fuzzy controller consists of an input module, an inference engine, and an output module. The input module is used to map the system inputs, given in numeric sensor readings and ambiguous human concepts, into appropriate fuzzy logic values that can support word computing [5]. The inference engine, the kernel of a fuzzy controller, processes word variables to identify the appropriate control actions for the plant under a set of specific conditions. These control actions are derived from perception-based control rules. The output module generates a numerical (*i.e.*, machine-executable) control signal based on the activated rules and the conversion method selected. Since they represent a control strategy that is capable of mimicking human behavior and nonlinear systems, fuzzy controllers have been successfully used to control vehicle steering systems [6]. The basic functional components of a fuzzy controller are knowledge-based control rules defined in natural language, which are used to rep-

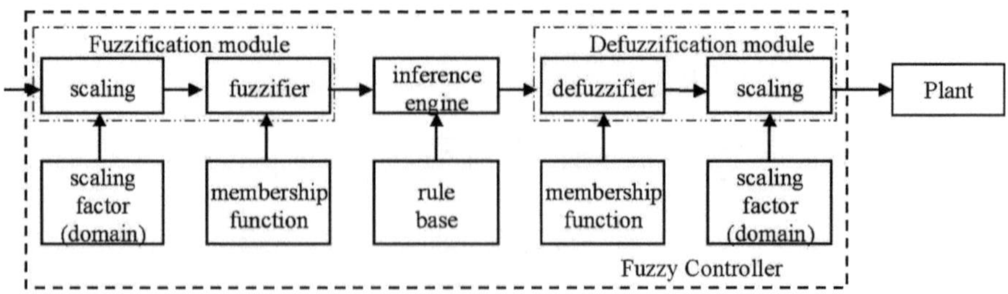

Figure 7.20 Overall structure of a fuzzy controller

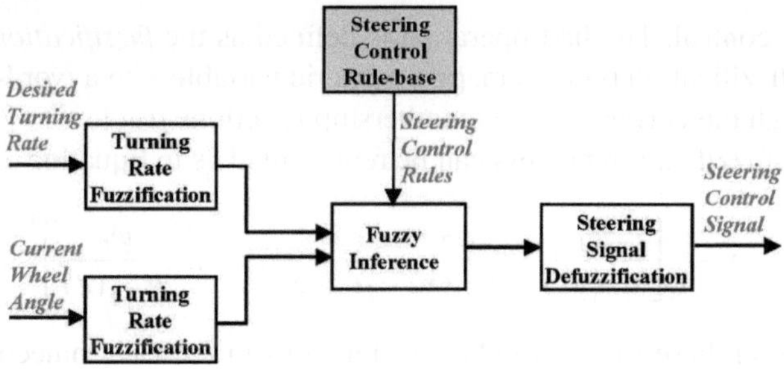

Figure 7.21 Fuzzy steering controller for off-road vehicles

resent typical control strategies of a human expert for various predefined circumstances.

The block diagram of Figure 7.21 shows the information flow in a fuzzy controller that is designed to steer an off-road vehicle. The decision-making information flow defines the fundamental control laws of the controller and provides the basis for all of the steering control rules. The final objective of this fuzzy steering controller is to steer the vehicle along a desired pathway. In real life, human drivers steer according to the magnitude of the offset between the vehicle's direction of travel and the desired course. Human perception of the vehicle's response helps to decide if the guiding action is understeered, properly steered, or oversteered. To mimic human reasoning, inputs to the fuzzy steering controller may include the desired heading angle and the current steering action. The control laws of the fuzzy controller model the appropriate steering actions for a different combination of input variables, and follow a format such as "if desired δ is *hard-right-turn* AND actual state is *minor-understeer*, THEN action is *minor-steer-to-right.*" This sample control rule uses two input word values: *hard-right-turn* to describe the intensity of the desired turning rate and *minor-understeer* to evaluate the perceived results, and one output word value: *minor-steer-to-right* to establish the required steering action. This two-input/one-output rule can be represented in a generic form as a conjunctive rule, without any loss of generality, as given in Equation 7.13, where X and Y are the state variables, Z is the control variable, and A, B, and C are the linguistic values of the state/control variables X, Y, and Z in their universe of discourse:

$$\text{IF} \quad X = A \quad \& \quad Y = B \quad \text{THEN} \quad Z = C \tag{7.13}$$

As can be seen in the example above, the linguistic descriptions of the fuzzy variables, such as *hard-right-turn* and *minor-steer-to-right*, contain a certain degree of imprecision within their corresponding universes of discourse. To make these word values computable, fuzzy logic controllers employ *fuzzy membership functions* to quantify these variables in their universe of discourse. The quantification process consists of two operations: first, the conversion of numeric state variables into word-valued inputs allowing for word computations; second, the conversion of word-valued control actions into the numeric control signals needed to implement the automatic control. The first operation is defined as the *fuzzification process*. In general, the fuzzification process maps a numeric variable x to a word-valued variable w_i through the corresponding membership functions $\mu_{W_i}(x) = [0,1]$. Mathematically, the fuzzification process can be represented as in Equation 7.14:

$$F : x \rightarrow \left[\frac{w_1}{\mu_{w_1}(x)}, \frac{w_2}{\mu_{w_2}(x)}, \frac{w_3}{\mu_{w_3}(x)}, \ldots, \frac{w_n}{\mu_{w_n}(x)} \right]^{\mathrm{T}}. \tag{7.14}$$

Implementing these fuzzy-based control actions in the electromechanical steering actuator of an intelligent vehicle requires the conversion of the fuzzy steering actions into a crisp real-valued control signal through the *defuzzification process*. The defuzzification process, therefore, converts two or more fuzzy-valued control actions into one crisp real-valued implementation signal. Although there are many defuzzification strategies that can be used for different applications, the *center of area* (COA) method is recommended for calculating the real-valued control signals of steering controllers like the one represented in Figure 7.21, because this method averages the domains of the selected steering actions and thus inherently reduces potential noise, improving robustness and accuracy. On the whole, the design process of a fuzzy controller is a trial-and-error procedure that heavily relies on the experience and accumulated knowledge of the designer.

The design of a steering controller, in one way or another, requires a certain amount of knowledge of the dynamic behavior of the actuating mechanism. This knowledge, often modeled as the plant of the steering system, can be obtained by *mathematically modeling* the dynamic system or, alternatively, by conducting a practical *system identification* protocol. Because the modeling and identification of a steering system is very dependent on the vehicle of interest, only a generic overview focusing on conventional agricultural tractors with electrohydraulic steering systems will be presented here. One fundamental feature of an off-road vehicle's steering system is its nonlinear dependency on the external load, which leads to deadbands, asymmetric flow gains, hysteresis, saturation, and time delays. This situation results in serious difficulties when designing high-performance steering controllers. A model of the steering system is helpful to analyze the system dynamics, predict key system states, optimize system parameters, and design an efficient steering controller. The description of a complete model for a steering system usually includes the steering linkage kinematics, the hydraulic actuator, and the EH control valve dynamics.

Kinematic analysis of the steering linkage implies the derivation of the conversion equations that relate the displacement of the rod of the steering cylinder to the steering angle of the front wheels. This relationship often provides the feedback for a closed-loop controller, and even the connection between the external loads on the wheels and the actuating forces of the steering mechanism. Conventional steering linkages for off-road vehicles comprise one or two hydraulic cylinders, a trapezoidal bar linkage with one degree of freedom, two symmetric kingpin axles, and the front wheels. An algebraic analysis is commonly carried out to describe the nonlinear geometric relationship deduced from the spatial steering linkage. The displacement of a steering actuator is normally chosen as the control variable, and the corresponding wheel angles are expressed as mathematical functions of the actuator displacement. In order to simplify the model, it is common practice to assume that all components are rigid bodies and that any gaps in the joints are negligible. As they are constrained by the bar linkage mechanism, the wheel angles for both sides of the vehicle are always different; that is, the wheel angle on the inner side is always larger than that on the outer side during the turning maneuver. To further simplify the modeling process, the Ackerman angle θ, calculated as the average of both wheel angles, is often used to represent the control steering angle for the vehicle. The steering linkage naturally behaves nonlinearly, and the Ackerman angle can be approximately expressed as a second-order function of the actuating cylinder displacement ΔX, as indicated by Equation 7.15 [7]. It is worth pointing out that the curve deduced to estimate the Ackerman angle can be linearized if the steering angle is small.

$$\theta = a \cdot \Delta X^2 + b \cdot \Delta X \tag{7.15}$$

One practical way of modeling the hydraulic steering system is to divide the entire system into two sections: the EH valve subsystem and the cylinder subsystem. The former normally includes the electronic driver submodel and the valve–actuator submodel. The electronic driver submodel is used to represent the dynamic relationship between the voltage control signal and the corresponding displacement of the valve spool. The valve–actuator submodel sets the relationship between the orifice opening of the valve, which is determined by the valve spool position, and the displacement of the steering actuator. Three universal laws of physics – flow continuity, the conservation of momentum, and the conservation of energy – can be applied to develop this submodel [8]. Being a typical flow control system, the plant output is the rate of change in the angle of the vehicle's front wheels with respect to the input steering control signal. Because the hydraulic steering system is clearly nonlinear, model linearization is desirable in order to simplify the model. Linearization is commonly based on the three assumptions of no friction, no hydraulic leakage, and fluid incompressibility. The steering system model is actually a higher-order model, but – because the natural frequency of the valve is always much higher than that of the steering system – it is reasonable to simplify the plant model into a first-order model for velocity control or a second-order model for position control, as indicated by Equation 7.16, where τ and k are system-based constants [7].

$$G_P(s) = \frac{\dot{\theta}(s)}{U(s)} = \frac{k}{\tau \cdot s + 1}; \quad G_P(s) = \frac{\theta(s)}{U(s)} = \frac{k}{s \cdot (\tau \cdot s + 1)} \tag{7.16}$$

System identification techniques aim to build a mathematical model of a dynamic system based on measured plant data rather than by applying equations and the laws of physics. This goal can be accomplished by gathering fundamental information about the system of interest from measured data, which is then analyzed to build an *empirical model*. In other words, the objective is to identify the *input–output relationship* between the system variables, as measured input–output data pairs are used

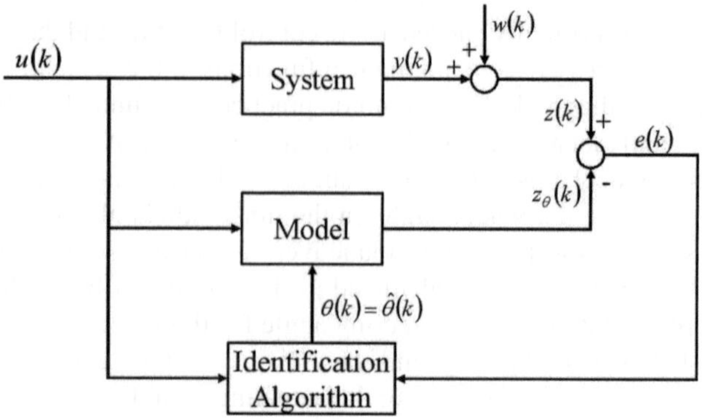

Figure 7.22 Principles of empirical system identification

to determine the quantitative relationship between the system variables. For this reason, it is important to collect a complete set of input–output data pairs that can support reliable system identification. Generally speaking, there are two main system identification methods: the frequency domain method and the time domain method. The *frequency domain method* provides essential information about the natural frequency, the damping ratio, and the stability of the system. The *time domain method* can perform either offline or online system identification, with the offline approach generally being more accurate than the online approach. Both methods, however, share similar working principles. As shown in Figure 7.22, the basic procedure is first to compare the actual output from the studied plant with the output estimated by the model for the same input (this may or may not take external disturbances into consideration), and then to apply the error between the actual and estimated outputs in order to adjust the model parameters according to a specific algorithm that attempts to minimize the error. A recursive procedure is normally applied to gradually optimize the empirical model, which can follow two general philosophies: nonparametric and parametric. *Nonparametric models* are easy to obtain, but they are often presented in graphs or tables, which complicates their direct application in simulations. Many well-known methods such as step response, pulse response, transient analysis, frequency analysis, correlation analysis and spectral analysis can be used for nonparametric model identification. The nonparametric model approach can be applied to identify almost any complicated system. In comparison, the *parametric model* needs to build the basic structure of the model first, and is therefore more difficult to obtain initially. The general procedure employed to design a parametric model for a dynamic system includes these four steps: 1) selection of an input signal; 2) construction of a model structure; 3) estimation of model parameters; and 4) validation of the model. After the empirical model has been identified, it can be used many times in simulation and control to design new applications. Methods commonly used for parametric model identification include (but are not limited to): the prediction-error identification method, linear regression and least-squares methods, maximum likelihood estimation methods, the recursive least-squares method, and fuzzy-neuro methods. Two particular system identification algorithms, namely the autoregression (AR) model and the autoregression moving average (ARMA) model, are widely applied in linear systems identification.

7.5 Electrohydraulic Valve Behavior According to the Displacement–Frequency Demands of the Steering Cylinder

After the disappointment suffered by early researchers in the area of artificial intelligence due to the failure of the General Problem Solver [9], it became apparent that intelligent systems should be focused and specialized to achieve a reasonable level of performance. In alignment with this philosophy, the control strategy for the EH valve that generates turns should take into account the nature of the commands expected when guiding an agricultural vehicle automatically. Using a semiautonomous behavioral approach, the vehicle will be autosteered within the crop rows or tree lanes – *i.e.*, for most of the working time of the vehicle – while turning maneuvers performed in the headlands will be left for the operator to control. Most of crop and tree rows are approximately straight or possess slight curvature; consequently, autoguided vehicles will require small angular corrections and an appropriate correction rate for prompt steering actuation. It is imperative, therefore, to understand the response of the EH valve to these kinds of control commands. The characteristic curves described in Section 7.3 and represented in Figures 7.12 and 7.13 serve this purpose well, although a comprehensive set of potential inputs must be tried in a complete hydraulic system. This section describes the characterization of an EH valve according to the displacement–frequency demands of the steering rod. The valve was tested in a hardware-in-the-loop simulator that complied with the structure of the block diagram depicted in Figure 7.11.

The characteristic curve obtained for the EH valve mounted on the simulator (Sauer-Danfoss, Ames, IA, USA) when the excitation command was a sinusoidal signal with an amplitude of 25 mm and a frequency of 1 Hz (Figure 7.13) was quite different from the general curve represented in Figure 7.12. Different steering situations are likely to cause changes to the shape of the characteristic curve; the key is to quantify these changes so that they can contribute to the design of the controller, thus enabling optimum vehicle performance (Section 7.6). Sinusoidal signals were selected as the patterns that most resemble steering during smooth driving. Operators exert continuous corrections around the center line of the vehicle's path. The amplitude of the input command (the magnitude of the lateral correction) and the frequency of the sinusoidal signal (the rate of correction) were the parameters studied in order to characterize the EH valve.

An amplitude of 25 mm is a reasonable trade-off to model autosteering commands when following a straight track. The actuation frequency is a trickier parameter to cope with, but understanding it is central to the design of a steering controller. Let us analyze what occurs when the frequency and amplitude are al-

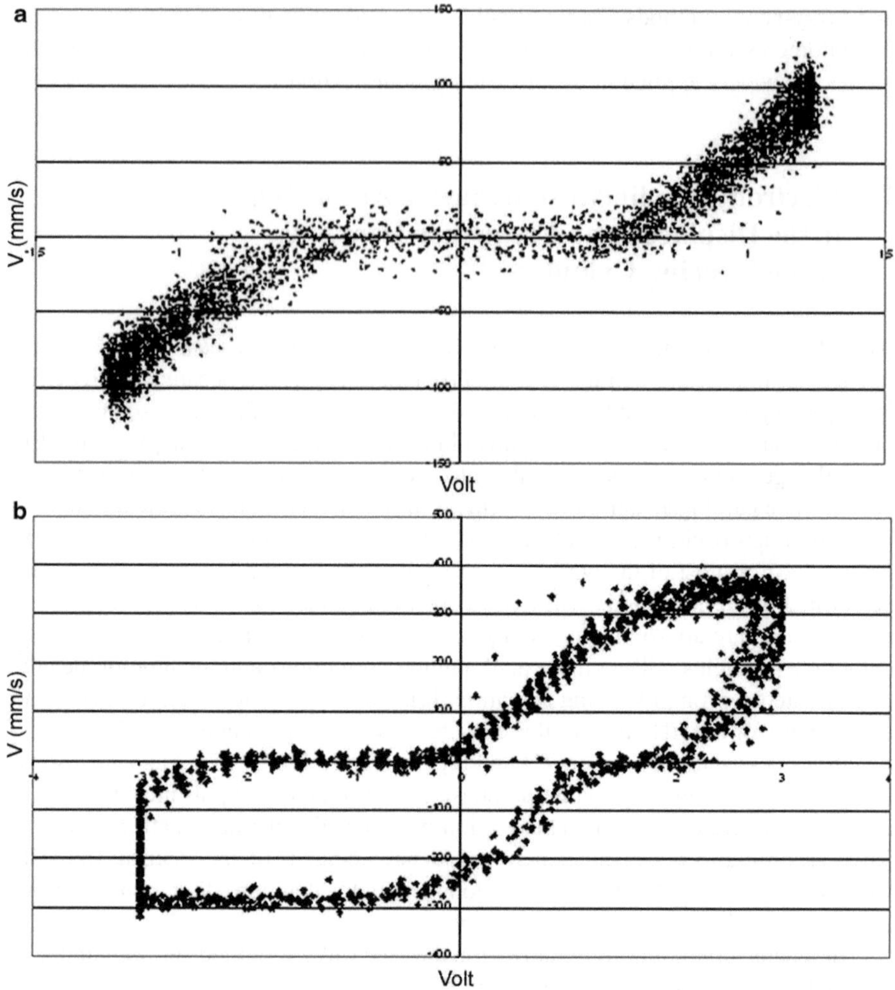

Figure 7.23 Responses to varying frequencies and amplitudes: (**a**) 0.1 Hz and 150 mm; (**b**) 2 Hz and 25 mm

tered for the same hydraulic system. Figure 7.23a shows the results of moving the rod very slowly (0.1 Hz) but making large displacements (15 cm), and Figure 7.23b shows what happens when the frequency is pushed up to 2 Hz and the amplitude is lowered to 25 mm.

The most noticeable difference between these figures is the *change of shape* provoked by altering the actuating frequency and amplitude. This can also be seen when comparing the figures with Figure 7.13 (which represents 1 Hz and 25 mm). Although the actuation frequency will probably depend on the traveling speed, a value of 0.1 Hz is not very realistic. However, the situation portrayed in Figure 7.23b is quite probable in the field. Doubling the frequency for the same amplitude of 25 mm resulted in a loss of symmetry (as shown in Figure 7.23b) due to the differential velocities in extension and retraction. The steering cylinder in the simulator was differential, just like the majority of cylinders mounted in off-road vehicles. Figure 7.23b shows the phenomenon of *voltage saturation* for both negative (−3 V) and positive (+3 V) voltages, but it also shows that the maximum rod speeds were reached: −300 mm/s in retraction and 400 mm/s in extension. Therefore, the frequency of the system shaped the characteristic curve of the EH valve, but it is important to determine the extent to which the curve can evolve. An extreme situation for the system tested (and thus for a common steering linkage) was simulated by keeping the amplitude at 25 mm but elevating the frequency to 4 Hz; that is, correcting steering four times per second. The curve's profile evolved such that the loop adopted a completely open, rectangular-like shape that was horizontally bounded by the maximum and minimum voltages and vertically limited by the top speeds during extension and retraction. The severe frequency demand of 4 Hz resulted in the delay of the rod with respect to the input command, despite the rod reaching its top speeds.

The multiple forms adopted by the EH valve curve depending on the requested working conditions (*i.e.*, amplitude and frequency) led to the *categorization of characteristic curves into four types*. A series of experiments [10] demonstrated that intermediate stages between two categorized conditions gave transitional curves between these two extreme cases, and both amplitude and frequency influenced the shape of the characteristic curve. Given that the effects of both parameters were coupled, valve curves were classified according to the product of the amplitude and the frequency. In fact, this product is essential in order to establish quantitative limits for the four types of curves found with the simulator. Figure 7.24 shows the results for the 24 tests performed, where each type of curve (I, II, III, and IV) is indicated

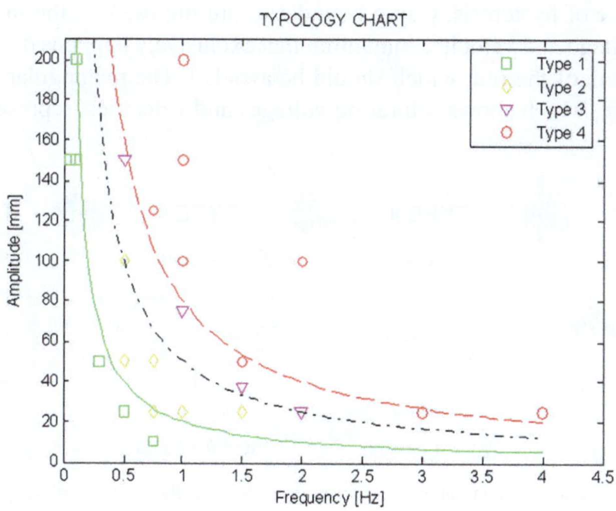

Figure 7.24 Frequency–amplitude chart for a hydraulic steering cylinder

and localized in a two-dimensional plot with the frequency (in Hz) measured on the abscissa and the amplitude (in mm) represented on the ordinate.

If the product of both variables – conveniently represented by Cartesian coordinates (x, y) in Figure 7.24 – is approximately constant, the locus of the points represented by the coordinates x and y follows a *rectangular hyperbola* of form $x \cdot y = K$. Furthermore, for the particular case studied in Figure 7.24, it is possible to define three rectangular hyperbolas that delimit the estimated domain of each type of curve. The function defining these four types of curve is given in Equation 7.17, and the mathematical representations of these hyperbolas are shown in Figure 7.24.

$$\text{Type:} \begin{bmatrix} \text{I} \\ \text{II} \\ \text{III} \\ \text{IV} \end{bmatrix} \rightarrow \begin{bmatrix} 0 < A \cdot f \leq 20 \\ 20 < A \cdot f \leq 50 \\ 50 < A \cdot f \leq 80 \\ A \cdot f > 80 \end{bmatrix} \tag{7.17}$$

The four types of curve defined by Equation 7.17 are schematized in Figure 7.25, where the progressive evolution from Type I to Type IV is apparent. The metamorphosis suffered by the curves of EH valves proves that the properties of the controller input signal actually shape the geometry of the curve, which is crucial to modeling and understanding the behavior of the EH valve.

Type I curves possess a *deadband* but lack the undesirable phenomena of *hysteresis* and *voltage saturation*. In this situation, the cylinder rod moves at speeds well below the most extreme values and, although the deadband needs to be properly compensated for, this type of curve is favorable for achieving effective control of the steering linkage. The profile representing Type II follows a double loop caused by hysteresis. It has a deadband but there are no signs of voltage or velocity saturation. This lack of saturation makes Type II a potential candidate for designing purposes. Type III curves force the loop to open; it becomes a unique ring that clearly shows the occurrence of hysteresis. Curve boundaries are imposed by the maximum values of the voltage and velocity, something that exclusively happens during the most extreme motion of the rod, which should be avoided. The rectangular shape of the Type IV curve, which shows saturating voltages and velocities, represents the least

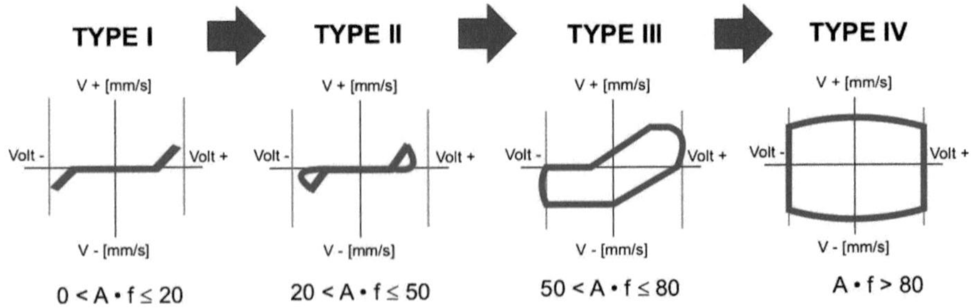

Figure 7.25 Evolution of characteristic curves according to the product of amplitude and frequency

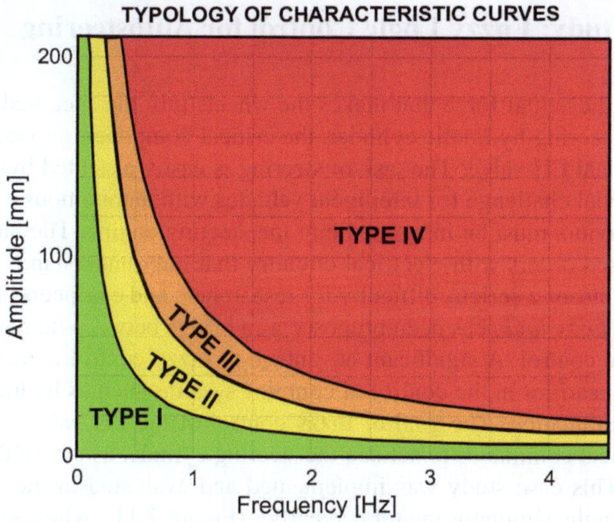

Figure 7.26 Typology chart for characteristic curves of EH valves

desirable of all of the cases analyzed, because the valve must work continuously under an intense flow of commands to keep up with the demanding input signal, which the valve can barely follow.

The dual-parameter classification of EH valves based on the amplitude and frequency magnitudes of the input signal provides a practical way to support the design of the controller. To facilitate its use, the graph of Figure 7.24 has been simplified in the *typology chart* of Figure 7.26. The chart places any pair of values for the amplitude and frequency of the signal in a *type area*. Generally speaking, areas marked as Type III and IV should be avoided, whereas Type I and II locations indicate promising conditions for optimum control. When there is no need for extreme rod velocities, the retraction and extension speeds will be approximately the same, thus yielding symmetrical curves like those represented by Types I and II. As autosteered vehicles face the same demands when turning left and right, the characteristic curves governing steering will have to be symmetrical – hence the preference for Types I and II.

While the differences between the shapes of the curves were explained as a consequence of the input signal above, the design of the hydraulic circuit on the simulator also influences the results. Thus, for instance, the fact that the cylinder is differential or the EH valve is proportional cannot be overlooked. In fact, this application gives an idea of how to characterize a hydraulic steering system in terms of procedure and philosophy, but a particular design would probably lead to different curve profiles. Furthermore, the design of each particular controller will also introduce distinct features into the curve shape, as demonstrated in the next section. Therefore, this section illustrates the complexity of characterizing hydraulic steering systems for automatic guidance, and suggests a methodology to cope with it. A more complete description of this procedure can be read in [10].

7.6 Case Study: Fuzzy Logic Control for Autosteering

Even though the actual force that makes the wheels turn the specified angle is exerted by the steering hydraulic cylinder, the critical component in steering control is the directional EH valve. The task of steering is easily executed by humans, but it is a non-trivial challenge for intelligent vehicles with autonomous navigation, as multiple behaviors must be integrated into the steering engine. The harmonization of human-based orders with electrical circuitry that only understands on–off electronic signals poses a serious difficulty for researchers and engineers. To cope with this problem, *fuzzy logic* has been proposed as a bridge between human knowledge and electronic control. A significant advantage of fuzzy control is its capability to handle non-linearities in the control of complex systems such as hydraulic steering for vehicle automation. The goal of fuzzy control of the EH valve is the generation of electrical commands to actuate the steering cylinder as smoothly and stably as possible. This case study was implemented and evaluated in the hardware-in-the-loop hydraulic simulator mentioned earlier (Figure 7.11), where small steering displacements and frequencies of below 4 Hz emulated in-field autosteering.

7.6.1 Selection of Variables: Fuzzification

The ultimate purpose of the feedback-based navigation controller of the vehicle is to calculate the steering commands, in the form of input voltage, that make the EH valve follow the desired path accurately. Fuzzy logic was selected to achieve this due to its unique ability to combine mathematical equations with knowledge-based linguistic variables. The first step in the construction of a fuzzy system is to define variables that determine the input–output functions. In a typical hydraulic system, such as the one represented in Figure 1.6, the controller output sent by the central processor is connected to the solenoid that actuates the valve; therefore, the *output variable* of the fuzzy controller is the *voltage signal* associated with the solenoids. Using the output variable as the input voltage to the EH valve is fairly common; however, the input variables selected for the fuzzy core rend to be more varied. For this study case, two *input variables* were considered: the *position error* (ε) of the steering cylinder rod defined by Equation 7.9, and the *rod velocity* (V) given by Equation 7.10. In general, fuzzy control of hydraulic systems tends to follow a proportional–derivative design, which utilizes the position error and its variation (derivative) with respect to time as input variables. Given that the velocity of the rod can easily be determined in the simulator through Equation 7.10, and bearing in mind that valve performance is evaluated via the characteristic curves that relate input voltage to rod velocity, selecting the rod velocity instead of the derivative of the error as an input variable results in a more intuitive and practical design. The structure of the fuzzy control system proposed for this case study is illustrated in Figure 7.27. The "curve selector" precedes the fuzzification phase, as it dictates

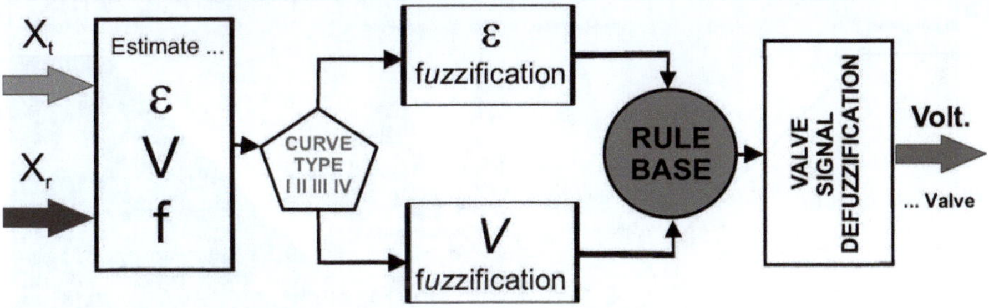

Figure 7.27 Fuzzy logic control system for hydraulic automatic steering

the philosophy of actuation and is based on the four types of curves defined in Figure 7.25. The curve selector embodies the typological chart of Figure 7.26.

According to the block diagram of Figure 7.27, the navigation engine containing the fuzzy algorithm receives two entries of diverse origin: the desired position of the rod X_t, corresponding to a specific wheel angle estimated by the perception–localization sensors, and the actual rod position X_r measured by the wheel sensor angle (potentiometer in the simulator; encoder in many robotized vehicles). As the controller was evaluated in a laboratory simulator, X_t was emulated by a computer in the form of sine waves with amplitudes and frequencies similar to those found while autosteering. The availability of X_t and X_r for every loop allows the two input variables ε and V to be calculated through the direct application of Equations 7.9 and 7.10, respectively. The frequency of the steering actuation needs to be estimated as well in order to select a curve from the typology chart of Figure 7.26. The *fuzzification* of the input variables ε and V involves assigning a category to the numeric values found after applying Equations 7.9 and 7.10. This operation is done using membership functions whose shapes and properties depend on the specific curve determined by the curve selector. For straight crop-track autosteering, it is reasonable to limit the rod stroke to 50 mm. Figure 7.28a shows the five levels (or categories) that were proposed to classify the position error ε: high negative (HN), low negative (LN), zero (Z), low positive (LP), and high positive (HP). The same linguistic levels (HN, LN, Z, LP, and HP) were defined for the membership functions of the rod velocity V, as shown in Figure 7.28b. Note that, even though the simulator top speeds for the rod were −300 mm/s in retraction and 400 mm/s in extension, the membership function for V is basically limited to $\pm$ 200 mm/s, as these are the typical maximum speeds obtained for Type I and II curves, which provide more stable control. The mapping of each input variable to the five classes resulted in *25 logic rules*, which provided the core of the expert system designed.

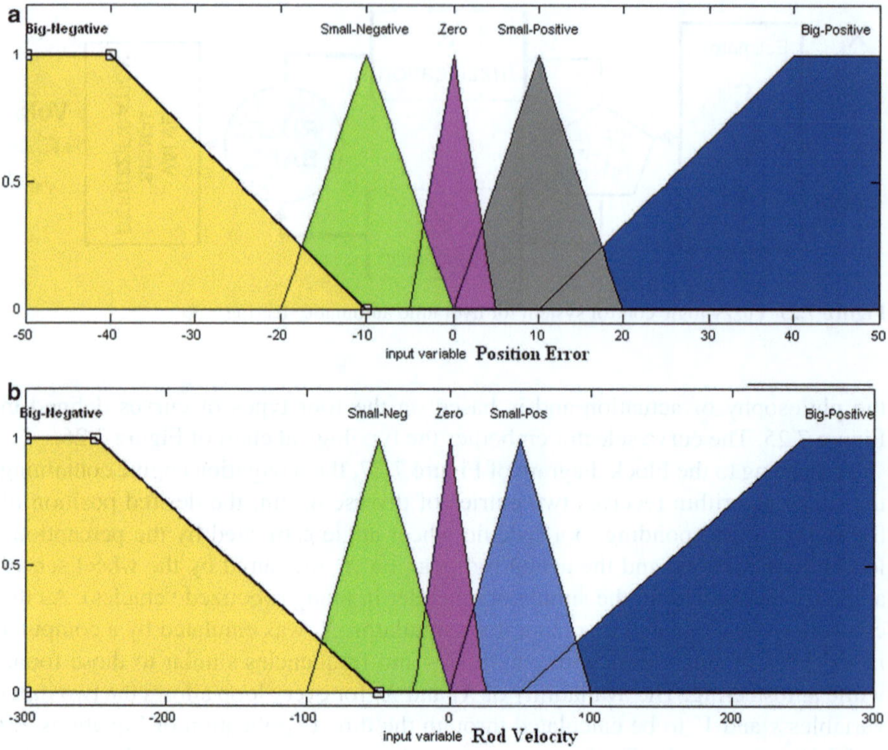

Figure 7.28 Membership functions for input variables of a fuzzy steering controller: (**a**) the position error ε; (**b**) the rod velocity V

7.6.2 Fuzzy Inference System

The *inference engine* of the fuzzy system consists of a set of rules, the *rule base*, which embody the core control strategy through knowledge-based instructions. The inference engine receives linguistic descriptions from the *fuzzification* stage, such as "large position error and low negative speed," and outputs linguistic variables as well, such as "low negative voltage." The final step in the fuzzy process is the conversion of the output variable (voltage) from a linguistic category to a numerical magnitude that will energize the solenoids of the EH valve to achieve the desired rod position X_t. This operation is denoted the *defuzzification* stage. The selection of position error and rod velocity as input variables introduces important changes in the general purpose loop given in Figure 7.11. The new block diagram for the proposed fuzzy control system is depicted in Figure 7.29. The sign of the velocity is set by Equation 7.10, and as a result, the rod moves from left to right when the estimated velocity V is positive, and *vice versa*.

The practical way of implementing experience-based inference rules is through conditional statements such as "IF Velocity is Low and Positive and Error is Low

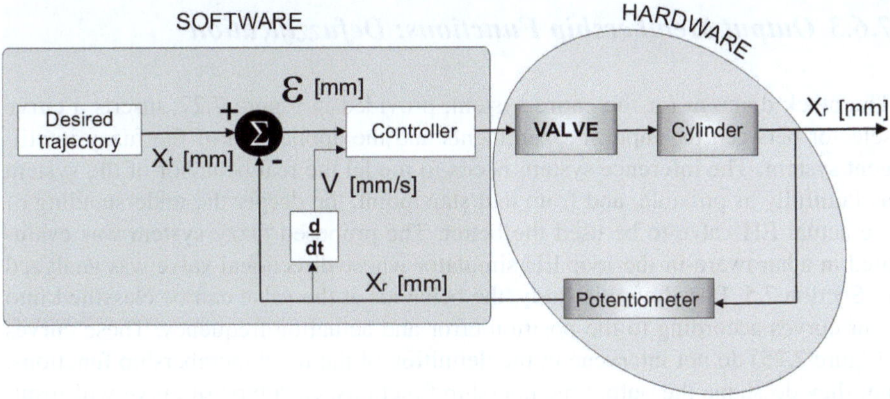

Figure 7.29 Block diagram for a fuzzy system that takes position error and rod speed as input variables

and Negative, THEN Voltage is Low and Negative." Given that the membership functions for both input variables have been grouped into five levels or categories, the total number of conditional propositions that form the base rule is 25 (5 × 5), as illustrated in the table of Figure 7.30. This procedure is also called an *expert system* because it is supposedly based on the experience of experts, or at least tries to mimic such a degree of expertise. The output of the rule base is a command voltage that belongs to one of the following five linguistic categories: *high negative* (HN), *low negative* (LN), *zero* (Z), *low positive* (LP), and *high positive* (HP). The defuzzification operation described in the next section establishes what HN voltage actually means and provides a unique number for it. This is the voltage that will actuate the EH valve.

Velocity of rod [V] \ Position Error [ε]	High Negative	Low Negative	Zero	Low Positive	High Positive
High Negative	HN	LN	LP	HP	HP
Low Negative	HN	LN	Z	LP	HP
Zero	HN	LN	Z	LP	HP
Low Positive	HN	LN	Z	LP	HP
High Positive	HN	HN	LN	LP	HP

Figure 7.30 Rule base, or inference engine, for the proposed EH control fuzzy algorithm

7.6.3 Output Membership Functions: Defuzzification

The block diagram for the entire system, provided in Figure 7.27, inserts a curve selector between the input measurements and the application of the fuzzy intelligent system. The inference system needs to model the real behavior of the system as faithfully as possible, and from that standpoint, the deeper the understanding of the actual EH valve to be used the better. The proposed fuzzy system was evaluated in a hardware-in-the-loop EH simulator whose directional valve was analyzed in Section 7.5. Based on this study, the behavior of the valve can be classified into four curves according to the position error and actuation frequency. These curves (Figure 7.25) do not intervene in the definition of the input membership functions, but they do shape the output membership functions; each type of curve will result in a particular set of functions. As a matter of fact, Type I curves follow several approaches with two, three, and five levels, because they provide the most commendable performance; Type II and Type III curves implement a five-level approach; and Type IV uses only three levels. A detailed description of the membership functions used for the output voltage can be found in [11]. The use of three levels of linguistic variables requires a simplification of the base rule, which can be carried out by means of the following criteria: HN and LN merge into one class, N (negative voltage), and HP and LP constitute class P (positive voltage). A further reduction into two levels can be performed according to the same procedure. The final operation to convert the linguistic variables (HN, HP, Z, *etc.*) into crisp voltages is executed on the selected (output) membership function when specific mathematical operations are applied, such as the center of gravity defuzzification method.

7.6.4 System Evaluation

The EH simulator described in Section 7.5 provided a means to evaluate the fuzzy algorithm proposed in this case study. In order to check its behavior, a simplistic proportional controller was run in the simulator under the same excitation inputs (sinusoidal signals emulating autosteering). Visual inspection of the rod position curve is not sufficient for a complete assessment of the controller response, so it is necessary to define a numerical parameter. Thus, the *error index* (EI) provides a quantitative evaluation of the controller performance by summing the differences between the desired rod trajectory X_{t} and the actual trajectory X_{r} for a given interval of time delimited by $[t_0, t_f]$, where $f+1$ represents the number of cycles evaluated. The EI is just the area trapped between the theoretical and real trajectory curves computed at discrete intervals and divided (normalized) by the time elapsed during the study. The calculation process can be followed through Equations 7.18 and 7.19, and a graphical representation of the error index is provided in Figure 7.31. Note that an intermediate variable, the *error sum* (ES), has been defined to ease the error

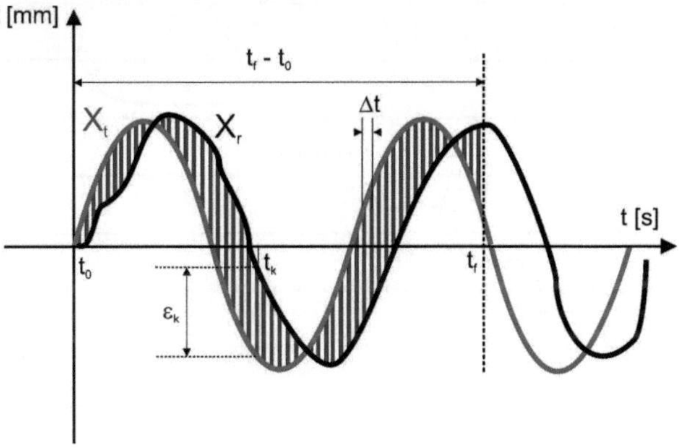

Figure 7.31 Graphical visualization of the concept of the error index (EI)

calculations; this represents the integral of the errors.

$$\mathrm{ES} = \sum_{i=0}^{i=f} |\varepsilon_i| \tag{7.18}$$

$$\mathrm{EI} = \frac{\mathrm{ES}}{t_f - t_0} \tag{7.19}$$

The total time considered for the calculation of EI is given by the expression $t_f - t_0$, so the total number of points studied (sampled) can easily be obtained as the quotient $(t_f - t_0)/\Delta t$, where Δt is the sampling interval in seconds. EI remains small as long as the trajectory followed by the rod sticks close to the input trajectory, but it starts to increase as soon as the rod cannot track the theoretical positions imposed by the signal generator of the simulator in a timely manner.

A typical attribute of the characteristic curves of EH valves, as shown in the Type I, II, and III curves of Figure 7.25, is the existence of a *deadband*. The presence of a deadband is a disadvantage for valve control: when a deadband is present, small voltages do not actuate the rod, but when the voltage exceeds a certain threshold and the rod initiates its motion, the position error ε has already grown excessively and the controller overreacts by displacing the rod at high speed to compensate for the error. In these circumstances, the rod often exceeds the target position and needs to reverse its motion unstably, resulting in fluctuating patterns. When the rod retracts, however, the errors ε_i are negative and the voltages tend to be negative, producing smoother motion. The effects of the deadband can be palliated by introducing the correction of Equation 7.20, where V_{al} is the voltage estimated by the fuzzy algorithm after the defuzzification phase, and V_{EH} is the voltage sent to the solenoid of the EH valve.

$$\begin{Bmatrix} \text{If} \quad V_{\mathrm{al}} > 0.05\ V & \Rightarrow V_{\mathrm{EH}} = 0.5 + V_{\mathrm{al}} \\ \text{If} \quad V_{\mathrm{al}} < -0.05\ V & \Rightarrow V_{\mathrm{EH}} = V_{\mathrm{al}} - 0.5 \end{Bmatrix} \tag{7.20}$$

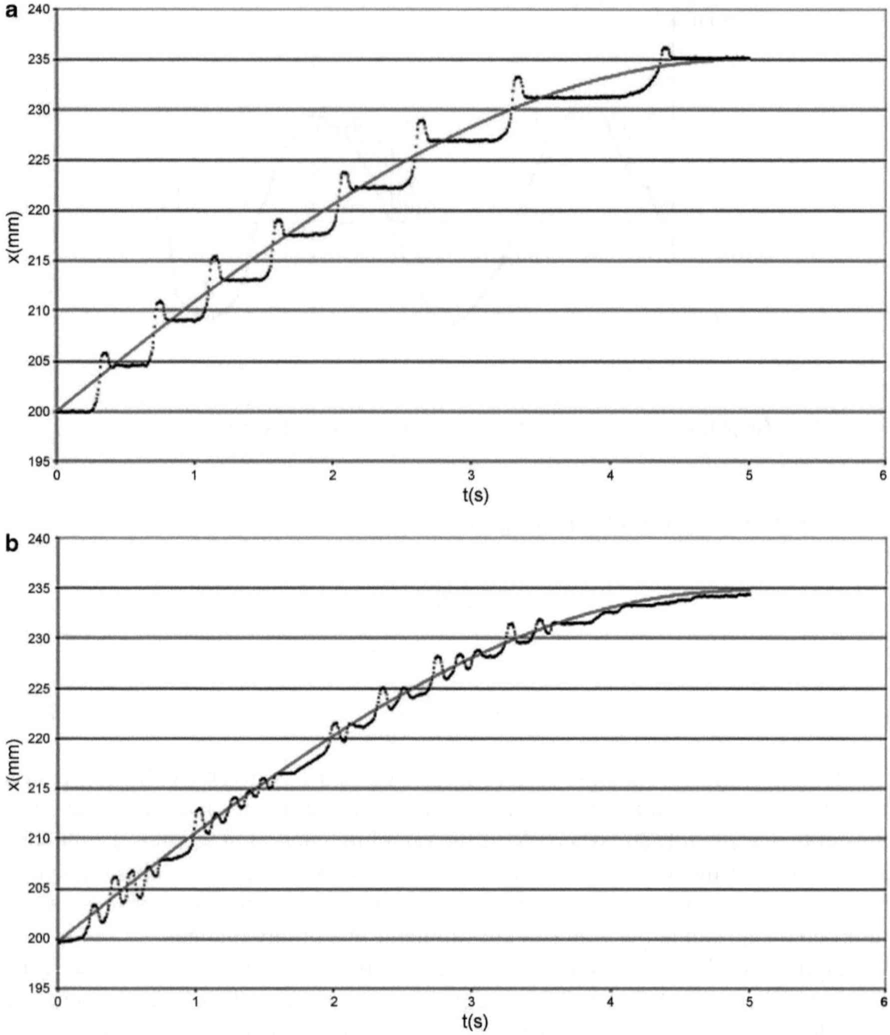

Figure 7.32 Rod positions and input commands (**a**) with and (**b**) without deadband compensation (rod stroke: 35 mm, frequency: 0.05 Hz)

The logic introduced by Equation 7.20 into the control algorithm improves results considerably. Our comparison parameter, the EI, reaches a value of 391 mm/s when the system is not deadband compensated; the input amplitude is 35 mm, and the actuating frequency is 0.05 Hz. The proportional controller gave an EI of 280 mm/s for the same input signal. When Equation 7.20 was inserted into the algorithm, the EI decreased to 135 mm/s. Figure 7.32 shows the system's response before (a) and after (b) deadband compensation for an extension stroke of 35 mm and frequency of 0.05 Hz.

The proposed fuzzy controller is designed to provide small autosteering corrections at frequencies of between 0.5 and 2 Hz. The performance of the EH valve depends on the properties of the commanding signal as well as the characteristics of the controller. A complete description of this case study is included in [11].

7.7 Safe Design of Automatic Steering

The current approach that is used to design intelligent navigation systems for implementation in off-road vehicles follows the hypothesis of semiautonomy. This philosophy expects the operator to remain in the cabin, even if most of the driving is performed automatically. The necessary coexistence of automatic and manual driving must be taken into account when designing the steering control system, in order to ensure that reliability will be high. The original default steering mechanism must work properly, so the manipulation of oil pipes must be studied carefully to make sure that the pressure will never drop when steering is performed manually. Even in automatic mode, the operator may need to grab the steering wheel in the event of an emergency. Similarly, the solenoids of the EH valve should only be energized if the automatic mode is connected and there is an operator onboard. All of these considerations, aside from programming reliable software and selecting consistent hardware, are of great importance when devising the automatic steering system of an intelligent vehicle.

References

1. Zhang Q (2009) Basics of hydraulic systems. CRC Press, Boca Raton
2. Zhang Q (1999) Hydraulic linear actuator velocity control using a feedforward-plus-PID control. Int J Flex Autom Integr Manuf 7(3):277–292
3. Dong Z, Zhang Q, Han S (2002) Control of an electrohydraulic steering system using a PID controller with a nonlinear compensation algorithm. Proc SPIE 4715:87–95
4. Ellis G (2000) Control system design guide, 2nd edn. Prentice-Hall, San Diego
5. Zadeh LA (1996) Fuzzy logic: computing with words. IEEE Trans Fuzzy Syst 4(2):103–111
6. Zhang Q (2003) A generic fuzzy electrohydraulic steering controller for off-road vehicles. J Automob Eng 217(9):791–799
7. Qiu H, Zhang Q, Reid JF, Wu D (2001) Modeling and simulation of an electro-hydraulic steering system. Int J Vehicle Design 17(2–3):259–265
8. Qiu H, Zhang Q (2003) Feedforward-plus-proportional-integral-derivative controller for an off-road vehicle electrohydraulic steering system. J Automob Eng 217(5):375–382
9. McCorduck P (2004) Machines who think. AK Peters, Natick
10. Rovira-Más F, Zhang Q, Hansen AH (2007) Dynamic behavior of an electrohydraulic valve: typology of characteristic curves. Mechatronics 17:551–561
11. Rovira-Más F, Zhang Q (2008) Fuzzy logic control of an electrohydraulic valve for auto steering off-road vehicles. J Automob Eng 222:917–934

Chapter 8
Design of Intelligent Systems

8.1 Basic Tasks Executed by Off-road Vehicles: System Complexity and Sensor Coordination

The design of intelligent off-road vehicles must fulfill practical needs; it is therefore important to determine what we require from intelligent equipment. Although needs change with time and location, it is possible to identify a set of basic competences for generic vehicles that perform tasks on agricultural fields. These aptitudes can be sorted into three fundamental layers: safety, user information, and machine actuation. Even though the three layers are essential for the correct deployment of off-road intelligent vehicles, they take different priorities: safety is most important, and actuation often requires sensory information before mechanical execution. The following paragraphs describe some of the typical items within each layer in detail, and Figure 8.1 depicts a schematic representation of the three-layer task division concept.

The *safety layer* represents the loop with the highest priority; this continuously checks that the vehicle is not at risk. Dangerous situations may occur on steep slopes – rollovers need to be avoided, both longitudinally and laterally, thus ensuring what is termed static stability. The vehicle also has to be dynamically stable, avoiding side slip laterally and showing good aptitude when braking longitudinally. Autoguidance abilities result in demanding security constraints, such as the ability to avoid objects in the path of the vehicle, negotiate negative obstacles (holes) or waterways, and cope with unsupervised guidance when the operator recklessly leaves the cabin or accidentally falls asleep. Finally, the usual engine and transmission failures, electrical power cuts, sensor malfunctions, and software crashes may compromise the proper execution of the task. The safety layer must provide fast and accurate responses to prevent these potential problems before engaging the intelligent machine in a hazardous or even fatal situation.

The *user information layer* is in charge of supplying all kinds of useful data to the operator of the machine, or, in the case of remote-controlled operations, to the wireless-linked operating station. Apart from traditional engine and vehicle checks,

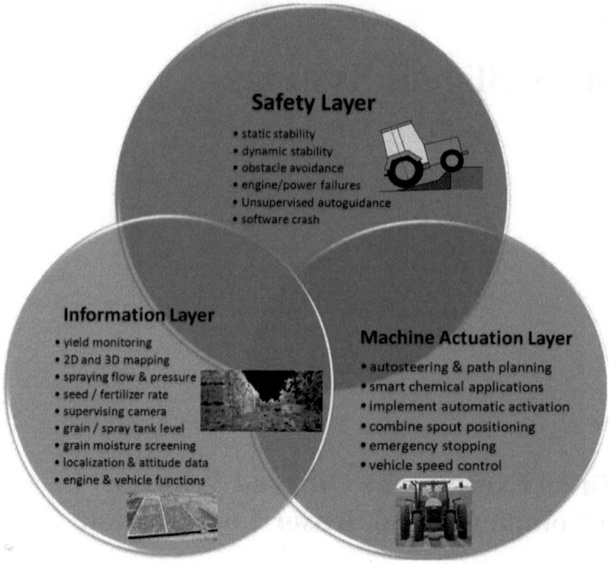

Figure 8.1 Basic tasks assigned to off-road intelligent vehicles in agricultural fields

such as those for coolant temperature, oil level, or forward speed, new functions may be added to these special vehicles in order to augment their "intelligence." Among these, we can include yield monitoring, positioning and attitude data, advanced engine functions such as real-time fuel consumption or tire pressure, map construction (either 2D or 3D), spraying flow and pressure checks, variable seeding or fertilizer rate monitoring, grain or chemical tank levels, positions of combine filling spouts, supervising cameras for demanding maneuvers, loading truck status, and grain moisture screening. In addition to passing this information on to the user in the cabin (or remote station), some data will also be transferred to the main onboard computer to pass to the decision-making routines of the actuation layer. Some of the data facilitated by the user information layer will be conveniently saved for further analysis and storage in historical record files, which are sometimes needed to compose precision farming maps.

The *machine actuation layer* is responsible for the actions automatically undertaken by the vehicle. These are prompted by the results of the computer reasoning that takes place in the vehicle's processors. A small vehicle may have just one central computer, but given the complexity of these systems, as proven by the multiplicity of potential tasks mentioned above, intelligent vehicles usually feature several interconnected processors. In general, several actions are expected to occur at a given time, so it is critical to establish the right frequencies for all of the processors. Commercial systems already incorporate some automatisms, while others still need to be standardized and widely disseminated. Many last-generation combine harvesters incorporated adaptive heads and threshing cylinder speed control. Future vehicles will be even more complex, since they will perform automatic steering and path planning, execute smart environmentally friendly chemical applications, activate implements when needed without human intervention, accurately position grain spouts as soon as loading trucks are ready for loading, and automatically control the velocity of the vehicle (allowing for prompt emergency stops when appropriate).

The diversity of the sensors required by the vehicle to fulfill all of the needs of the three layers makes it crucial to optimize the selection and implementation of the sensors, so that the technology on the vehicle is as functional as possible. This implies perfect coordination among sensors such that the same sensor can serve different layers and, when possible, various applications within a layer. For example, a stereoscopic camera can provide the user information layer with a 3D terrain map, assist in the planning of a trajectory for automatic navigation in the actuation layer, and trigger high-priority obstacle avoidance warnings within the safety layer. This intertwining of devices requires order and planning based on a well-established architecture from both hardware and software standpoints. One critical aspect of the latter is the "frequency budget." Each sensor typically operates at a specific data rate, so GPS receivers often output strings at 1 or 5 Hz, video streaming often occurs at 30 frames/s, and fiber-optic gyroscopes may reach 100 Hz. In light of this diversity, what is the recommended frequency of the main loop? Conventional wisdom sets the minimum frequency for off-road autonomous vehicles at 10 Hz [1], given that the normal velocities of these vehicles are moderate. It is therefore obvious that GPS rates of 1 Hz are not suitable for most of the tasks currently assigned to intelligent vehicles, and this problem will become even worse in the future as more tasks are added. Excessive data streaming is not always advantageous because, despite allowing for redundancy, it may result in saturation and computer overflow. Numerical data are not usually problematic, but digital images, especially if they have large resolutions or support 3D information, often lead to slow loops. In conclusion, each intelligent vehicle will require particular settings that suit its main purpose, although there is an archetypical general architecture that follows the proposed three-layer configuration (with the safety layer taking priority) and makes rational use of coordinated sensors and actuators working at a minimum rate of 10 Hz.

8.2 Sensor Fusion and Human-in-the-loop Approaches to Complex Behavior

Intelligence is typically associated with complexity, or more accurately complex behavior. For instance, the chart in Figure 1.2 proves that brain power has increased as the complexity of life has risen from bacteria to primates. Robotic vehicles, which are designed to aid humans in habitual tasks, inherit this high level of sophistication. Managing complexity in autonomous vehicles is not a trivial matter; sensor failures, signal corruption, intricate environments with changing illumination, electromagnetic or magnetic interference, power outages, and real-time constraints are common challenges for intelligent vehicles. Team Tormenta [1] fielded an autonomous jeep in the DARPA Grand Challenge 2005. In order to endow the vehicle with an operational system while maintaining its fundamental capabilities, the team had to limit complexity by establishing a logical progression of actions through the main control loop, which was augmented with secondary time-consuming threads. Velocity requirements for the vehicle made it necessary to sample the sensors at least ten times per second, as well as to discard information that had already been processed about the world model. A divide and conquer approach – probably unavoidable under such circumstances – led the team to partition the system into *sensing*, *planning*, and *control*, which were implemented in two processors communicated via an Ethernet.

An obvious outcome of environmental and task complexity is the need for *sensor fusion* and *redundancy*. Humans and animals naturally combine the information from multiple senses to navigate and execute basic operations. Lacking the powerful brains of living creatures, robotic vehicles require sophisticated hardware and software in order to allow them to adapt to changing environments and accomplish exigent missions in a safe and efficient way. Two different approaches have become essential characteristics of intelligent vehicles: combining local information with global localization to enhance autonomous navigation, and integrating inertial systems with GNSS for vehicle automation.

The synergetic effect of merging *local perception* with *global localization* is especially attractive for agricultural scenarios. Satellite signals are readily available in open fields but less so in tight arrangements of rows; this and the possibility of unexpected obstacles make local sensing indispensable. At present, the only reliable source of global positioning is the Navstar GPS; however, local perception can be achieved with a variety of sensors, such as ultrasonic sensors, lidars, or machine vision. Despite the options available for local perception, different sensors have different capabilities, and conventional wisdom heralds computer vision as being the richest source of neighborhood awareness. *Case Study II* of Chapter 4 discussed the overall implementation of GPS–vision fusion to guide a tractor automatically; this section looks deeper at on the algorithms and control strategies behind this approach. Before deciding on the particular methodology to use to merge information from alternative sources, it is essential to analyze the nature of this information to make sure that coherent data are fused. The steering controller shown in Figure 4.32 allows two different inputs: *offset error* and *heading error*. These two errors together form the *position error vector*, which can be independently estimated using either the GPS or an imaging sensor; the question is: which one should be chosen at a particular time? In reality, there is no need to make this choice mutually exclusive: both sets of data can take be used to get the final estimate. A mixing ratio can be calculated based on the instantaneous quality of each estimate. This *signal quality mixing ratio* is expressed in Equation 8.1, where $\boldsymbol{O_c}$ is the output position error vector sent to the steering controller, M_v is the mixing ratio matrix for the vision sensor, $\boldsymbol{E_v}$ is the position error vector estimated by the vision system, M_g is the mixing ratio matrix for the GPS, and $\boldsymbol{E_g}$ is the position error vector output by the GPS receiver. Notice that these definitions of the matrices M_v and M_g lead to Equation 8.2, where I is the identity matrix. As well as incorporating the offset and heading errors, the position error vector can be augmented with an estimate of the curvature obtained from the GPS receiver. This definition of the position error vector makes I a three-dimensional matrix.

$$\boldsymbol{O_C} = \boldsymbol{M_v} \cdot \boldsymbol{E_v} + \boldsymbol{M_g} \cdot \boldsymbol{E_g} \tag{8.1}$$

$$I_{3x3} = \boldsymbol{M_v} + \boldsymbol{M_g} \tag{8.2}$$

The generation of the mixing ratio matrices $\boldsymbol{M_v}$ and $\boldsymbol{M_g}$ is a delicate step in this procedure. In *Case Study II* (Chapter 4), the quality of the data gathered with machine vision depended on a confidence index that was based on how closely a customized template matched the crop rows detected in the images [2]. GPS quality, on the other hand, was a combination of the number of satellites in the solution and other quality indicators provided by the GPS receiver, such as DOP, HDOP, VDOP, *etc*. Once the quality indicators have been determined, Equation 8.1 can be expanded in detail according to Equation 8.3, where the definition of each component can be found in Table 8.1. Note that the vision algorithm implemented in CS II cannot estimate the curvature of the trajectory, so there is no κ_v.

$$\begin{bmatrix} \delta \\ \phi \\ \kappa \end{bmatrix} = \begin{bmatrix} \alpha_\delta & 0 & 0 \\ 0 & \alpha_\phi & 0 \\ 0 & 0 & \alpha_\kappa \end{bmatrix} \cdot \begin{bmatrix} \delta_v \\ \phi_v \\ 0 \end{bmatrix} + \begin{bmatrix} 1-\alpha_\delta & 0 & 0 \\ 0 & 1-\alpha_\phi & 0 \\ 0 & 0 & 1-\alpha_\kappa \end{bmatrix} \cdot \begin{bmatrix} \delta_g \\ \phi_g \\ \kappa_g \end{bmatrix} \tag{8.3}$$

Given that δ_v, ϕ_v, ϕ_g, δ_g, and κ_g are determined by the perception and localization sensors, the objective of the algorithm, according to the fusion strategy defined by Equation 8.1, is to resolve the *mixing ratio vector* $(\alpha_\delta, \alpha_\phi, \alpha_\kappa)$. This vector represents the best combination (*i.e.*, the optimum proportions) of data from different sensors, and can be deduced using different approaches. Even though there are plenty of ideas regarding the best way to perform sensor combination, the novelty of these techniques means that there is no generalized solution (*i.e.*, no *paradigm* in the sense defined by Kuhn [3]; that is, no well-established and unique scientific route).

Table 8.1 Quality indicators incorporated into the matrices $\boldsymbol{M_v}$ and $\boldsymbol{M_g}$

Parameter	Definition
δ	Offset error sent to the steering controller
ϕ	Heading error sent to the steering controller
κ	Curvature estimate sent to the steering controller
α_δ	Signal quality mixing ratio for the offset error
α_ϕ	Signal quality mixing ratio for the heading error
α_κ	Signal quality mixing ratio for the curvature estimate
δ_v	Offset error provided by machine vision
ϕ_v	Heading error given by machine vision
δ_g	Offset error from the GPS receiver
ϕ_g	Heading error supplied by the GPS receiver
κ_g	GPS-based curvature estimate

		QUALITY OF MACHINE VISION DATA		
		GOOD	FAIR	POOR
QUALITY OF GPS DATA	GOOD	α_δ = large α_ϕ = medium α_κ = 0	α_δ = medium α_ϕ = medium α_κ = 0	α_δ = small α_ϕ = small α_κ = 0
	FAIR	α_δ = large α_ϕ = large α_κ = 0	α_δ = large α_ϕ = medium α_κ = 0	α_δ = medium α_ϕ = small α_κ = 0
	POOR	α_δ = 1 α_ϕ = large α_κ = large	α_δ = large α_ϕ = medium α_κ = large	α_δ = medium α_ϕ = medium α_κ = large

Figure 8.2 Rule base for crop rows traced following a slight curvature

The sensor fusion procedure developed in [4, 5], for instance, uses *fuzzy logic* to find the coefficients of the mixing ratio vector. The problem of having to decide the proportions in which the different data sets should be merged is well suited to fuzzy set theory, where the variables are defined linguistically and converted to values in the range 0–1 using *membership functions* and *logic rules*. In this particular case, three different sets of rules were devised for the following three situations: straight rows, gentle curvature rows, and safety rules. The curvature membership function is needed to determine whether the rows are aligned or curved, and therefore apply the appropriate rule base. The complete approach can be read in [5], and as an example, the *rule base* for the first situation (straight rows) is provided in Figure 8.2. This set of rules establishes a relationship between the quality of the signal output by the sensors and the values of the components of the mixing ratio vector $(\alpha_\delta, \alpha_\phi, \alpha_\kappa)$. These rules show a preponderance of machine vision estimates over those given by the GPS, because the perception system assembled in [5] worked accurately for quasi-straight lines.

The entries of the rule base defined by Figure 8.2 set five values for the coefficients of the mixing ratio vector: {0, *small*, *medium*, *large*, 1}. The numerical magnitudes corresponding to the linguistic variables *good*, *fair*, and *poor* are calculated from the input membership functions, such as the GPS quality function of Figure 8.3. Likewise, the intuitive attributes *small*, *medium*, and *large* are mapped to actual numbers with the output membership functions, such as the *offset mixing ratio* α_δ of Figure 8.4. Finally, the *inference engine* (or rule base) of Figure 8.2 holds expert knowledge stating that "if the vision data is *good* but the GPS estimates are *poor* the offset mixing ratio is *large* when the vehicle travels within straight rows." Figures 8.3 and 8.4 both represent triangular membership functions. At this point, we could have a pertinent debate on the shapes of membership functions, but in practice most applications end up implementing triangular and trapezoidal functions due to their simplicity and the good results they yield. The actual execution of the rule base of Figure 8.2 in *Case Study II* led to associations of this kind: when the

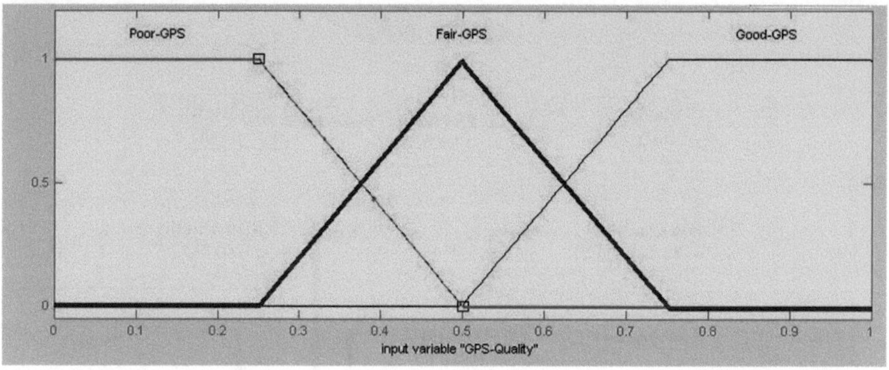

Figure 8.3 Fuzzy logic (input) membership functions for GPS quality estimates

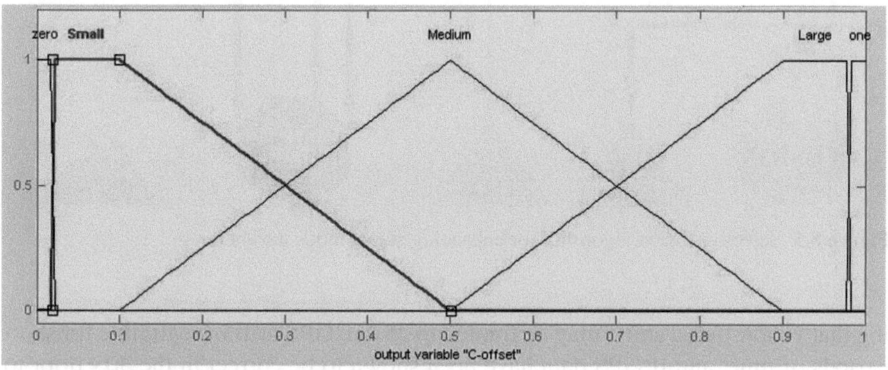

Figure 8.4 Fuzzy logic (output) membership functions for the offset mixing ratio

GPS quality is 88% and the quality of the vision system is (simultaneously) 75%, the corresponding mixing ratio vector is (0.9, 0.5, 0).

The fuzzy-based reasoning algorithm developed above assumes good GPS performance whenever its quality indices remain high. However, a close examination of how these indices were defined reveals potential issues with *guess rows* (see Figure 4.3), *bias errors*, and receiver *drift*. As a matter of fact, field experiments have demonstrated that after GPS signal dropouts (and subsequent receiver initialization) there is a jump (bias) in positioning coordinates; after 15 min of continuous work, drift is significant; and a guess row cannot be identified via satellite signals, regardless of the number of satellites and high-precision indices. Therefore, a more sophisticated logic procedure is required to increase reliability and enhance robustness. The idea of *selective fusion* depends on the capability of highly accurate vision positioning to correct for GPS bias and drift. These new rules precede the execution of the fuzzy model previously discussed, and incorporate new operational modes: *GPS-only*, *vision-only*, *no guidance*, and *fuzzy fusion*. The following set of rules form the basis of this approach: the information supplied by a local sensor is restricted to its field of view, so large positioning corrections cannot be assured, mean-

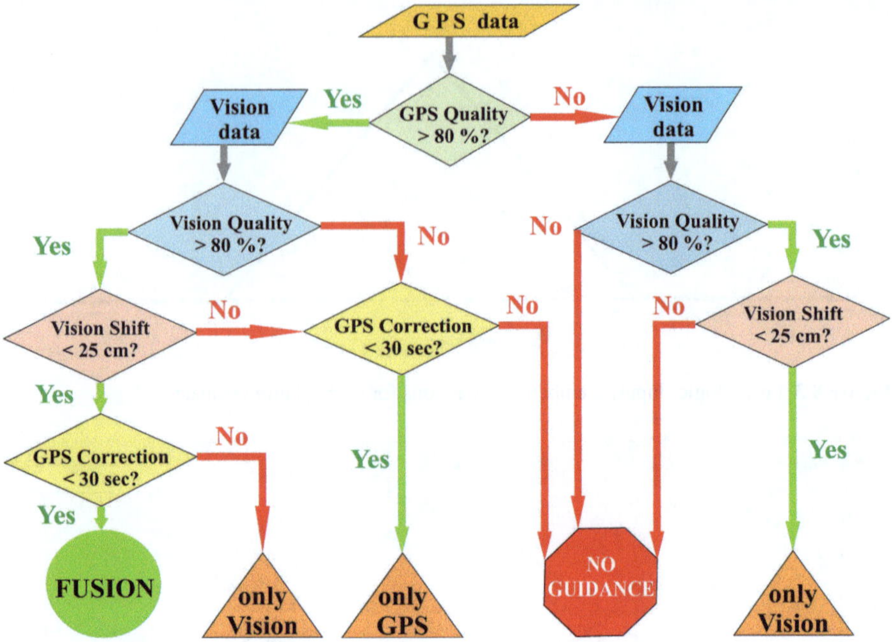

Figure 8.5 Selective fusion algorithm for enhancing autonomous navigation

ing that vision-based shifts may be limited to 25 cm; GPS drift is negligible for short periods of time, and if GPS data have been shown to be correct in the 30 s prior to a vision sensor failure, global information can be utilized to audit visual perception; the GPS-only mode is activated when the GPS receiver shows high quality and vision has been good in the last 30 s, even though it is poor at present; the vision-only mode starts up when there are high-quality vision data and relatively small shifts, but the GPS quality is low or its output has not recently been checked and rectified by the vision system; fuzzy fusion only takes place with high-quality estimates on both sides and favorable conditions in terms of low-vision shifts and permanent GPS consistency; and finally, the system will adopt the no guidance mode in the case of unreliable sensor data, ceasing the motion of the vehicle for security reasons. All of this *artificial thinking* is mapped out graphically in the chart of Figure 8.5.

The integration of inertial navigation systems (INS) with GNSS is one of the most – if not *the* most – widely used technique for achieving a reliable navigation solution. This sensor integration can be performed via two alternative approaches: *artificial intelligence* (AI) methods and the *Kalman filter* (KF). On the whole, during medium-to-long GPS outages, an AI module performs better than a KF module, but KF modules tend to give better results during short GPS signal losses. Even though most current INS/GPS integration modules rely on the Kalman filter, KF-based fusion techniques entail serious difficulties, such as the complexity involved in deducing an accurate dynamic model, or the uncertainty associated with the estimation of relevant system parameters (observability) and covariance matrices. Present AI-based techniques for INS/GPS integration, on the other hand, make use of a variety of methods, such as artificial neural networks, fuzzy logic, and hybrid neuro-fuzzy algorithms [6]. AI modules need to be trained before they can reach an acceptable level of reliability. For example, the initial weights employed in neural networks are usually random, and a certain number of iterations are needed to get to the correct initial weights before the actual navigation task. As illustrated in Figures 8.2–8.5, fuzzy logic systems seem "more rational," as many initial parameters are set during the design phase.

Section 3.5 gives an overview of INS/GPS integration with the Kalman filter. *Case Study II* in Chapter 4 presents a practical application of vision–GPS sensor fusion carried out with two alternative methodologies: fuzzy logic and Kalman filtering. The particulars of the former were covered in previous paragraphs of this section, and the workings of the latter are discussed in the following lines. The state vector for the generalized expression of the KF given in Equation 3.1 does not incorporate any camera-based vehicle-fixed local data. However, the schematic of Figure 4.34 considers the localization of the target point expressed in local coordinates as the vehicle state. Under this new definition of state, both the state vector $\boldsymbol{x_k}$ and the measurement vector $\boldsymbol{z_k}$ are augmented with the addition of the local (vehicle-fixed) coordinates of the target point determined via computer vision. Equation 8.4 specifies the components of the state and measurement vectors. Recall that the target point at time step k is defined as the instantaneous point along the trajectory towards which the vehicle is directed to fulfill a navigation mission.

$$\boldsymbol{x_k} = \begin{bmatrix} X_k \\ Y_k \\ U_k \\ V_k \\ \varphi_k \\ X_k^P \\ Y_k^P \end{bmatrix} = \begin{bmatrix} X \\ Y \\ U \\ V \\ \varphi \\ X^P \\ Y^P \end{bmatrix}_k ; \qquad \boldsymbol{z_k} = \begin{bmatrix} x_{\text{GPS}} \\ y_{\text{GPS}} \\ \varphi_{\text{GPS}} \\ \text{vel}_{\text{GPS}} \\ x_{\text{vision}}^P \\ y_{\text{vision}}^P \end{bmatrix}_k \tag{8.4}$$

The measurement vector $\boldsymbol{z_k}$ and the vehicle state vector $\boldsymbol{x_k}$ are related by the recursive expression of Equation 8.5, where H_k is the *measurement sensitivity matrix* and $\boldsymbol{v_k}$ is the 6×1 *measurement noise vector* with known covariance. The first four components of $\boldsymbol{z_k}$ are measurements provided by the GPS receiver (the vehicle's position, orientation, and forward velocity); the remaining parameters (x_{vision}^p, y_{vision}^p) represent the position of the target point P (Figure 4.34) estimated with the onboard video camera. Given that all of the variables in Equation 8.5 occur at time step k, there is no need to retain the subscript k when deducing matrix H_k.

$$\boldsymbol{z_k} = H_k \cdot \boldsymbol{x_k} + \boldsymbol{v_k} \tag{8.5}$$

The inference of the measurement sensitivity matrix H_k is an important step in the implementation of the Kalman filter, especially considering the unusual combination of different types of sensors. The GPS receiver directly provides estimates for the three vehicle states (x and y positions, and heading) indicated in Equations 8.6–8.8. The next objective is to express the elements of the measurement vector as a function of the components of the state vector; these functions constitute the elements of H_k.

$$x_{\mathrm{GPS}} = X \tag{8.6}$$
$$y_{\mathrm{GPS}} = Y \tag{8.7}$$
$$\varphi_{\mathrm{GPS}} = \varphi \tag{8.8}$$

The decomposition of the observed forward velocity $\mathrm{vel}_{\mathrm{GPS}}$ into its two normal components U and V (related to the local tangent plane, LTP) follows the conventions of Euclidean geometry, as stated in Equation 8.9. However, the expression of local coordinates (x^P_{vision}, y^P_{vision}) as a function of LTP global coordinates requires a more involved route. The first stage of this process casts the LTP global coordinates of the target point P (X^P, Y^P) as a function of the vehicle's LTP global coordinates (X, Y), the target-point vehicle-fixed local coordinates (x^P_{vision}, y^P_{vision}), and the vehicle's LTP heading φ. Equations 8.10 and 8.11 show this step in mathematical form.

$$\mathrm{vel}_{\mathrm{GPS}} = \sqrt{U^2 + V^2} \tag{8.9}$$

$$X^P = X + x^P_{\mathrm{vision}} \cdot \cos\varphi + y^P_{\mathrm{vision}} \cdot \sin\varphi \tag{8.10}$$
$$Y^P = Y - x^P_{\mathrm{vision}} \cdot \sin\varphi + y^P_{\mathrm{vision}} \cdot \cos\varphi \tag{8.11}$$

Equations 8.10 and 8.11 can be reorganized to yield Equations 8.12 and 8.13, which are intermediate steps in solving for x^P_{vision} and y^P_{vision}.

$$X^P - X = x^P_{\mathrm{vision}} \cdot \cos\varphi + y^P_{\mathrm{vision}} \cdot \sin\varphi \tag{8.12}$$
$$Y^P - Y = -x^P_{\mathrm{vision}} \cdot \sin\varphi + y^P_{\mathrm{vision}} \cdot \cos\varphi \tag{8.13}$$

The linear combination of Equations 8.12 and 8.13, as suggested by Equations 8.14 and 8.15, leads to Equations 8.16 and 8.17, which in turn prepares the way for the definitive expression of the local coordinates of the target point P in terms of a global frame, as given in Equations 8.18 and 8.19.

$$[\text{Equation 8.12}] \cdot \sin\varphi + [\text{Equation 8.13}] \cdot \cos\varphi \tag{8.14}$$
$$[\text{Equation 8.12}] \cdot (-\cos\varphi) + [\text{Equation 8.13}] \cdot \sin\varphi \tag{8.15}$$

$$\left[X^P - X\right] \cdot \sin\varphi + \left[Y^P - Y\right] \cdot \cos\varphi = y^P_{\mathrm{vision}} \cdot \sin^2\varphi + y^P_{\mathrm{vision}} \cdot \cos^2\varphi \tag{8.16}$$
$$\left[X - X^P\right] \cdot \cos\varphi + \left[Y^P - Y\right] \cdot \sin\varphi = -x^P_{\mathrm{vision}} \cdot \sin^2\varphi - x^P_{\mathrm{vision}} \cdot \cos^2\varphi \tag{8.17}$$

$$y^P_{\mathrm{vision}} = X^P \cdot \sin\varphi - X \cdot \sin\varphi + Y^P \cdot \cos\varphi - Y \cdot \cos\varphi \tag{8.18}$$
$$x^P_{\mathrm{vision}} = X^P \cdot \cos\varphi - X \cdot \cos\varphi - Y^P \cdot \sin\varphi + Y \cdot \sin\varphi \tag{8.19}$$

The equalities found in Equations 8.6–8.19 provide all of the information needed to compose the measurement sensitivity matrix H_k, whose final expression is given in Equation 8.20. After expanding Equation 3.1 to include the vision estimates, and applying matrix H_k according to Equation 8.20, the only matrices that are still to be determined are the noise covariance matrices Q and R. As discussed in Section 3.5, the *process noise covariance matrix* Q models uncertainty in the dynamic process. Given the complexity involved in deducing this model accurately from direct observation, the conventional procedure for obtaining such a model has been an iterative process for in-field tuning. The *measurement noise covariance matrix* R, on the other hand, is related to the noise expected in sensor readings, which is typically provided by sensor manufacturers. For this vision–GPS integration, the covariance matrix R of Equation 3.3 would need to be augmented with the accuracy of the vision estimates for the horizontal and vertical coordinates, which in the practical implementation described in CS II was about 30 cm. A description of the entire project can be found in [7].

$$H_k = \begin{bmatrix} 1 & 0 & 0 & 0 & 0 & 0 & 0 \\ 0 & 1 & 0 & 0 & 0 & 0 & 0 \\ 0 & 0 & 0 & 0 & 1 & 0 & 0 \\ 0 & 0 & \frac{U}{\sqrt{U^2+V^2}} & \frac{V}{\sqrt{U^2+V^2}} & 0 & 0 & 0 \\ -\cos\varphi & \sin\varphi & 0 & 0 & 0 & \cos\varphi & -\sin\varphi \\ -\sin\varphi & -\cos\varphi & 0 & 0 & 0 & \sin\varphi & \cos\varphi \end{bmatrix} \tag{8.20}$$

The advantages of control systems based on fuzzy logic rather than other conventional methods are based on the ability of fuzzy sets to adapt to situations that are typically handed by people, as human knowledge can be incorporated into the system's intelligence through a set of heuristic rules. This *modus operandi* presents clear advantages when systems respond nonlinearly and with uncertainty. A further step that brings the system's response closer to the human response is to incorporate humans into the design loop using the *human-in-the-loop* (HIL) interactive control system design tool [8]. The HIL design tool provides a way to include a human operator in the system design process. It is implemented in a virtual environment that allows a new design to be tested and improved before building a prototype. This configuration closes the loop for qualitative open-loop designs by creating multiple user-defined modular components that are either removable or replaceable. The HIL design process has been applied to create an adaptive steering controller that enhances maneuverability in a front-end loader regardless of operator behavior [8]. The steering control rules of the model were selected with fuzzy logic.

8.3 Navigation Strategies and Path-planning Algorithms

There is no unified theory that establishes an unequivocal relationship between the coordinates of the desired location (*target point*) of an autonomous vehicle with

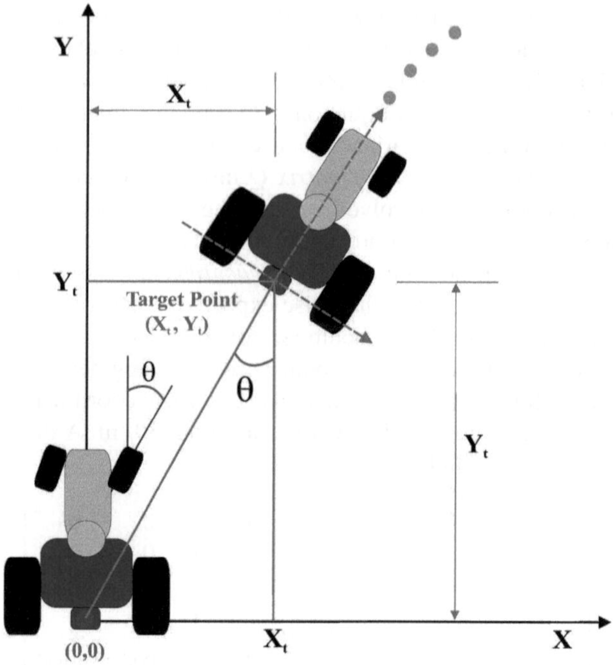

Figure 8.6 Basic methodology used to estimate steering commands

Ackerman steering and the optimum *steering angle* that must be turned by the front wheels of the vehicle. Therefore, a number of approaches have been proposed to solve this fundamental problem of field robotics and automatic navigation. A straightforward method of calculating the steering angle needed to direct a tractor from its actual position to the computer-selected target point is shown in Equation 8.21 [9], where the vehicle-fixed coordinates of the target point are (X_t, Y_t) and the steering angle is θ, as graphically represented in Figure 8.6. Notice that the origin of the coordinates moves with the vehicle and is arbitrarily located at the tractor's drawbar. The further the tractor is from the desired position, the greater the correction. It seems quite intuitive that the steering angle θ must be a function of this deviation, and Equation 8.21 identifies both parameters in a very simplistic way; indeed, a tractor was automatically driven at up to 11 km/h when this approach was followed [9]. Note that negative values of X yield negative steering angles (*i.e.*, left turns). This determination of the steering angle implicitly considers offsets and heading angles, which are combined without distinction.

$$\theta = \arctan\left(\frac{X_t}{Y_t}\right) \tag{8.21}$$

The path constructed with the method associated with Equation 8.21 and Figure 8.6 does not account for curvature changes, variable forward speed, or other errors associated with vehicle dynamics, including the steering system and tire slip. Therefore, a more complex solution may be needed to increase reliability. Sophistication can be introduced by substituting a unique target point for a matrix of several locations representing a desired trajectory in vehicle space. This particular idea has been implemented in an automated tractor whose computer vision system generates the *trajectory path* (originally comprising five points) in image space before it is transformed to vehicle space by means of a coordinate transformation matrix [10]. Similar to the previous case depicted in Figure 8.6, the origin is defined at the center of the rear axle; however, in addition to the calculation of the desired wheel angle, this application also determines a speed-dependent *look-ahead distance*. Curvature is detected by noting a change in direction for three successive points. The direction of the trajectory path can be oriented left or right; a curve is detected when the current direction changes from the previous direction. The look-ahead distance determines the target point; the location of the point is used to get the desired turning angle of the wheels. However, this time the steering angle is given as a function of the vehicle's wheel base (*i.e.*, the distance between rear and front axles), and the radius of curvature for the circular trajectory. This procedure follows the philosophy of SAE Standard J695 [11], as implemented in [12], and can be schematized as in Figure 8.7. If the target point is represented by coordinates (X_t, Y_t) and the wheel base is W_b, the approximate Ackerman steering angle θ can be calculated with Equation 8.22. The radius of curvature R_{curv} can be deduced directly from Figure 8.7, as shown in Equation 8.23. When the angle θ of this approach (Equation 8.22) is compared with that given by the simplistic approach (Equation 8.21), we can see that

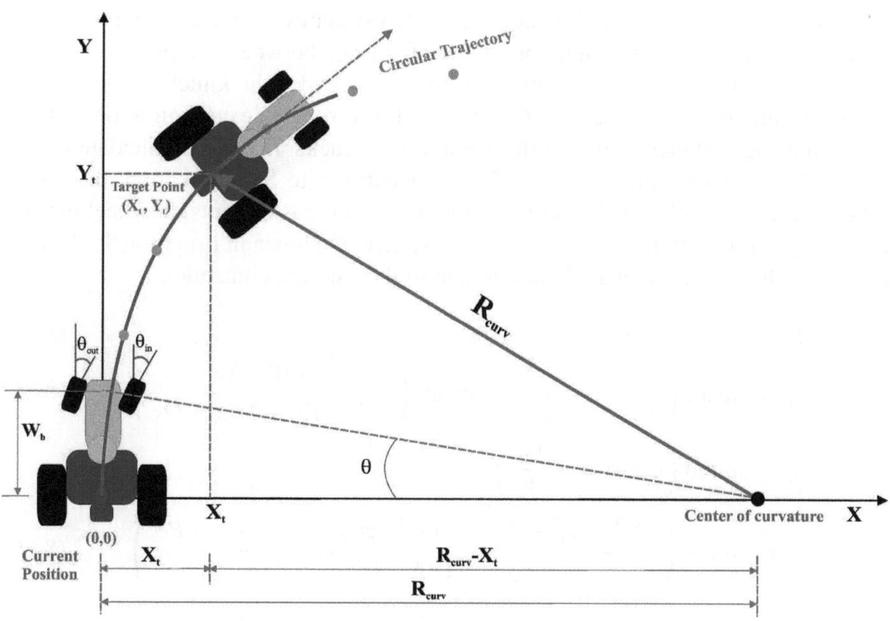

Figure 8.7 Geometry used to estimate steering commands with SAE Standard J695

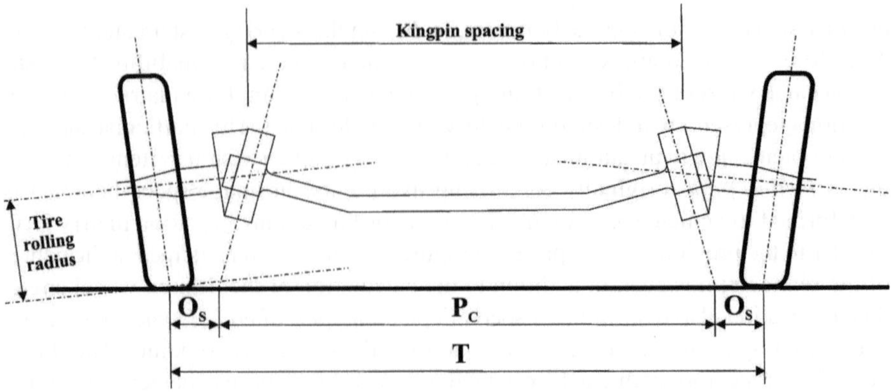

Figure 8.8 Key parameters of front-axle configuration vehicles

the former yields shallower turns than the basic method.

$$\theta = \frac{\theta_{\text{in}} + \theta_{\text{out}}}{2} \approx \arctan\left(\frac{W_{\text{b}}}{R_{\text{curv}}}\right) = \arctan\left(\frac{2 \cdot W_{\text{b}} \cdot X_{\text{t}}}{X_{\text{t}}^2 + Y_{\text{t}}^2}\right) \tag{8.22}$$

$$R_{\text{curv}}^2 = Y_{\text{t}}^2 + (R_{\text{curv}} - X_{\text{t}})^2 \quad \rightarrow \quad R_{\text{curv}} = \frac{X_{\text{t}}^2 + Y_{\text{t}}^2}{2 \cdot X_{\text{t}}} \tag{8.23}$$

The approximation for the Ackerman steering angle θ given in Equation 8.22 has been effectively used to autosteer a large tractor in the field [12]. However, SAE Standard J695 [11] yields more accurate estimates for the inside and outside wheel turning angles as a function of the offset O_S between the pivot center and the middle of the tire track, and the distance P_C between the knuckle pivot centers at the ground. Figure 8.8 depicts O_S and P_C for a front-axle configuration vehicle. Note that the distance between the front wheel tracks T can be calculated from O_S and P_C using Equation 8.24. Finally, according to Standard J695, the inside wheel angle θ_{in} is obtained with Equation 8.26, whereas the outside wheel turning angle θ_{out} is given by Equation 8.25; their average, θ, shown in Equation 8.27, is the corresponding Ackerman angle that is sent to the steering controller.

$$T = P_C + 2 \cdot O_S \tag{8.24}$$

$$\theta_{\text{out}} = \arcsin\left(\frac{W_{\text{b}}}{R_{\text{curv}} - O_S}\right) = \arcsin\left(\frac{2 \cdot W_{\text{b}} \cdot X_{\text{t}}}{X_{\text{t}}^2 + Y_{\text{t}}^2 - 2 \cdot X_{\text{t}} \cdot O_S}\right) \tag{8.25}$$

$$\begin{aligned} \theta_{\text{in}} &= \arctan\left(\cot \theta_{\text{out}} - \frac{P_C}{W_{\text{b}}}\right) \\ &= \arctan\left(\frac{\left(X_{\text{t}}^2 + Y_{\text{t}}^2 - 2 \cdot X_{\text{t}} \cdot O_S\right) \cdot \cos \theta_{\text{out}} - 2 \cdot X_{\text{t}} \cdot P_C}{2 \cdot X_{\text{t}} \cdot W_{\text{b}}}\right) \end{aligned} \tag{8.26}$$

$$\theta = \frac{1}{2}\left[\arcsin\left(\frac{2 \cdot W_b \cdot X_t}{X_t^2 + Y_t^2 - 2 \cdot X_t \cdot O_S}\right) + \arctan\left(\frac{\left(X_t^2 + Y_t^2 - 2 \cdot X_t \cdot O_S\right) \cdot \cos\theta_{out} - 2 \cdot X_t \cdot P_C}{2 \cdot X_t \cdot W_b}\right)\right] \quad (8.27)$$

The navigation strategies discussed so far respond to the real-world need for an autonomous agricultural machine to drive between narrowly separated rows of vegetation, so the *a priori* simplicity of the solution should never be seen in a derogative way. As a matter of fact, both of these strategies [9, 12] were used to impressively effect to autoguide two > 200 HP tractors, each weighing several tonnes (University of Illinois, 2000–2001). However, it is sometimes necessary to direct the automated vehicle toward a predefined goal (*i.e.*, a sequence of targets are known beforehand), rather than expecting the vehicle to determine its targets as it moves ("on the fly"). This was the case for the jeep that participated in the Grand Challenge competition [1], where the basic navigation strategy was to follow an initial waypoint list provided by the Department of Defense (DARPA). The strategy used in [1] was to keep the car as close to the center of the corridor as possible, since the center should contain the fewest obstacles and the smoothest terrain. In essence, the vehicle followed the original waypoint list, and when it came to within a fixed distance of the point (8 m in the actual race), the next waypoint in the list became the current waypoint. The navigation software determined the steering command that minimized the distance between the vehicle and an ideal piecewise linear path obtained by drawing lines between each pair of consecutive waypoints. The advantages of steering along a path rather than towards a target point were the increased accuracy obtained by tracing curves defined by close waypoints and smoother vehicle motion, as the rate of change in heading did not depend on the distance to the target point.

The easy availability of GPS information has resulted in several research projects where an autonomous vehicle must reach a goal from a starting point and conduct a farming mission on the way. While this has made for eye-catching demonstrations, actual requirements in the field are quite different. At present, autonomous off-road vehicles must provide the level of autonomy and intelligence that we previously associated with *semiautonomous vehicles*; that is, autopiloting tasks in the field with the operator positioned in the cabin for security and monitoring purposes. Does this mean that all of the conventional techniques for planning and navigation developed within field robotics cannot be applied to off-road agricultural equipment? Of course not: they will be essential in the next section. Recall that there are two key competencies required for mobile robot navigation [13]: *path planning* (identifying the trajectory that needs to be followed by the vehicle); and *obstacle avoidance* (re-planning while traveling to avoid collisions). The trajectory planning techniques discussed above will direct the vehicle toward a moving target point or a waypoint-defined ideal path, but unexpected obstacles may appear in the way and interfere with the projected trajectory. In these circumstances, safety takes priority and the autonomous vehicle must be endowed with the proper intelligent safeguarding instruments to cope with the dynamism of real-world environments and resolve any problem satisfactorily before any risks can arise. The following section discusses some classic approaches to robot navigation in the presence of obstacles that are randomly distributed around partially known surroundings.

8.4 Safeguarding and Obstacle Avoidance

An efficient planner incorporates new information gathered by onboard sensors in real time as the vehicle navigates; therefore, unanticipated new information may require significant changes in the initial strategic plans [13]. At the end of the 1960s, there was neither the underlying theory nor an adequate conceptual framework to solve the general problem of determining a path through a navigation graph containing obstacles and free space. This situation stimulated Hart, Nilsson, and Raphael to propose the pioneering A* algorithm [14], which provides an optimal heuristic approach to determining the path of minimum cost from a start location to a goal. The A* algorithm uses special knowledge of the problem after the environment has been modeled through a two-dimensional graph. Such graphs consist of *nodes* representing vehicle locations and line segments between adjacent nodes called *arcs*. The arcs have costs associated with them in such a way that the optimal path is the path with the smallest cost. One major problem with the practical implementation of the A* algorithm is the determination of the *evaluation function* that estimates the costs of potential paths. Such a cost typically comprises two terms: the actual cost of the path from the starting point to the current point, and an estimate of the cost of an optimal path from the current point to the preferred goal. A useful characteristic of this method is that it should minimize the cost of traversing rather than the number of steps needed to find the optimal path.

Even though there is, generally speaking, a certain amount of *a priori* knowledge of the environment that an agricultural or forestry vehicle may be exposed to, such as the existence or absence of rows, the spacing between them, and the types of boundaries around the field, obsolete data and unanticipated objects can negatively affect the success of an automated task. The problem of planning paths through partially known dynamic environments has been efficiently solved by the D* algorithm [15]. The D* algorithm is an adaptation of the A* algorithm (with "D" standing for *dynamic*) to partially known environments. The novelty of the D* approach is the concept that arc cost parameters can change during the problem-solving process. For environments represented by a large number of cells (over 10,000), the CPU runtime of the D* algorithm is significantly lower than those of other optimal replanners. This computational efficiency is crucial for autonomous vehicles that need to make steering decisions in real time. There are many other path-planning algorithms that can be employed to avoid obstacles, such as visibility graphs, Voronoi diagrams, potential fields, the bug algorithm, or curvature velocity techniques [13], but none of those solutions have proven to be as efficient as the D* algorithm for safeguarding intelligent off-road vehicles in real time.

General obstacle-avoidance techniques that work acceptably inside laboratories and within controlled environments need to be adapted and tuned before they can be implemented in real vehicles operating outdoors. The aforementioned autonomous jeep participating in the Grand Challenge [1] was guided towards an ideal path defined by a set of waypoints. When onboard perception sensors detected obstacles within 1.7 m of the path, the vehicle avoided the obstacles by inserting two temporary waypoints, thus forming a bridge to the side of the obstacle while still minimizing the distance to the ideal path and reducing the amount of steering necessary. This strategy was significantly different from grid path-planning algorithms, as it impeded abrupt changes in the jeep direction. Steering was executed by a PD controller outputting commands proportional to the distance of the vehicle to the ideal waypoint line. *Case Study IV* in Chapter 5 considered an obstacle detection and avoidance perception system that was based on stereoscopic vision [16]. The path planner implemented on a utility vehicle was based on the A* algorithm [14], and introduced a heuristic function that depended on the Euclidean distance to a predefined target point. The goal, selected by the operator, was within the field of view (a rectangle 8 m wide and 20 m long) of a stereo camera. Figure 5.55 shows three challenges faced by the vehicle's safeguarding engine, where the target point is marked by the red dot and is located between 10 and 18 m from the vehicle.

The detection of an object does not necessarily have to lead to its perfect identification; autonomous vehicles often only need to check whether the terrain ahead is traversable. This idea can be further developed by using *traversability*-based navigation grids to scrutinize a vehicle's surroundings. Whether an object is traversable depends not only on the size and dimensions of an obstacle but also on the type and the design of the vehicle. Fruit trees are typically non-traversable, but crop rows and grass swaths depend on the free height of the piece of equipment used. An assessment of traversability requires the implementation of a local perception unit onboard the autonomous vehicle. Stereoscopic vision, through its three-dimensional perceptive capabilities, offers excellent information for the practical application of traversability analysis [17]. Figure 8.9a shows a *frontal traversability map* (FTM)

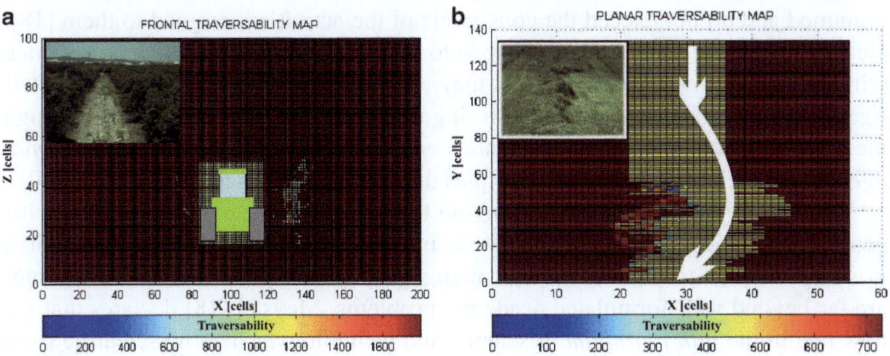

Figure 8.9 Traversability analyses of agricultural scenes: (**a**) frontal map (FTM) for orchards; (**b**) planar map (PTM) of a windrow

generated from the front view of an orchard scene captured with a stereovision sensor, where the big square in the center of the grid marks the dimensions of the vehicle, and the other filled cells indicate the positions of the trees on both sides of the vehicle. Figure 8.9b, in contrast, provides a *planar traversability map* (PTM) inferred from the top view of a cut grass swath. The vertical band depicted in the image has the width of the vehicle, and given that the windrow was high enough to be considered non-traversable, the vehicle had to be guided around the swath according to the path traced by the white arrow.

8.5 Complete Intelligent System Design

The period of time following the Renaissance in the sixteenth century is regarded as the *Age of Scientific Revolution*; comparatively, we could denote the twenty-first century as the *Age of Computational Reasoning*, considering that artificial intelligence is now reaching the long-desired degree of maturity that should enable machines to perform actions based on their own initiative. Many appliances, industrial machines, and commercial vehicles currently incorporate a certain degree of intelligent automation. However, new ideas tend to be driven by the need to resolve a practical problem along with fresh scientific discoveries. Therefore, this book has been assembled around practical applications as a framework for the nascence and development of *intelligent off-road vehicles*. Thomas Kuhn [3] concluded that a new theory is always announced together with its application to a range of natural phenomena; without them it would not even be a candidate for acceptance, because the process of learning a theory depends upon the study of its applications. The analysis of all of the applications (and their solutions) presented here provide the basis for proposing the *intelligent system design* of future vehicles. Meystel claimed in 1991 that there is no definition of an intelligent machine as a technological entity (in fact, he believes that the words *intelligent* and *intelligence* have been depreciated), and although thousands of so-called automated guided vehicles are in operation all over the world, they cannot be considered *autonomous robots*, since they are programmed and do not exceed the constraints of the activities assigned to them [18]. Here, *autonomous* robots are understood to be intelligent machines that can operate with no human involvement; that is, they are endowed with the property of intelligent control. However, rather than trying to replicate human intelligence through the *laws of thought*, it is more advantageous to consider the vehicle as a *rational agent* controlled by the general principles that comprise artificial intelligence [19].

When *designing* the *reasoning engine* that must be integrated into the intelligent vehicle (and done so as naturally as in other mechanical or electric designs), it is important to consider that the challenges that these special vehicles encounter are far beyond well-formulated academic problems. Meystel [18] declares that we now live in the *fifth paradigm* of science and technology, where no scientific truth can be clearly stated and no combination of experimental conditions can be reproduced. As a result, rather than trying to objectively reproduce and test situations, our efforts should be directed towards complying with a vaguely formulated list of requirements; errors should not be considered negatively – instead of avoiding them, we should take them into account. The artificial intelligence transferred to the vehicles – machine-based rationality – must be considered in a wider, more complete sense in order to cope with such a diverse list of requirements (tracking, positioning, obstacle avoidance, mapping, perceiving, recording, monitoring, self-awareness, and so forth). *Sensory systems* such as vision and sonar cannot deliver perfectly reliable information about the environment around the vehicle; therefore, *reasoning* and *planning systems* must be able to handle *uncertainty*, which is a key problem in field robotics [19]. This complexity demands optimal integration of *hardware* and *software*, and of *global* and *local localization*. Off-road environments are notoriously difficult to model, not only due to the intrinsic difficulties of partially known terrains, but also because they change with time because outdoor conditions are often unpredictable. In this regard, *heuristics* provide a certain degree of attachment to the reality of the field, a kind of "grounding" that allows the vehicle to handle unpredictability. *Expert systems* and *semantic networks* have been proven efficient for the navigation of agricultural semiautonomous vehicles, especially when combined with *reactive control* according to the actuation level desired. *Cell decomposition* techniques for *path planning* that are further augmented with *obstacle detection* abilities and *multiple perception* (monocular vision, stereo vision, laser, thermal mapping, *etc.*) are probably safe assets for future agricultural robots.

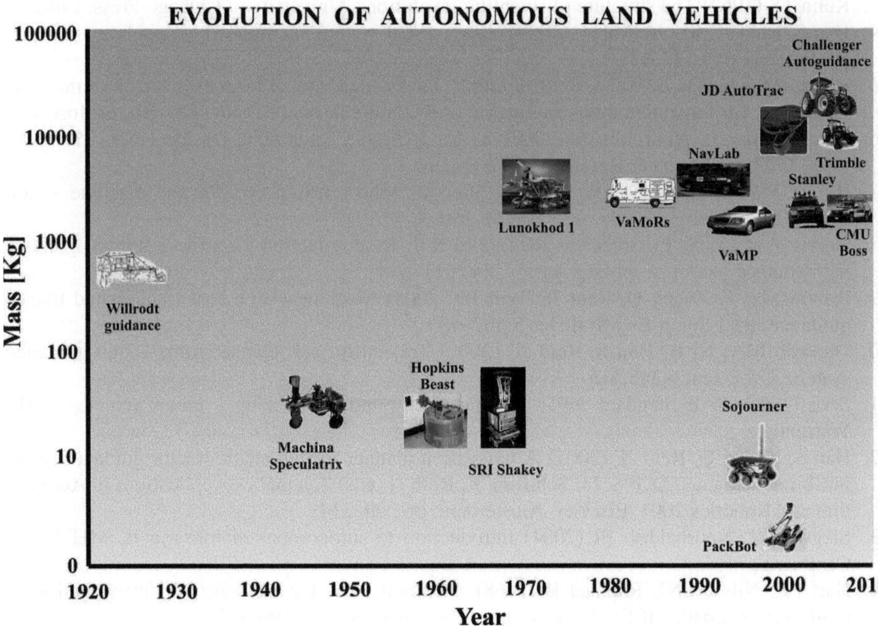

Figure 8.10 Evolutionary chart of some autonomous land vehicles

Emergent behavior – the behavior that emerges through the interplay of a *simple controller* and a *complex environment* [19] – is another option to bear in mind. Figure 8.10 illustrates, as an epilogue, how autonomous land vehicles have evolved over the last few decades, plotting year against vehicle weight. Off-road agricultural and forestry equipment, due to its size, weight, and power, poses extra challenges to automation, control, and safety – challenges that make applications within this discipline even more exciting. In the chart of Figure 8.10, after a shared robotic evolutionary trend up to 1980, note that there are two opposing tendencies: one towards small robots, and the other towards heavy autonomous off-road machines. Present day off-road agricultural equipment includes some of the largest and most advanced automated vehicles, which prvide an optimal solution to many current commercial applications in modern farms. The old dream of Willrodt to automatically operate farm machines is finally becoming a reality over 80 years later. We hope that this technological trend continues in the future, in order to reduce drudgery and improve working life in the field.

References

1. Bebel JC, Raskob BL, Parker AC, Bebel DJ (2006) Managing complexity in an autonomous vehicle. Proc Position Locat Navig Symp 356–365
2. Wei J, Han S, Rovira-Más, F, Reid JF (2005) A method to obtain vehicle guidance parameters from crop row images (ASABE Publ. no. 051155). ASABE, St. Joseph
3. Kuhn TS (1962) The structure of scientific revolutions. University of Chicago Press, Chicago
4. Han S, Reid JF, Rovira-Más F (2009) Vision-aided system and method for guiding a vehicle. US Patent 7,610,123 B2
5. Rovira-Más F, Han S, Wei J, Reid JF (2005) Fuzzy logic model for sensor fusion of machine vision and GPS in autonomous navigation (ASABE Publ. no. 051156). ASABE, St. Joseph
6. Chanthalansy L, Noureldin A (2009) AI for INS/GPS integration (Inside GNSS, Sept/Oct 4(5)). Gibbons Media & Research LLC, Eugene
7. Rovira-Más F, Han S (2006) Kalman filter for sensor fusion of GPS and machine vision (ASABE Publ. no. 063034). ASABE, St. Joseph
8. Norris WR, Zhang Q, Sreenivas RS (2006) Rule-base reduction for a fuzzy human operator performance model. Appl Eng Agric 22(4):611–618
9. Rovira-Más F, Zhang Q, Reid JF, Will JD (2003) Machine vision based automated tractor guidance. Int J Smart Eng Syst Des 5:467–480
10. Dickson MA, Ni B, Han S, Reid JF (2002) Trajectory path planner for a vision guidance system. US Patent 6,385,515
11. SAE (1998) SAE Standard J695: Turning ability and off-tracking – motor vehicles. SAE, Warrendale
12. Han S, Zhang Q, Reid JF (2001) A navigation planner for automatic tractor guidance using machine vision and DGPS. In: Schilling K, Roth H (eds) Telematics Applications in Automation and Robotics 2001. Elsevier, Amsterdam, pp 209–214
13. Siegwart R, Nourbakhsh IR (2004) Introduction to autonomous mobile robots. MIT Press, Cambridge
14. Hart PE, Nilsson NJ, Raphael B (1968) A formal basis for the heuristic determination of minimum cost paths. IEEE Trans Syst Sci Cybernet SSC-4 2:100–107
15. Stentz A (1994) Optimal and efficient path planning for partially-known environments. Proc IEEE Int Conf Robot Automat 4:3310–3317

16. Rovira-Más F, Reid JF, Zhang Q (2006) Stereovision data processing with 3D density maps for agricultural vehicles. Trans ASABE 49(4):1213–1222
17. Rovira-Más F (2009) 3D vision solutions for robotic vehicles navigating in common agricultural scenarios. In: Proc IFAC Bio-Robotics Conf IV, Champaign, IL, USA, 10–11 Sept 2009
18. Meystel A (1991) Autonomous mobile robots: vehicles with cognitive control. World Scientific, Singapore
19. Russell S, Norvig P (2002) Artificial intelligence: a modern approach. Prentice Hall, Upper Saddle River

Index

K

L

M

N

O

MIX
Papier aus verantwortungsvollen Quellen
Paper from responsible sources
FSC® C105338

If you have any concerns about our products,
you can contact us on
ProductSafety@springernature.com

In case Publisher is established outside the EU,
the EU authorized representative is:
Springer Nature Customer Service Center GmbH
Europaplatz 3, 69115 Heidelberg, Germany

Printed by Libri Plureos GmbH
in Hamburg, Germany